Communications in Computer and Information Science 2654

Rationale

The CCIS series is devoted to the publication of proceedings of computer science conferences. Its aim is to efficiently disseminate original research results in informatics in printed and electronic form. While the focus is on publication of peer-reviewed full papers presenting mature work, inclusion of reviewed short papers reporting on work in progress is welcome, too. Besides globally relevant meetings with internationally representative program committees guaranteeing a strict peer-reviewing and paper selection process, conferences run by societies or of high regional or national relevance are also considered for publication.

Topics

The topical scope of CCIS spans the entire spectrum of informatics ranging from foundational topics in the theory of computing to information and communications science and technology and a broad variety of interdisciplinary application fields.

Information for Volume Editors and Authors

Publication in CCIS is free of charge. No royalties are paid, however, we offer registered conference participants temporary free access to the online version of the conference proceedings on SpringerLink (http://link.springer.com) by means of an http referrer from the conference website and/or a number of complimentary printed copies, as specified in the official acceptance email of the event.

CCIS proceedings can be published in time for distribution at conferences or as post-proceedings, and delivered in the form of printed books and/or electronically as USBs and/or e-content licenses for accessing proceedings at SpringerLink. Furthermore, CCIS proceedings are included in the CCIS electronic book series hosted in the SpringerLink digital library at http://link.springer.com/bookseries/7899. Conferences publishing in CCIS are allowed to use Online Conference Service (OCS) for managing the whole proceedings lifecycle (from submission and reviewing to preparing for publication) free of charge.

Publication process

The language of publication is exclusively English. Authors publishing in CCIS have to sign the Springer CCIS copyright transfer form, however, they are free to use their material published in CCIS for substantially changed, more elaborate subsequent publications elsewhere. For the preparation of the camera-ready papers/files, authors have to strictly adhere to the Springer CCIS Authors' Instructions and are strongly encouraged to use the CCIS LaTeX style files or templates.

Abstracting/Indexing

CCIS is abstracted/indexed in DBLP, Google Scholar, EI-Compendex, Mathematical Reviews, SCImago, Scopus. CCIS volumes are also submitted for the inclusion in ISI Proceedings.

How to start

To start the evaluation of your proposal for inclusion in the CCIS series, please send an e-mail to ccis@springer.com

Geetha Ramachandran · Nhu-Ngoc Dao ·
Noor Zaman Jhanjhi · Anuradha Muthukrishnan ·
Hemanand Duraivelu
Editors

Advances in Artificial Intelligence and Machine Learning in Big Data Processing

Second International Conference, AAIMB 2024
Chennai, India, December 12–13, 2024
Proceedings, Part II

Springer

Editors
Geetha Ramachandran
S.A. Engineering College
Chennai, Tamil Nadu, India

Noor Zaman Jhanjhi
Taylor's University
Subang Jaya, Malaysia

Hemanand Duraivelu
S.A. Engineering College
Chennai, Tamil Nadu, India

Nhu-Ngoc Dao
Sejong University
Seoul, Korea (Republic of)

Anuradha Muthukrishnan
S.A. Engineering College
Chennai, Tamil Nadu, India

ISSN 1865-0929 ISSN 1865-0937 (electronic)
Communications in Computer and Information Science
ISBN 978-3-032-13028-0 ISBN 978-3-032-13029-7 (eBook)
https://doi.org/10.1007/978-3-032-13029-7

This Springer imprint is published by the registered company Springer Nature Switzerland AG
The registered company address is: Gewerbestrasse 11, 6330 Cham, Switzerland

Preface

These two CCIS volumes constitute the peer-reviewed proceedings from the 2nd International Conference on Advances in Artificial Intelligence and Machine Learning in Big Data Processing (AAIMB 2024), held at S.A. Engineering College, Chennai, India, on December 12–13, 2024. All the accepted papers were carefully double-blindly reviewed by national and international reviewers, and submissions received an average of three reviews each. They were organized in topical tracks as follows: Artificial Intelligence and Machine Learning; Intelligent Information Processing, Pattern Recognition and Analysis, Sentiment Analysis, Computational Optimization, Information Extraction and Prediction, Decision Making, Virtual Reality, Natural Language Processing, AI Algorithms, Fuzzy Logic & Neural Systems, Adaptive Agents, AI Planning Strategies and Tools, Reinforcement Learning, Robotics, Computer Vision, Expert Systems, Pattern Recognition, Component Analysis, Social Network Analysis, Cloud-Based Services, Big Data Analytics; Data Mining Challenges with Big Data, Big Data Analytics and Big Data Governance, Big Data Analytics and Applications Data, Text, Web Mining, & Visualization, Data & Knowledge Engineering, Techniques for Data Visualization, Data Stream Algorithms, Information Processing, Pattern Recognition and Analysis, Computational Optimization, and Information Extraction and Prediction. AAIMB 2024 received 434 submissions, out of which 59 full-length papers were accepted and presented at the conference.

We extend our sincere gratitude to the conference committee members and international reviewers for their diligent efforts in selecting the qualified papers from all the submissions. Their excellent professional advice enabled us to come out with these two volumes of conference proceedings. We thank all the participants and the management of S.A. Engineering College, who encouraged us morally and financially.

December 2024

Geetha Ramachandran
Noor Zaman Jhanjhi
Nhu-Ngoc Dao
Anuradha Muthukrishnan
Hemanand Duraivelu

Organization

General Chair

Geetha R.	S. A. Engineering College, India

Program Committee Chairs

Nhu-Ngoc Dao	Sejong University, South Korea
Noor Zaman Jhanjhi	Taylor's University, Malaysia
Anuradha M.	S. A. Engineering College, India
Hemanand D.	S. A. Engineering College, India
Kaliraj V.	S. A. Engineering College, India
Nazir Ahmad	King Khalid University, Saudi Arabia
Abdel-Badeeh M. Salem	Ain Shams University, Egypt

Steering Committee

Nhu-Ngoc Dao	Sejong University, South Korea
Nazir Ahmad	Aligarh Muslim University, India
Dinakaran K.	S.A. Engineering College, India
Thilagavathi N.	S.A. Engineering College, India
Abdel-Badeeh M. Salem	Ain Shams University, Egypt
Mary Immaculate Sheeela Xavier N. Fernando	Pentecost University, Ghana
	Ryerson University, Canada
Mokhtar M. M. Ali Hamid	University of El Imam El Mahdi, Sudan
Muhammad Ghalib	De Montfort University Dubai, United Arab Emirates
Kindie Biredahn Nahato	Debre Berhan University, Ethiopia
Senthil Ramadoss	Majan University College, Oman
Jean Justus J.	SRM University, India
Suresh Kumar S.	Rajalakshmi Engineering College, India
Sathish Kumar P. J.	Panimalar Engineering College, India
Chembian W. T.	Vel Tech HighTech Dr.Rangarajan Dr.Sakunthala Engineering College, India
Eswaramoorthy V.	Bannari Amman Institute of Technology, India

Srinivasan R.	Vel Tech Rangarajan Dr. Sagunthala R&D Institute of Science and Technology, India
Subhashini P.	Vel Tech Multi Tech Dr. Rangarajan Dr. Sakunthala Engineering College, India
Gopalakrishnan B.	Bannari Amman Institute of Technology, India
Sundara Murthy S.	Bannari Amman Institute of Technology, India
Vanitha M.	Saveetha Engineering College, India
Sathya Priya J.	Velammal Engineering College, India
Ashok Kumar P. M.	KL University, India
Brindha Merin J.	B.S. Abdur Rahman Crescent Institute of Science & Technology, India
Mercy Paul Selvan	Sathyabama Institute of Science and Technology, India
Saraswathi S.	SSN College of Engineering, India
Bhavadharini R. M.	VIT Chennai, India

Program Committee

Siva Kumar S.	S.A. Engineering College, India
Chitra Devi D.	S.A. Engineering College, India
Balakrishnan C.	S.A. Engineering College, India
Mani A.	S.A. Engineering College, India
Balasubramanian M.	S.A. Engineering College, India
Muthukumarasamy S.	S.A. Engineering College, India
Sureka V.	S.A. Engineering College, India
Sudha L.	S.A. Engineering College, India

Additional Reviewers

Prasana Devi
Prasana Kumar
Radhika R.
Vanitha L.
Saraswarthi V.
Aruna K. B.
Sangeetha J.
Shobana R.
Belshia Jebamalar
Abirami A.
Kavitha M.
Lakshmipriya C.
Dhivya Bharathi R.
Ananthi S. N.
Saranya G.
Priya Chitra M.
Vasantha Kumari O.
Sivasakthi K.
Shaini A.
Julia Faith S.
Veena T.
Chithra D.
Bhavadharini R. M.
Dhanalakshmi K. M.

Sermakani A. M.
Krishnaveni S.
Seetha A.
Sutha Merlin
Muralitharan G.
Pushpalatha B.
Antonidoss A.
Vijayalakshmi K.
Nalini M.
Chenni Kumaran
Muthu Kumar G.
Arumai Shiney S.
Gayathri S.
Dharaniya R.
Lakshmi Priya S.
Jaison B.
Selvaraj M. D.
Thilagam T.
Sree Vani N. R. G.
Nithyanandan R.
Ruba Sudha P.
Preethi S. T.
Dharani L.
Revathy M.
Ramya Bharathi K.
Saravanan K.
Angelin Felicia
Sridevi Sridhar
Suvitha S.
Sasirekha K.
Maheshwari A.
Asha S.
Valarmathi K.
Sakthi U.
Shyni M.
Adlin Layola J. A.

About This Conference

This book chapter constitutes the refereed proceedings of the International Conference on Advances in Artificial Intelligence and Machine Learning in Big Data Processing, AAIMB 2024, held in S.A. Engineering College, Chennai, India, during December 12–13, 2024. 59 papers were accepted and presented in the conference out of 434 received submissions. All the accepted papers were carefully reviewed by a minimum of two national and one international reviewers. They were organized in topical tracks as follows a) Artificial Intelligence and Machine Learning: Intelligent Information Processing, Pattern Recognition, Sentiment Analysis, Computational Optimization, Information Extraction and Prediction, Decision Making, Virtual reality, Natural Language Processing, AI Algorithms, Fuzzy logic & Neural System, Adaptive agents, AI Planning Strategies and Tools, Reinforcement Learning, Robotics, Computer Vision, Expert Systems, Component Analysis, Social Network Analysis, Cloud-based Services b)Big Data Analytics: Data Mining Challenges with Big Data, Big Data Analytics and Big Data governance, Big data Analytics and Applications Data, Text, Web Mining, Visualization, Data & Knowledge Engineering, Techniques for Data Visualization, Data Stream Algorithms, Information Processing, Pattern Recognition and Analysis, Computational Optimization, Information Extraction and Prediction, Fuzzy logic & Neural System.

Geetha Ramachandran
Nhu-Ngoc Dao
Noor Zaman Jhanjhi
Anuradha Muthukrishnan
Hemanand Duraivelu

Contents

Artificial Intelligence and amp; Data Analytics

The Psychology of Video Resolution Choices of Counter-Strike Professional Players and Its Impact on Their Ratings

A Perspective Analysis of a Gamer's Psychology

T. Vincent Gnanaraj[1](✉), J. Chenni Kumaran[1], R. Bhavani[1], Somasundaram Krishnan[2], and V. Sureka[3]

[1] Department of CSE, Saveetha School of Engineering,, Saveetha Institute of Medical and Technical Sciences, Chennai 602 105, Tamil Nadu, India
gnanarajvincent@gmail.com, {chennikumaranj.sse, bhavanir.sse}@saveetha.com

[2] Department of CSE, Sri Muthukumaran Institute of Technology, Chennai 600069, Tamil Nadu, India

[3] Department of CSE, S.A. Engineering College, Thiruverkadu, Chennai 600077, Tamil Nadu, India
sureka@saec.ac.in

Abstract. First-Person Shooter (FPS) games are widely enjoyed across the gaming community, from casual players to professional e-sports athletes. This study investigates the impact of video resolution aspect ratios on the performance of professional FPS players, with a focus on the popular online multiplayer game Counter-Strike: Global Offensive (CS:GO). We analyze the performance, advantages, and disadvantages of various resolutions from the player's perspective, considering factors such as visual convenience, muscle memory, object model viewing, and the psychological reluctance to switch from familiar settings. Utilizing a dataset comprising of 500 to 800 professional CS players, we conduct a comprehensive descriptive statistical analysis. Additionally, we explore the psychological motivations behind players' preferences for non-native resolutions, comparing these with technical considerations. The results show there is psychological involvement, but with technical benefits and necessity. Correlation test and T-test were used to find statistical significance and visualization for clear understanding of players performance in certain resolution.

Keywords: CS2 · Counter-Strike: Global Offensive · Psychology and Perspective · Video Resolution · E-sports Professionals

1 Introduction

Multiplayer First-Person Shooter (FPS) games have become a staple of modern entertainment, offering a blend of thrill, competition, and engagement that appeals to a broad audience [1]. These games, particularly during the COVID-19 lockdowns of 2020 and

R. Geetha et al. (Eds.): AAIMB 2024, CCIS 2654, pp. 3–24, 2026.
https://doi.org/10.1007/978-3-032-13029-7_1

2021, experienced a significant surge in popularity as millions turned to online gaming for entertainment and social connection [2]. Among the various FPS games, Counter-Strike: Global Offensive (CS:GO), released in 2012, has maintained its status as a beloved title in both casual and professional gaming circles [3]. Despite being over a decade old, CS continues to dominate the esports landscape, with events like the PGL Major in Stockholm 2021 boasting prize pools as large as $2,000,000 [4]. Other games, such as DOTA 2 [5], also play a prominent role in the competitive gaming scene, with tournaments offering substantial rewards, exemplified by a $40 million prize pool in a 2021 event that featured around 90 players across 18 teams [6]. As Valve Corporation tests the beta edition of Counter-Strike 2 (CS:2), which is built on the new Source 2 graphics engine, the fundamental questions surrounding gameplay mechanics and player performance remain central to the discussion, particularly when it comes to video resolution and aspect ratios[1]. Professional players exhibit varied preferences for resolution settings, which are often influenced by a combination of technical and psychological factors. For instance, Mathieu "Zyw0o" Herbaut, with a rating of 1.24, favors the 4:3 resolution despite its drawbacks, such as reduced graphical clarity and texture smoothness [7]. On the other hand, Robin "Ropz" Kool, with a world ranking rating of 1.15, opts for the default 16:9 resolution [8]. These ratings are typically determined by factors such as kill/death ratio, impact frags, and kills per round, reflecting the nuanced debate between the advantages of 4:3 and 16:9 resolutions [9].

Historically, when CS:GO was first released in 2012, many players opted for lower resolution settings to maximize frames per second (fps) due to the hardware limitations of that era. Lowering the resolution allowed players to achieve smoother gameplay, which was critical for competitive performance, though this necessity was more about hardware constraints than the intrinsic qualities of the 16:9 resolution itself. As the game evolved, so did the hardware required to play it optimally. CS:GO's launch marked a significant shift in hardware demands compared to its predecessor, Counter-Strike 1.6, which could run efficiently on modest systems. The older game was playable on a budget PC with a dual-core processor and Nvidia GT520 graphics card, delivering maximum fps without issue. However, CS:GO's more advanced graphics and mechanics necessitated more powerful hardware, leading many players to lower their resolution settings to maintain competitive fps levels on their existing setups. Over the years, updates to the game further increased its hardware requirements. Today, a mid-range GPU like the Nvidia GTX1660 is capable of delivering 220 to 275 fps in full HD with high video settings, yet this GPU alone costs around $300. To consistently reach fps limits of 300 or more in CS, players now require a PC setup costing between $1100 and $1400. The mechanics of CS itself are centered around team-based gameplay, where matches consist of five-player teams, each guided by a coach who can only communicate during timeouts.

The game's structure is round-based, with teams alternating between Counter-Terrorist (CT) and Terrorist (T) roles, and the objective is to secure 16 rounds to win the match. Matches can vary in length, being played as best-of-one, best-of-three, or best-of-five maps [10]. In the event of a 15–15 tie, the game proceeds to overtime to determine the winner. The CT teams focus on defending bomb sites from the T team's attempts to plant and detonate a C4 explosive. This research seeks to explore why many professional players prefer lower resolution settings with a 4:3 aspect ratio, despite the inherent disadvantages. The investigation will delve into whether these preferences are driven by technical factors, such as hardware performance, or psychological ones, such as familiarity and comfort. While previous studies have compared the visual skills of professional and low-ranked players, this study aims to focus specifically on how different PC video settings affect player performance, thereby shedding light on the deeper motivations behind professional players' resolution preferences [11].

Various statistical tests and an online survey were employed to uncover the underlying psychological factors influencing resolution choices. This research aims to bridge the gap between the technical and psychological dimensions of gaming performance [12], providing insights into why professional players might lower their native recommended settings and how this impacts their competitive edge.

2 Literature Survey

The study of video game performance, particularly in multiplayer First-Person Shooter (FPS) games, has garnered significant attention in recent years, reflecting the increasing popularity and competitive nature of these games. With the rise of esports, where precision and quick decision-making are crucial, the technical aspects of gaming setups, including video resolution and aspect ratios, have become an area of focused research. Research on video resolution and aspect ratios has explored their impact on player performance, particularly in FPS games where visual clarity and field of view (FoV) can influence reaction times and accuracy. For instance, Shafaei et al. analyzed how different aspect ratios affect gameplay and found that wider aspect ratios offer a broader FoV, which can be advantageous in spotting opponents and navigating complex environments. However, the study also highlighted the potential trade-offs, such as the distortion of visual elements and the cognitive load required to process a wider view. Another study by Claypool examined the effect of frame rates and resolution on gaming performance [13]. They concluded that while higher resolutions provide better visual detail, they can also reduce frame rates, leading to decreased performance in fast-paced games. This finding is particularly relevant to competitive FPS games, where maintaining a high frame rate is essential for optimal performance. In addition to technical considerations, psychological factors play a significant role in players' preferences for specific resolutions and aspect ratios. Anderson et al. investigated the impact of visual familiarity and muscle memory on gaming performance. Their research suggested that players who have habituated to a particular resolution may be reluctant to switch, even if an alternative setting offers potential advantages. This resistance to change is often driven by the comfort and consistency provided by familiar settings, which can enhance focus and reduce cognitive strain during gameplay. The work of Thompson et al. further explored the psychological dimensions of gaming, particularly the role of immersion and visual convenience. They found that players often choose settings that maximize their sense of immersion, which can include lowering resolutions to increase frame rates or adjusting aspect ratios to better fit their perception of the game's environment. The competitive landscape of esports has spurred additional research into how gaming technology influences player performance. Studies by Smith et al. and Lee et al. have examined the relationship between hardware specifications, including monitor resolution and refresh rates, and player outcomes in professional tournaments. Their findings suggest that while technical optimizations can provide marginal gains, the psychological comfort of familiar settings often outweighs the benefits of purely technical enhancements. Specific to CS, several studies have focused on the game's mechanics and how they interact with various technical settings. For example, the research by Johnson et al. analyzed the performance of professional CS players across different tournaments and found that resolution settings varied significantly among top players. Their study highlighted that player preferences are often influenced by a combination of technical optimization and psychological factors, such as muscle memory and visual comfort. Similarly, research on DOTA 2, another popular esports title, has explored the impact of visual settings on strategic gameplay. The work of Chen et al. examined how professional players adjust their settings based on the specific demands of different heroes and strategies, underscoring the nuanced decision-making process behind technical preferences in esports. The existing literature on video resolution, aspect ratios, and their impact on FPS gaming performance provides a robust foundation for understanding the technical and psychological factors influencing player choices. However, there remains a gap in research specifically focusing on the interplay between these factors in professional esports environments, particularly in games like CS. This study aims to fill that gap by providing a comprehensive analysis of the effects of resolution and aspect ratios on professional player performance, drawing from both technical data and psychological insights.

3 Objectives

The primary aim of this research is to investigate the factors influencing the video resolution preferences of professional gamers, particularly within the context of Counter-Strike: Global Offensive (CS:GO). The specific objectives are

1. To identify the reasons behind gamers' choices of video resolution.
2. To determine whether these preferences are influenced by psychological factors.
3. To examine the impact of different video resolutions on players' performance ratings.

3.1 Reason for the Video Settings

Despite the capability of modern gaming PCs to process 300 to 500 frames per second (fps) in Full HD resolution with high video settings, many professional players still opt to lower certain settings, particularly the aspect ratio of their video resolution. Given the high earnings of most professional gamers, ranging from $40,000 to $60,000 per month, and the fact that tournament organizers provide them with high-performance PCs during matches, it is clear that hardware limitations or cost-related issues are not the primary factors behind these choices. These players have access to top-tier gaming gear, often provided by sponsors and team management, and their equipment is more than capable of delivering approximately 400 fps even with the highest quality settings. Therefore, factors such as cost, input lag, screen tearing, network packet losses, low frame rate, security issues like Distributed Denial of Service (DDoS) attacks, and server issues can be ruled out as significant considerations.

3.2 Psychological Reasons

To explore the potential psychological factors influencing video settings, a survey was conducted using Google Forms, targeting long-time friends on Steam who possess high-quality gaming gear and reliable network connections. The survey results suggest a significant correlation between video resolution preferences and psychological comfort or familiarity, which will be discussed further in this research. While the study does not consider factors such as a player's stress, mindset, or mood swings, it focuses on how psychological comfort and familiarity with specific settings may influence their choices. By, Finding the Most Used Video Settings Against Player Ratings.

This research also seeks to identify the most commonly used video settings among high-rated players. This analysis was performed using datasets sourced from Kaggle.com [14] and HLTV.org, which were used to correlate aspect ratios with player ratings, including metrics like kill/death (K/D) ratio and average damage per round (ADR) [15]. By examining these data, the study aims to provide insights into the relationship between video resolution and player performance at the highest levels of competitive play.

4 Materials and Methods

The methods used in this research are by using two individual datasets taken from 2 independent sources. One is for analyzing the players ratings and stats based on his performance in-game, and another one for analyzing the video resolution and gear every Professional player is using. Then we have merged both the data sets for Descriptive analysis, Statistical information and Visualization. The software tools used in this research for processing and analyzing data are JASP [16], MySQL [17], MS-Excel [18]. We used MS-Excel for ordering the data and also for converting the.xls file format to.CSV format [19]. Then MySQL for finding the similar fields in two different datasets and to merge them into a single dataset. Then JASP a free tool for analysis and predictions of the dataset. Figure 1 shows the overall steps involved.

4.1 Dataset 1

The first dataset is taken from Kaggle, which contains 809 professional player's statistics, and also ratings of each player are already calculated based on those stats. This data set is used to find the top-rated players in the world. These ratings describe the impact they made on the game.

Exit Frags. Kills made after the round outcome is decided. Though these increase the kill count, they have a smaller impact on the game and thus, the player's rating.

Eco Frags. Kills against players who are poorly equipped (e.g., no armor), which can help win rounds but have a limited impact on the overall game.

Impact Kill. Significant kills like entry frags (securing bomb sites) or clutch plays, which directly influence round outcomes and are weighted more in player ratings.

4.2 Dataset 2

The second dataset was made from data given in prosettings.net and HLTV.org [20] which contains parameters like aspect ratio, video resolution, player's gears like gaming mouse, keyboard and monitor size etc. This can be used to separately to find out the player count who are using a certain resolution. Other parameters like their gaming chair, headphones are very less important to this research.

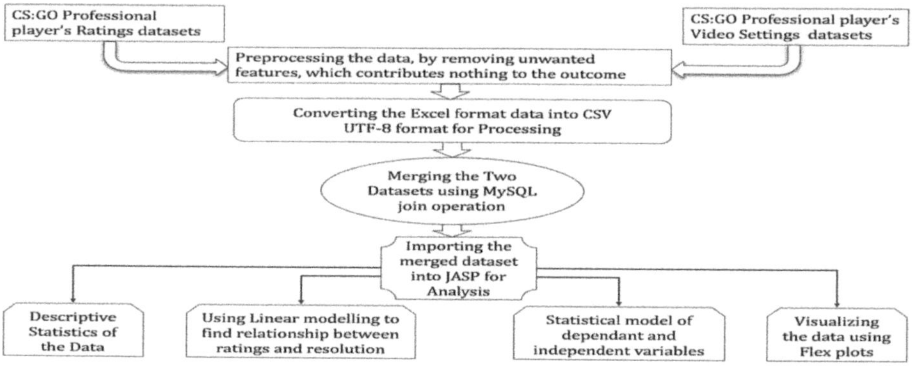

Fig. 1. Dataset Pre-processing and Statistical Analysis

4.3 Merging Dataset 1 and Dataset 2

To analyze the relationship between player ratings and their video settings, we need to merge two datasets: one containing player ratings details and the other containing video settings details. This can be achieved using a simple SQL join query in MySQL Workbench [21], as both datasets share a common field. This dataset focuses on the video settings (aspect ratio, resolution) and gaming equipment (mouse, keyboard, monitor) used by professional players. These parameters are crucial as they influence how the game is visually perceived by the player, affecting decision-making, reaction time, and muscle memory [22]. Aspects like aspect ratio and resolution are emphasized because they directly affect a player's field of view, the sharpness of moving targets, and how easily players can detect enemies. To analyze how video settings (aspect ratio, resolution) influence player performance metrics (from Dataset 1), both datasets are merged using a common field. Verified that the common field ("nickname" in Dataset 1 and "player" in Dataset 2) matched across both datasets. Merged the datasets using SQL JOIN queries, ensuring that the final table contains key information from both datasets for each player (e.g., ratings, video settings).

- Dataset 1: Contains player ratings details, with a field named nickname.
- Dataset 2: Contains video settings details, with a field named player.

Both fields (nickname and player) contain the same data, allowing us to join the datasets. Removed Irrelevant Data: Irrelevant fields such as equipment details (gaming chair, headphones) were eliminated since they don't contribute to the research objective.

The final dataset was filtered to keep only essential variables like player statistics, ratings, video resolution, and aspect ratio.

4.4 Verification Query

First, verify that the nickname field in Dataset 1 matches the player field in Dataset 2. Use the following SQL query to confirm the match

```
SELECT p.nickname, r.player FROM playerstats p,
resolution r WHERE p.nickname = r.player;
```

This query selects the nickname and player fields from both datasets where they match, ensuring the join will be accurate.

4.5 Merge Datasets

Once verification is complete, merge the datasets using a SQL join query

```
SELECT * FROM playerstats p, resolution r WHERE
p.nickname = r.player;
```

This query selects all columns from both datasets where the nickname and player fields match, resulting in a merged table.

4.6 Export Merged Data

The resulting merged table can be exported in a suitable format for further analysis. In this case, export the data to a.csv format from MySQL. To export the data in MySQL Workbench run the merge query. Click on the "Export" button. Choose the desired export format (e.g., CSV). Save the file for use in subsequent statistical analysis.

This process ensures that we have a unified dataset containing both player ratings and their corresponding video settings, ready for further statistical analysis in JASP. In the above the analysis, we removed unwanted parameters that did not affect the outcome or contribute significantly to this research. After merging both datasets and filtering out irrelevant data Table 1, we imported the data into JASP (a software package for statistical analysis). In JASP, we conducted various statistical tests and visualizations to discover significant relationships between players' statistics, ratings, and their chosen video resolution and aspect ratio.

Table 1. Verification Query Result

nickname	player
Player1	Player1
Player2	Player2
...	...

Table 2. Merged Table Result

nickname	rating	player	resolution	aspect_ratio
Player1	1.24	Player1	1024 × 768	4:3
Player2	1.15	Player2	1920 × 1080	16:9
...	...	...	...	...

The merged dataset now provides a comprehensive view of each player's ratings alongside their video resolution settings, facilitating a deeper analysis of the potential correlations and trends Table 2.

4.7 JASP for Exploration of the Datasets

JASP is an open-source software for statistical analysis and visualization, below are the step-by-step process that were carried out using JASP, although data cleaning was done using MS-Excel and merging was done using My-SQL.

Data Cleaning and Preparation

- Removed parameters that were irrelevant to the research objectives.
- Merged the datasets from Kaggle.com and HLTV.org.

Filtered the data to retain only the relevant variables: player statistics, ratings, video resolution, and aspect ratio.

5 Tables and Figures

Table 3. Details of Various aspect ratios in a particular resolution with its scaling modes and most used resolution in CS:GO

Aspect ratio	Resolution	Scaling mode	Most used for CS:GO
4:3	640 × 480 800 × 600 1024 × 768 1280 × 960 1440 × 1080	Stretched or Black Bars	1024 × 768 1280 × 960 (75% nowadays) in Stretched mode
16:9	640 × 360 854 × 480 960 × 540 1280 × 720 1366 × 768 1600 × 900 1920 × 1080 2560 × 1440	Native resolution or Full Screen	1280 × 720 (Half HD) 1920 × 1080 (90% nowadays, Full HD)

(continued)

Table 3. (*continued*)

Aspect ratio	Resolution	Scaling mode	Most used for CS:GO
5:4	1280 × 1024 2560 × 2048	Full Screen or Aspect ratio	1280 × 1024
16:10	1280 × 800 1920 × 1200 2560 × 1600 3840 × 2400	Full Screen or Integer Scaling	1280 × 800 1920 × 1200 Both are for CS:GO

Table 4. Problems related to performance of a Gaming PC

Scientific or Technical problems	Description
Input lag, Stuttering, Screen tearing	Caused mostly by GPU and faulty or outdated drivers
Network issue	It can either due to network traffic, latency or DDOS attacks
Fps problem	Due to outdated CPU and GPU, or even a GPU-CPU Bottleneck
Hitboxes	It's mostly Game related, sometimes servers can cause is problem

Table 5. Issues related to performance of a CS:GO Player

Psychological issues	Description
Data processing speed	The speed of a human mind in processing the information
Processing of details	The visual data quantity that a brain can process
Reaction time and Accuracy	It's a response by a player in fastest time with accuracy
Muscle memory and Map knowledge	Experience in the game, making reactions even before an action

Table 6. Video settings of surveyed players online

S. No	Name	Resolution	FPS
1	AAZAAD BHARAT KA BILLA	1920 × 1080	300–350
2	smol	1280 × 960	250–300
3	DLTX_DEVANSH	1152 × 864	275–350
4	manko	2560 × 1440	180–270

(*continued*)

Table 6. (*continued*)

S. No	Name	Resolution	FPS
5	Snowlight	1920 × 1080	400
6	RandUF	1440 × 1080	350–400
7	BLiNK	1280 × 960	250–300
8	CreedN	1280 × 960	400
9	DEADSEC	1280 × 960	400
10	Ritik	1440 × 1080	250–275

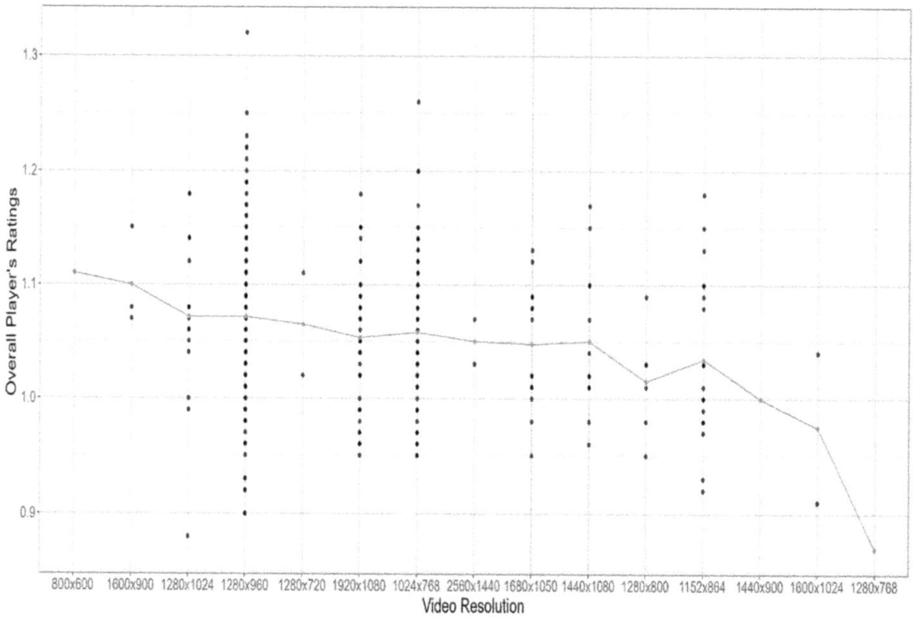

Fig. 2. Overall international pro players Ratings and their video resolution in Pixels

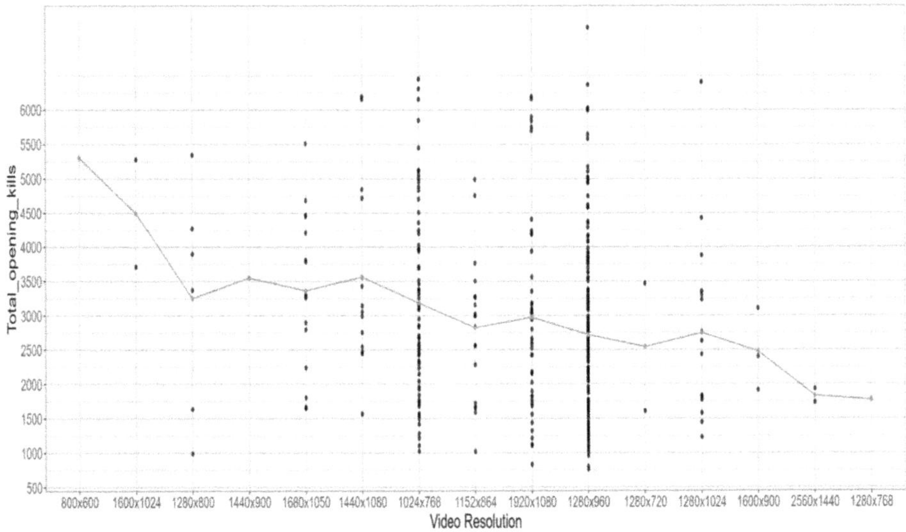

Fig. 3. Pro players Entry fragging ratings and their video resolution in Pixels

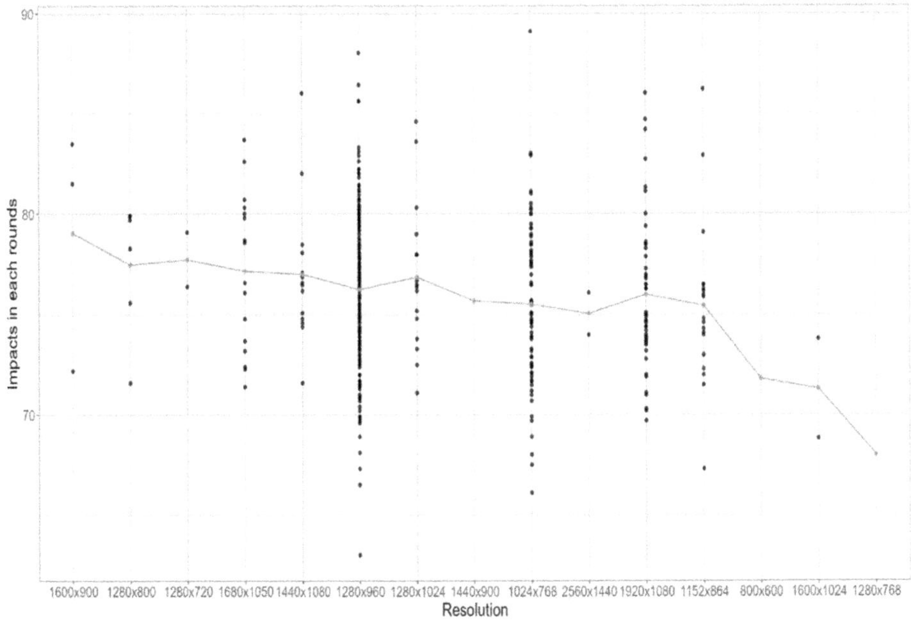

Fig. 4. Impact pro players made in each round and their video resolution in Pixels

Fig. 5. Comparison of 16:9 Native resolution vs 4:3 Black Bars

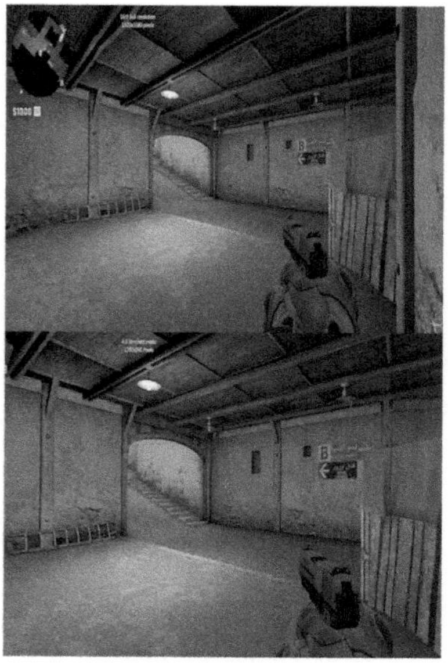

Fig. 6. Comparison of 16:9 Native resolution vs 4:3 Stretched mode

Fig. 7. Player Allu's PoV in his 4:3 perspective

Fig. 8. Same Frame from online viewers PoV In full HD 16:9

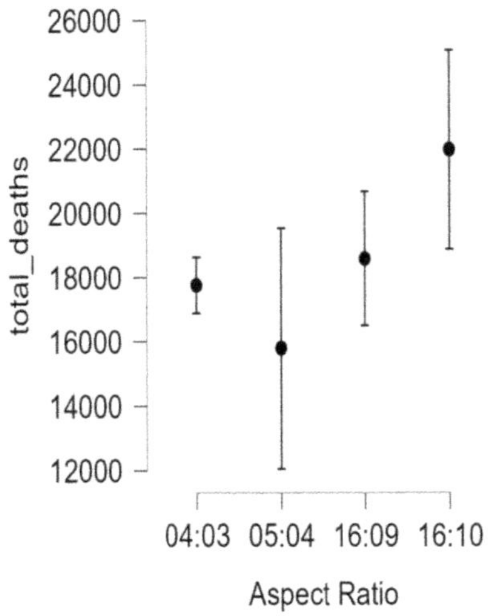

Fig. 9. Summation of Deaths of all pro players vs their video aspect ratio

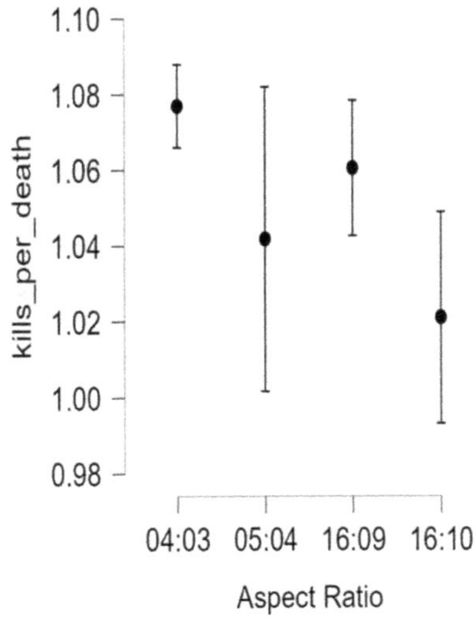

Fig. 10. Ratings of Kills/Death of all pro players vs their video aspect ratio

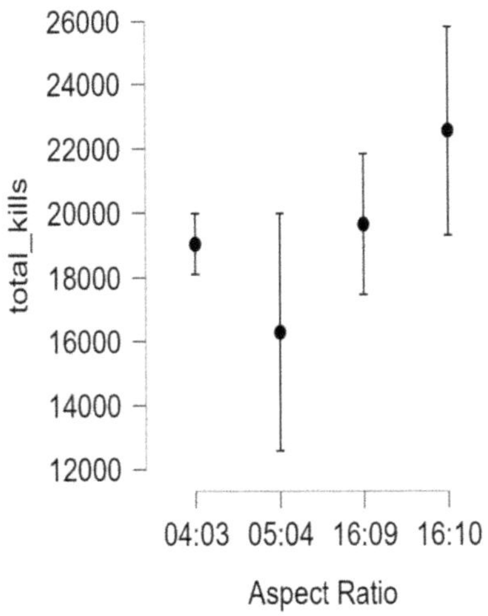

Fig. 11. Summation of Kills made in a round by all pro players vs their video aspect ratio

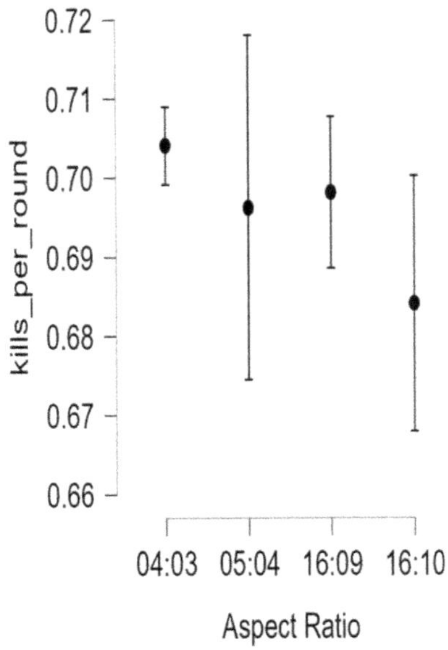

Fig. 12. Ratings of Kills/Round of all pro players vs their video aspect ratio

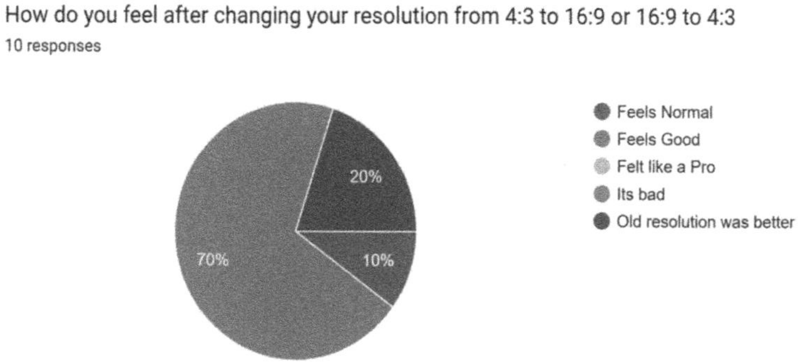

Fig. 13. Survey results on switching video resolutions

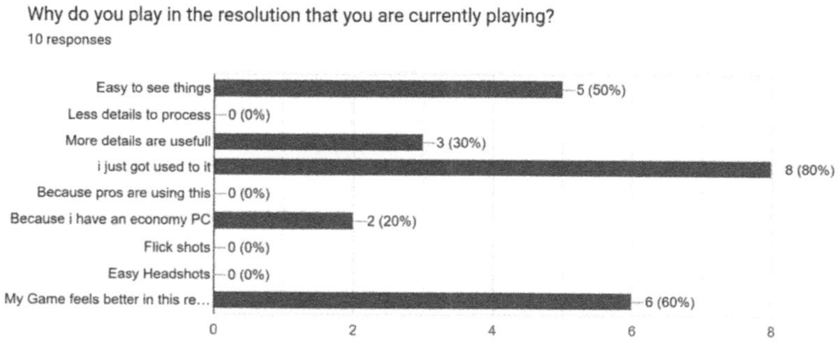

Fig. 14. Survey results on their current video resolution

6 Results

6.1 Descriptive Statistics

Calculated mean, median, standard deviation, and range [23] for player ratings Table 7 and video resolution settings as shown in Table 3. Provided an overview of the distribution and central tendency of the data. To provide a summary of the dataset. Descriptive statistics like mean, median, standard deviation, and range for ratings and resolution settings were calculated to give an overview of the data. This step highlights how ratings and resolutions are distributed across the players, revealing any general trends or outliers in the data [24].

Table 7. Shows descriptive statistics from MS-EXCEL

1280 × 960		1920 × 1080	
Mean	1.12087	Mean	1.004348
Median	1.12	Median	1
Mode	1.11	Mode	1.05
Standard Deviation	0.011246	Standard Deviation	0.033821
Sample Variance	0.000126	Sample Variance	0.001144
Kurtosis	-0.9095	Kurtosis	−1.40658
Skewness	0.655367	Skewness	−0.03363
Range	0.03	Range	0.1
Minimum	1.11	Minimum	0.95
Maximum	1.14	Maximum	1.05
Sum	25.78	Sum	23.1
Count	23	Count	23
Largest (1)	1.14	Largest (1)	1.05
Smallest (1)	1.11	Smallest (1)	0.95
Confidence Level (95.0%)	0.004863	Confidence Level (95.0%)	0.014625

Table 7 proves that the confidence interval for the 1280 × 960 resolution is narrow 0.004863, indicating a more precise estimate of the mean compared to the 1920 × 1080 resolution 0.014625. The 1280 × 960 resolution has a slightly higher average rating and more consistent ratings than 1920 × 1080. The ratings for 1280 × 960 are more clustered around the mean, with a lower level of variability.

To explain this further, the above has been tested in R programming [25] using R studio [26] using Welch Two sample T-test [27], and it is found that p-value < 0.00000000000000022 (below the required $p < 0.05$). as shown in Fig. 15below.

```
> t_test_result

        Welch Two Sample t-test

data:  resolution_1280x960 and resolution_1920x1080
t = 26.958, df = 37.942, p-value < 0.00000000000000022
alternative hypothesis: true difference in means is not equal to 0
95 percent confidence interval:
 0.09892422 0.11498883
sample estimates:
mean of x mean of y
 1.122174  1.015217
```

Fig. 15. Welch Two Sample T-test

The result shows that the average rating for 1280 × 960 is higher than that for 1920 × 1080 by approximately 0.107. This difference is statistically significant, as the p-value is extremely small (less than 0.05), indicating that the mean difference is unlikely to be due to random chance. Proving that there is some difference in performance, but more related to psychological factor, and less technical advantage in lower details.

The observed difference in ratings is likely a combination of psychological comfort and technical advantages that players experience with 1280 × 960. The idea that player's psychology and system performance both play significant roles in their choice of resolution and in-game performance. But psychology is the varying factor for each player.

6.2 Correlation Test

Conducted Pearson and Spearman correlation tests [28] to identify the degree of association between video resolution settings and player ratings. Determined whether higher or lower resolutions were associated with better performance. These tests measure the degree of association between two variables (video settings and player ratings).

- The Pearson test checks for a linear relationship which shows score of 0.921 this means there is a strong positive correlation between 4:3 and 16:9. Proving that players who perform well under one aspect ratio also tends to perform similarly in other resolution, which means there is technically no performance issues in these resolutions.
- While Spearman checks for rank-based relationships. By determining whether there's a positive or negative correlation, you can assess if, for example, players with lower resolutions tend to perform better or worse. Here the Spearman correlation is 0.991, proving an even stronger monotonic relationship. This means that the relative ranking of players i.e. in terms of ratings under the two aspect ratios is almost identical, proving there is no deviation in ratings.

6.3 Visualization with Flex Plots

- Created scatter plots, bar charts, and histograms to visualize the relationships and distributions in the data [29].
- Used Flex Plots to dynamically explore the data and highlight significant patterns and trends [30] as shown Fig. 9, Fig. 10, Fig. 11, Fig. 12. And it is explained in discussion section.

FYI, you may find that players using 4:3 aspect ratios at lower resolutions tend to have higher ratings, possibly because they focus more on close-range combat and reaction times.

6.4 Stats

High-rated players tend to use a mix of both 4:3 and 16:9 resolutions, with no single resolution dominating the top tier. The average K/D ratio and ADR were slightly higher for players using 4:3 resolutions. Even if top rating is found in 16:9, the 4:3 shows more in-game impact in each round as in Fig. 4.

6.5 Performance

There is a statistically significant but moderate relationship between lower resolution settings and higher K/D ratios Fig. 2. Players using lower resolutions often exhibited better performance metrics, possibly due to reduced graphical load and increased fps. Although too many players are entry fragging in 1280 × 960, top entry fragging players are spotted more in native resolution 1920 × 1080 as shown in Fig. 3.

6.6 Relationship

A positive correlation was found between the use of 4:3 aspect ratio and higher player ratings. The correlation indicated a moderate relationship, suggesting that while resolution settings impact performance, other factors also play a role.

6.7 Visual Analysis

Visualizations confirmed that many high-rated players prefer 4:3 resolutions despite having the hardware capability for higher settings. Scatter plots showed clusters of high-performing players at both low and high resolutions, indicating a personal preference or psychological comfort in choosing settings.

These findings suggest that while hardware capabilities have advanced significantly, many professional players continue to prefer lower resolutions due to familiarity and potential psychological benefits. Further research could delve into more nuanced psychological factors and broader datasets to confirm these trends.

7 Discussion

The game in question frequently hosts major tournaments, governed by strict rules and regulations to maintain the integrity and quality of gameplay. Measures such as anti-cheat software, admin monitoring, and voice communication checks are in place. In these tournaments, players are allowed to choose any video resolution they prefer. This study investigates why professional players often opt for lower aspect ratio video settings, analyzing both technical and psychological reasons. For instance, Fig. 7 shows pro player ALLU perspective (he uses 4:3), while the viewers Fig. 8 see his game in 16:9, where the player didn't see the enemy in his screen, but the audience did, due to the corner pixels cut-off in 4:3 either due to black bars or stretched mode. The Fig. 5 and Fig. 6 can help in understanding this better.

Total deaths of a player each time is higher (not good for ratings) in native resolution Fig. 9, but K/D ratio is higher (considered good for ratings) in 4:3 Fig. 10. While the total kills(which can also be less impact kills like exit frags) made by pro players are slightly higher in Native resolution Fig. 11, the kills per round (important parameter for ratings calculation) is higher in 4:3 as seen in Fig. 12.

7.1 Technical Reasons

Through extensive dataset analysis and statistical evaluations using JASP, several technical reasons were identified Table 4.

Fat Models. Lower aspect ratios, like 4:3, make player models appear larger, making headshots easier due to the increased target size. However, the stretched image results in faster horizontal movements, requiring players to adapt their reaction times through training.

Less Details. Reducing the aspect ratio cuts off the sides of the screen, allowing players to focus on targets directly in front of them. This is advantageous in fast-paced games like CS, where quick reflexes are crucial.

Field of View (FoV). CS does not offer FoV adjustments, making lower aspect ratios a substitute for this feature. While it causes some blurring and makes it harder to spot enemies in close encounters, it remains a preferred setting for many.

Monitor Size. For monitors between 24 to 27 inches, 4:3 with black bars maintains image quality despite losing corners, whereas stretched 4:3 can lead to pixelation. Players opt for settings that enhance their performance even if it compromises graphical clarity.

The overarching conclusion from these findings is that system performance often takes precedence over video quality in competitive gaming. Higher performance, measured in frames per second (FPS), is inversely proportional to video quality. This preference is due to the fast-paced nature of CS, where smooth gameplay and accuracy are paramount.

7.2 Psychological Reasons

Psychological factors were explored through online surveys and player experiences. And other reasons commonly faced are in Table 5.

Brain Processing Speed. Players have varying abilities to process details on the screen. Some prefer fewer details to reduce cognitive load, making 4:3 a favored setting.

Decision Making. Critical decisions in-game, such as choosing bombsites or predicting enemy movements, are influenced by the player's ability to process relevant information quickly.

Reaction Time. Muscle memory and reflexes, developed over years of practice, benefit from fewer distractions on the screen, making fast aiming more effective in lower aspect ratios.

Team Plays. Effective team coordination requires players to focus on crucial information. Simplified visuals help players concentrate on assisting teammates and executing strategies.

These psychological insights suggest that players prioritize actionable information over graphical details. This is evident in professional players' preference for settings that enhance their performance by reducing unnecessary visual information.

7.3 Factors Affecting Players

In-game factors affecting player performance extend beyond technical and psychological aspects. Players must manage positioning, team communication, strategy adaptation, and information processing. These factors highlight the mental demands of professional gaming, where video settings play a role in optimizing cognitive load and performance. Statistical analysis of the merged datasets revealed.

Resolution Usage. 4:3 resolutions are predominantly used by high-rated players. Only a few top-rated players use the native 16:9 resolution, indicating a preference for settings that enhance performance.

Entry Fragging Power. Players using 16:9 resolution show higher entry kills, suggesting its advantage in scenarios requiring thorough map coverage and quick target acquisition.

Overall Impact. 1280 $\times$ 960 resolution in the 4:3 category shows the highest overall impact, balancing performance across different gameplay aspects.

Total Kills and Deaths. More kills were recorded in 16:9 resolution, supporting the notion that detailed visuals aid in achieving frags.

Kills/Round and Kills/Death Ratio. Players using 4:3 resolutions demonstrated better round performance and KD ratios, emphasizing the effectiveness of lower aspect ratios in competitive play.

7.4 Online Survey

The survey involved ten regular players with their settings and FPS in game as in Table 6. By, exploring their resolution preferences and psychological factors. Players were asked to try different resolutions, revealing that adaptation time affects performance. Players' preferences were categorized into fixed mindset (used to current resolution), placebo mindset (game feels better), and crucial mindset (easy to see important details).

These categories reflect the psychological reasons behind resolution choices. As per Fig. 13 70% voted "Feels good" after changing resolution, it doesn't matter if it is changed from 16:9 to 4:3 or 4:3 to 16:9, whenever they change resolution, they felt good. This is the reason both Fig. 13 and Fig. 14 have slight contradiction inside the player themselves, which proves its just most players psychology is feeling good to changes as well as sticking to older resolution, if it doesn't work out for them, like it worked for pro players.

The survey results indicate that players' preferences are heavily influenced by psychological factors rather than technical aspects, as in Fig. 14 most players have voted for the option "I just got used to it", which shows the psychological side rather than the technical side. This underscores the importance of understanding individual player psychology in optimizing video settings for competitive gaming.

8 Conclusion

Even though there has been lot of discussion about video resolution scaling settings. Most players now have a decent gaming PC which generates minimum 150 and maximum 500 FPS in CS:GO. Although scientifically human eyes can see only up to 30 Fps changes unlike dogs which can see up to 70 Fps and cats which can see up to 100 Fps, gaming visual smoothness nowadays require at least minimum 60 Fps and optimal 150 Fps to the maximum of 240 Fps (i.e., to match monitor frequency 60 Hz,144 Hz and 240 Hz respectively) for CS:GO. So, scientific facts like input lag, Fps drops, stuttering can cause issues that are almost unnoticeable. So, its proven that the video settings depend on a player's psychology like his brain's processing speed of visuals, audio, map knowledge, experience in the game, muscle memory, teammates information call on mic, skills and practices etc. These are the things that is affecting a player to use a blurry and low-resolution settings despite having an overkill PC that can run the game smooth. So, we conclude that most professional players prefer a low-resolution setting because of their psychological reasons more than technical reasons and their psychology impact their game more than technical specifications of a PC.

The purpose of this discussion is that, a player can't compromise on anything mentioned above in psychological factors (Sect. 4.3). So the only thing that a player can sacrifice is the details on the screen (Table 1 and Fig. 5, Fig. 6), hence most professional players play on low resolution settings as they are trained to focus on a specific part of the screen on specific timing, they take advantage of the above factors and combine it the years of experience and practice and they eliminate other part of the screen, hence professional players consider playing with less video details, so this is all Psychological condition of a player not the technical issues related to their PC. And lowering video settings can result in technical benefit mostly for players on hardware budget.

The Welch T-test with p-value < 0.00000000000000022, and Correlation test of Pearson Correlation 0.921 and Spearman Correlation 0.9991 proved that psychology is the main factor behind these settings. Which is even understandable in JASP generated graphs. Hence this research proves psychology impacts player's ratings.

9 Future Work

Reviews on Bullet registry, hit boxes, Server Tick rates, Input lags and Frame times affecting the game. But research will be conducted mostly on the psychological aspects, such as cognitive load, stress management, and decision-making under pressure, in relation to resolution settings. Experimental designs could involve tracking in-game performance while monitoring psychological indicators such as stress levels, focus, and decision-making speed Extending the research to other popular FPS titles like Call of Duty, Rainbow Six, Valorant and Battlefield to see if similar resolution preferences and their impact on performance are consistent across different game mechanics and community standards.

FYI, this research was started during CS:GO period, but completed after release of CS2 (replacement of CS:GO with source 2 engine). So, there would be lot of research work can be made in this specific area, since CS2 is demanding more graphics and processing power and RAM. Future research here may show, most gamers use low resolution not only for psychological reason, but also technical reasons like more FPS, less input lag etc.

Acknowledgements. We would like to thank Kaggle and HLTV for providing the required datasets, with player's gears and ratings respectively. Also, for software such as MS-Excel, MySQL and JASP for data processing and analysis. And MS-PowerPoint for architecture diagram design. And R Programming for Welch Two Sample T-test.

Declarations. We declare that there is no conflict of interest or competing interest, and no funding or grant received from any person or agencies. This, is original research was made by the 1st author, guided by the 2nd author, analyzed by the 3rd author, validated by the 4th author and formatted by the 5th author as mentioned in the title page.

References

1. Weber, R., Behr, K.M., Tamborini, R., Ritterfeld, U., Mathiak, K.: What do we really know about first-person-shooter games? An event-related, high-resolution content analysis. J. Comput. Med. Commun. **14** (2009). https://doi.org/10.1111/j.1083-6101.2009.01479.x
2. Yang, J., Wang, R., Cook, A., Fuller, R.: Gaming during the COVID-19 pandemic: examining its effect on loneliness & motivation, playing and gratification differences between competitive and recreational gamers. Telemat. Informat. Rep. **11** (2023). https://doi.org/10.1016/j.teler.2023.100093
3. Pluss, M.A., et al.: Examining the game-specific practice behaviors of professional and semi-professional esports players: a 52-week longitudinal study. Comput. Hum. Behav. **137** (2022). https://doi.org/10.1016/j.chb.2022.107421
4. Шинкарук, О., Анохін, Е., Юхно, Ю., Лут, І., Пінчук, В., Бондар, М.: Вплив глядацької аудиторії на популяризацію кіберспортивних дисциплін та проведення змагань. Теорія і методика фізичного виховання і спорту (2023). https://doi.org/10.32652/tmfvs.2023.2.86-94
5. Yudha, H.T., Utami, A.R.: The Effect of Online Game Dota 2 in Students' Vocabulary. Pustakailmu.id. 2 (2022)
6. Mike, S.: The International 9 "Dota 2" Tournament Prize Pool Breaks $30 Million
7. HLTV.org: Mathieu "ZywOo" Herbaut Counter-Strike Statistics | HLTV.org. https://www.hltv.org/stats/players/11893/zywoo. Accessed 17 Sept 2024
8. HLTV: Robin "ropz" Kool Counter-Strike Statistics | HLTV.org. https://www.hltv.org/stats/players/11816/ropz. Accessed 17 Sept 2024
9. Zamfir, E., Conde, M.V., Timofte, R.: Towards real-time 4K image super-resolution. In: IEEE Computer Society Conference on Computer Vision and Pattern Recognition Workshops (2023). https://doi.org/10.1109/CVPRW59228.2023.00155
10. Nazhif Rizani, M., Iida, H.: Analysis of counter-strike: global offensive. In: Proceedings of 2018 International Conference on Electrical Engineering and Computer Science, ICECOS 2018 (2019). https://doi.org/10.1109/ICECOS.2018.8605213
11. Fagerholt, E., Lorentzon, M.: Beyond the HUD. User Interfaces for Increased Player Immersion in FPS Games. Chalmers University (2009)
12. King, D.L., Delfabbro, P.H.: The cognitive psychology of Internet gaming disorder (2014). https://doi.org/10.1016/j.cpr.2014.03.006

13. Liu, S., Claypool, M., Kuwahara, A., Scovell, J., Sherman, J.: Lower is better? The effects of local latencies on competitive first-person shooter game players. In: Conference on Human Factors in Computing Systems - Proceedings (2021). https://doi.org/10.1145/3411764.3445245
14. Bojer, C.S., Meldgaard, J.P.: Kaggle forecasting competitions: an overlooked learning opportunity. Int. J. Forecast. **37** (2021). https://doi.org/10.1016/j.ijforecast.2020.07.007
15. Rusk, F., Ståhl, M.: A CA perspective on kills and deaths in counter-strike: global offensive video game play. Soc. Interact. Video Based Stud. Hum. Soc. **3** (2020). https://doi.org/10.7146/si.v3i2.117066
16. Love, J., et al.: JASP: graphical statistical software for common statistical designs. J. Stat. Softw. **88** (2019). https://doi.org/10.18637/jss.v088.i02
17. Manual, M.R.: MySQL 8.0 Reference Manual. MySQL 8.0 Reference Manual. 1 (2023)
18. Lehardy, R.K., Luczynski, K.C., Hood, S.A., McKeown, C.A.: Remote teaching of publication-quality, single-case graphs in Microsoft Excel. J. Appl. Behav. Anal. **54** (2021). https://doi.org/10.1002/jaba.805
19. How-To Geek: What Is a CSV File , and How Do I Open It ? Miscellaneous (2018)
20. Rejthar, J., Kotrba, V.: Confirmation of the validity of the HLTV ranking. ICGA J. **45** (2023). https://doi.org/10.3233/icg-230229
21. Sun, B., Zhao, S., Tian, G.: SQL queries over encrypted databases: a survey. Connect. Sci. **36** (2024). https://doi.org/10.1080/09540091.2024.2323059
22. DRKS00023197: mentalFIT: the effects of exercise on psychological processes a pilot study (2021). https://trialsearch.who.int/Trial2.aspx?TrialID=DRKS00023197
23. Wan, X., Wang, W., Liu, J., Tong, T.: Estimating the sample mean and standard deviation from the sample size, median, range and/or interquartile range. BMC Med. Res. Methodol. **14** (2014). https://doi.org/10.1186/1471-2288-14-135
24. Nassreddine, G., Younis, J., Falahi, T.: Detecting data outliers with machine learning. Al-Salam J. Eng. Technol. **2** (2023). https://doi.org/10.55145/ajest.2023.02.02.018
25. Lang, M.: Efficient R programming. J. Stat. Softw. **80** (2017). https://doi.org/10.18637/jss.v080.b03
26. Martin, G.: R studio. In: An Introduction to Programming with R (2021). https://doi.org/10.1007/978-3-030-69664-1_1
27. Thukral, S., Kovac, S., Paturu, M.: t-test. In: Translational Interventional Radiology (2023). https://doi.org/10.1016/B978-0-12-823026-8.00104-8
28. Lani, J.: Correlation (Pearson, Kendall, Spearman). Statistics Solutions (2010)
29. Hervé, M.: GrapheR: a multiplatform GUI for drawing customizable graphs in R. R J. **3** (2011). https://doi.org/10.32614/rj-2011-018
30. Fife, D.A., Longo, G., Correll, M., Tremoulet, P.D.: A graph for every analysis: map ping visuals onto common analyses using flexplot. Behav. Res. Methods **53** (2021). https://doi.org/10.3758/s13428-020-01520-2

A Simple Review of Artificial Intelligence and Machine Learning Applications in Petroleum Engineering

Alfitouri Ibrahim Jellah(✉), Taha A. Burgan, and Mohammed I. Shawish

Bani-Waleed University, Bani-Waleed, Libya
alfitouri.jellah@bwu.edu.ly

Abstract. Artificial intelligence (AI) is a transformative force in the oil and gas industry, reshaping the sector and attracting scholarly attention. This study reviews a substantial body of published work on AI applications in the oilfield industry. As AI evolves, it is expected to enhance operational efficiency, optimize resource management, and drive innovation.

Key trends include advanced predictive analytics for exploration, innovative drilling techniques, and improved safety measures, all contributing to a more sustainable energy sector. Machine learning, a subset of deep learning, utilizes *artificial neural network*s to model complex data. These techniques yield significant insights from large datasets without human intervention in this industry.

This review highlights various machine learning applications in petroleum engineering, focusing on reservoir, drilling, and production engineering. It also notes the growing integration of these technologies in the field.

Keywords: Machine Learning · Artificial Intelligence · Petroleum Engineering

1 Introduction

1.1 Overview

According to various studies, genetic algorithms (GA) are widely incorporated to optimize and improve matching speed and accuracy, and artificial neural networks (ANN) are widely employed for training and history matching in oil field development. As we all know, every phase of an oilfield's growth calls for a vast array of variables—typically, "big data". Artificial intelligence algorithms that tackle real-world engineering problems are essential to boost productivity while reducing expenses and risks [12].

R. Geetha et al. (Eds.): AAIMB 2024, CCIS 2654, pp. 25–39, 2026.
https://doi.org/10.1007/978-3-032-13029-7_2

2 Machine Learning Applications in Reservoir Engineering

An extreme learning machine was utilized in a study by Jianhua Cao et al. (2015) to assess reservoir parameters in a heterogeneous sandstone reservoir. This effort aims to assess the porosity and permeability of wells without relying on core data. This task has demonstrated considerable challenges when approached through traditional petrophysical techniques. Comparisons with back propagation (BP) and support vector machine (SVM) networks show that the robust ELM method provides precise and fast prediction results. The neural network was trained and validated using around a thousand samples from eight wells. The network model is utilized to assess the remaining porosity and permeability. In order to facilitate simple supervised learning for Single-hidden Layer Feedforward Neural Networks (SLFN), the Extreme Learning Machine (ELM) was introduced in this study. ELM performs better than standard neural networks regarding universal approximation capabilities and learning efficiency. Unlike the BP network, the ELM network's input weights and biases are assigned randomly; thus, they do not need to be changed during the training phase. The output weights of the network can be analytically calculated by finding the least squares solution. This allows the neural network to be built quickly and requires relatively little computational power. With \(M \)hidden nodes, a Single Layer Feedforward Network (SLFN) can be expressed mathematically as follows:

2.1 Maintaining the Integrity of the Specifications

$$j = 1, 2, \ldots, N,$$

Where: The *i*th hidden node is connected to the input nodes by the weight vector wi, to the output nodes by the weight vector βi, and its threshold is represented by bi. fi is the nonlinear activation function of the hidden node. It could be represented by an identity function, a sigmoid function, a Gaussian function, or various other polynomial functions. The optimally pruned extreme learning machine is an SLFN variant of the ELM method (OP-ELM). The OP-ELM method consists of the following three crucial steps:

1. First, create a standard ELM model with a significant number of neurons.
2. Using multiresponse sparse regression (MRSR), the neurons in the hidden layer are organized based on their contribution to explaining the ELM output linearly.

Due to the nonlinear nature of the problems this study addresses, the optimal kernel type for both OP-ELM and ELM will be selected through a comparative analysis of their respective levels of efficiency in the ensuing trials. The computational experiments flow diagram is displayed in the image below.

The first step in the experiment is preparing the input data, which includes measures of porosity and permeability acquired using cores and the original log data. There are two types of datasets: testing and training (Fig. 1).

Porosity and permeability are computed using five logs: GR, AC, DEN, CNL, and RT. These logs are physically connected to petrophysical characteristics. Consequently,

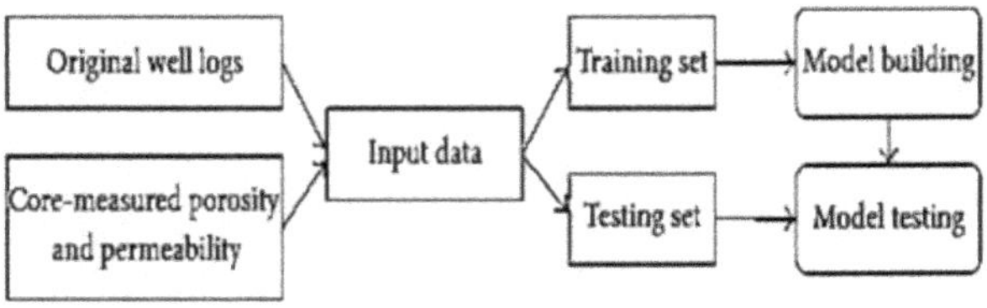

Fig. 1. Experimental Flow Diagram of the Estimation Experiments (Jianhua Cao et al. 2015).

the ELM network receives these five logs as input, with one log represented by each node. Porosity and permeability are the network neurons considered at the output layer. Figure 2 depicts the architecture of the network. Four thousand nine hundred fifty samples of Permian sandstone were obtained from the survey's eight wells; 90% of these samples were utilized as training data for the ELM network, while the remaining 10% served as test samples.—an analysis of the healthy eight porosity prediction performance of OP-ELM compared to BP and SVM techniques. Microcomputer tomography (CT) and the Lattice Boltzmann Method (LBM) measure fluid flow in continental carbonates. The fluid flow properties of these carbonates with multiscale pore networks are exceptional, and the method used to study them allows for the investigation of heterogeneity and anisotropy[1] (Table 1).

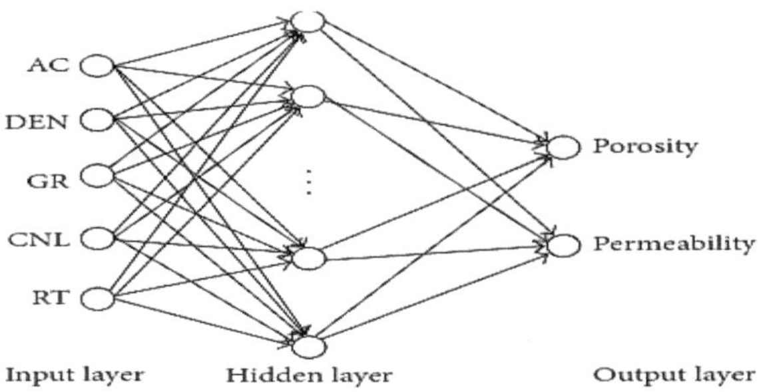

Fig. 2. The ELM network model has an input layer with five nodes and an output layer with two nodes: porosity and permeability. The hidden layer node count will be determined using the OP-ELM method.

Table 1. Results Porosity projection performance using OP-ELM versus BP and SVM methodologies for the eighth well. The comparison star is from the Permian reservoir (Jianhua Cao et al. (2015).

Algorithm	Accuracy	MAE	Training time (s)
BP	0.901	0.827	72
SVM	0.935	0.639	51
OP-ELM	0.956	0.205	23

Palabos is a software application that allows 3D pore network models to be used in single-phase flow simulations. It is used for part cle-based modeling of traditional computational fluid dynamics. Additionally, the simulation of permeability using multiple point geostatistics (MPS) was run on rock models. The findings for the three types of network predictors are presented using the same testing dataset. Three factors are employed in the comparisons: training time, accuracy, and MAE. The values are from averaging sample estimates from the eighth well. The Table shows the accuracy of the time in seconds and the mean absolute error (MAE) for each of the three processing methods. All three optimized network estimators worked well; however, OP-ELM produced the best results, with a mean absolute error of 0.205, an accuracy of 95.6%, and a quick learning time of 23 s.

Porous media's tortuosity and porosity regulate fluid flow. At larger scales, flow properties in continental carbonates are influenced by micro- and mesopores. MPS upscaling is necessary to address the volume-resolution problems with CT scanning equipment. The LBM-MPS technique can treat numerous lithologies with pore types, shapes, and networks. A three-dimensional approach is necessary because straightforward relationships of porosity-permeability only exist in simple carbonates. 3D fluid flow research provides the finest understanding of flow in porous media and is essential for reservoir modeling. Moreover, they provide the following summary:

1. Dealing with 3D datasets was essential to comprehending flow in complicated continental carbonates because 2D observations could not yield credible estimates of features that impact fluid flow in a rock, such as tortuosity and connectivity.
2. The subhorizontal and cascade facies related to pseudo-fenestral and inter-peloidal pores had the lowest permeabilities and porosities. The reed and water ll facies, which had reed molding and framework porosity, respectively, had far better properties. The resulting porous ties in the simulated and measured permeabilities easily surpassed 30% and 50 D.
3. The suggested multiple-point geostatistics (MPS) process is another tool for resolving typical difficulties in reservoir upscaling.
4. This scale-independent method has the added benefit of applying to 2D and 3D datasets. General averaging, often utilized in upscaling, is not employed in MPS.
5. The findings suggest that TI images can be used across multiple samples of the same facies type, improving the efficiency of conventional medical CT scanners. This method can also create detailed pore models when a deeper understanding of the network is needed.
6. Another study by Hongqing Song et al. (2020) demonstrates that specific machine-learning approaches have started to be progressively incorporated into the oil sector and have accomplished some success in both unconventional and conventional reservoirs due to the quick advancement of computer technology. This study proposes an alternate approach to reservoir vertical heterogeneity prediction using several deep neural networks and dynamic production data. The necessary data, which comprises dynamic production statistics calculated under various heterogeneous reservoir conditions, was gathered using a numerical simulation approach.
7. It is shown that machine learning approaches perform well in forecasting heterogeneity, with an RMSE of 12.71 mD, and successfully estimate reservoir permeability

through model validation. In addition, compared to BP and LSTM neural networks, the total AARD of the predicted result created by the CNN technique was fixed at 11.51%, indicating the most fantastic accuracy. Moreover, with a deviation of less than 10%, the permeability contrast, an essential metric for characterizing heterogeneity, may also be predicted precisely. Based on machine learning approaches, this work offered a possibility for vertical heterogeneity prediction in reservoirs.

8. The backpropagation technique combines the multilayer perceptron structure to form the BP neural network. It can now more accurately represent nonlinear mapping connections between variables since the backpropagation mechanism is continuously being refined. The two stages of a BP neural network's operation are forward propagation and backward propagation. Forward propagation is one neural network learning technique that produces predictions for vertical permeability by capturing the shifting features of dynamic production data. Ten hundred eighty neurons represent data on water cut, oil output, and water injection pressure, making up the model's input layer to capture the shifting characteristics of the data; the hidden layer comprises three layers, each with fifty neurons. Lastly, five neurons in the output layer match the five thin layers of vertical permeability of the simulated reservoir shown in Fig. 3's illustration (Table 2).

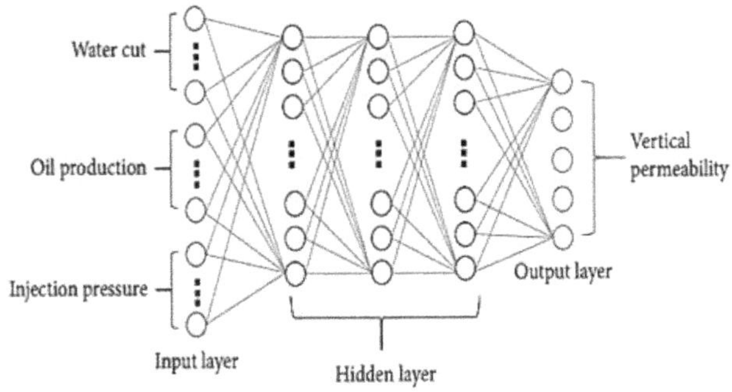

Fig. 3. Description of the network configuration of BP (Hongqing Song et al. 2020).

Table 2. An analysis of statistical errors was conducted on several neural networks.

Methods	ARD (%)	AARD (%)	RMSE (mD)	Training time	Prediction time
LSTM	25.66	43.25	51.04	15 min 41 s	0.8 s
BP	10.32	19.64	24.18	10 min 34 s	0.8 s
CNN	6.28	11.51	12.71	32 min 02 s	1.1 s

Based on the evaluation findings in Table, the AARD and ARD computed using CNN's estimation results are only 11.51% and 6.28%, respectively.

3. In addition, the RMSE may be maintained at 12.71 mD, the lowest value among LSTM and BP neural networks. The CNN model's unique data extraction technique is essential to reversing the reservoir's vertical permeability. In addition, the LS M model, which excels at handling time series issues, is held to high standards in this test because the study's input data is tied to time sequence. With an RMSE of 12. 1 mD, the machine learning model outperformed conventional methods regarding vertical permeability prediction accuracy, effectively estimating the reservoir's permeability as a whole instead of simply a single spot. Based on the model's valid action, the total AARD of the CNN technique's prediction result was managed at 11.51%, less in error calculation than the BP and LSTM networks. The three neural networks' combined prediction times are speedy, at less than a second. After carefully analyzing forecast time and accuracy, CNN is the optimal model.

A strong nonlinear relationship exists between logging parameters and porosity, as data collected from strata at varying depths indicates. These logging parameters reflect the properties of stratigraphic sediments in many geological periods. Porosity is a crucial factor for hydrocarbon storage because it captures the geological features of various historical eras, according to Zhenhua Zhang et al. (2021). It is essential to assess the reservoir's shale content and porosity using all available logging parameters to describe the reservoir accurately. In recent years, de p neural networks have been utilized to estimate shale content due to their enormous data structure mining capabilities. Besides its ability to process serialized input, the gated recurrent unit (GRU) neural network provides an additional advantage.

Thus, this paper suggests a method for forecasting porosity by combining several logging data using the GRU neural network. The logging parameters most pertinent to porosity were first chosen using a correlation measurement method per the Copula function. The GRU neural network finds a nonlinear mapping link between logging data and porosity parameters. One kind of neural network that analyses sequence data is the recurrent neural network (RNN). RNN creates links between the time steps and circularly assigns weights. An RNN structure with a single hidden layer is shown in Fig. 4. The RNN hidden layer is connected to both the output layer and the hidden layer nodes, in contrast to multilayer perceptron. In other words, the output layer and the hidden layer get the hidden layer's output. RNN establishes a linear relationship between datasets over time while reducing the number of parameters. Consequently, RNN offers clear benefits in handling time series and nonlinear problems.

Network is an important development. It effectively improves the network's memory capacity and resolves the problems of RNN gradient explosion and disappearance. Moreover, the LSTM network has an exterior and can remember more past data.

An internal "LSTM cell" circulation system, also known as self-circulation, is an integral part of the recurrent neural networks (RNNs) structure. As a result, LSTM m dels can apply nonlinearity to individual and combined elements.

Each unit is equipped with a gate control system that manages the flow of information using additional parameters. This configuration allows for a similar input and output mechanism in shared recirculating networks (Fig. 5).

Despite its complicated structure, many parameters, and sluggish convergence time, the LSTM network outperforms RNN in learning long-term dependencies

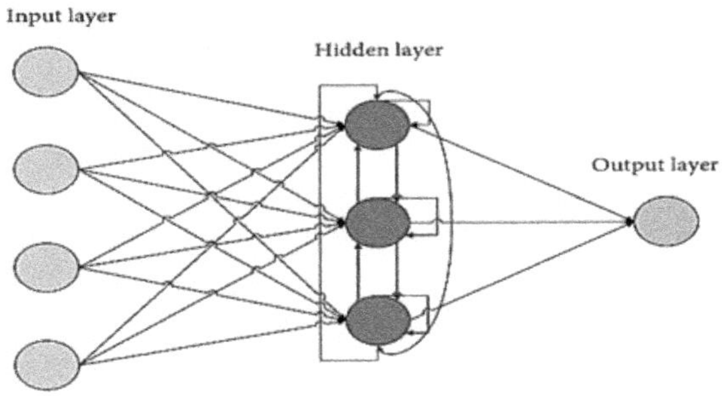

Fig. 4. RNN Structure with Single Hidden Layer (Zhenhua Zhang et al. 2021). The RNN's long—and short-term memory (LSTM)

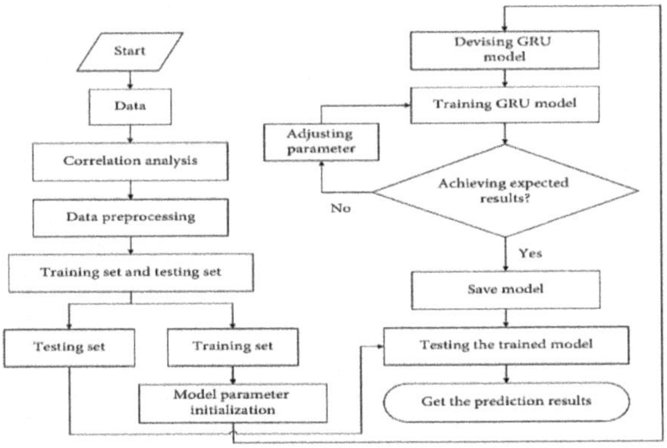

Fig. 5. The topology of the LSTM hidden layer.

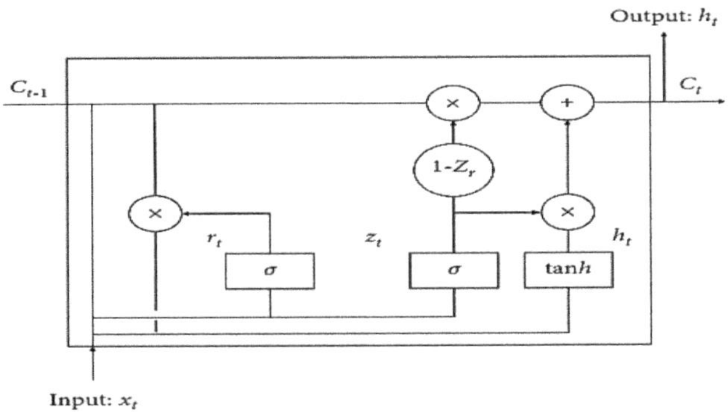

Fig. 6. CA-GRU-based Porosi y parameter prediction model (Zhenhua Zhang et al. 2021).

between sequence data. Figure 6 depicts the modeling process for the CA-GRU-based combination forecasting model.

The GRU and CA-GRU models are both capable of accurately predicting porosity. However, the CA-GRU model demonstrated greater precision, making it more effective for porosity predictions in this study. Additionally, when using optimal inputs, the CA-GRU model is a valuable tool for forecasting porosity, particularly in areas where high-precision porosity data is essential.

Logging parameters indicate the acoustic, discharge, and electrical sedimentary properties from various geological ages. The property that sets different formations apart is their porosity, which also has important time series properties. The high cost of using some technologies, including nuclear magnetic resonance and rock physical analysis, may be significantly decreased by directly predicting porosity using logging parameters. This can offer a curated and affordable basis for oil and gas exploration and production decision-making.

3 Machine Learning Applications in Drilling Engineering

In 1989, the first usage of neural networks in the E&P industry was for intelligent drill bit diagnostics, the interface of reservoir simulators, and the interpretation of good logs. Recognizing seismic arrangements, characterizing reservoirs, predicting porosity and permeability, determining PVT properties, estimating pressure drops in wells and pipes, optimizing production health, assessing sound performance, managing portfolios, and making informed decisions are key challenges in the oil and gas industry that needs to be addressed. To a certain extent of application and trend within the sector, Opeyemi Bello's (2015) study investigates and assesses the efficient implementation of artificial intelligence in drilling, one of the essential functions of the oil and gas business. Various papers and research projects will be compiled to highlight the boundaries of AI applications. This research aims to support the identification of neglected areas in the field and the ongoing development of this technique. The Table below illustrates his findings.

Various papers and research projects will be compiled to highlight the boundaries of AI applications. This research aims to support the identification of neglected areas in the field and the ongoing development of this technique. The Table below illustrates his findings (Table 3).

He also concluded that applying artificial neural networks (ANNs), genetic algorithms, fuzzy tools, and support vector machines in assessing drilling operations has been very encouraging, whether in modeled comparisons or test field trials. This has led to the need for focused development and research of AI tools for industrial use and trend-setting. Significant improvements in drilling may result from this, such as using more affordable rigs and DS, constant maximum feasible ROP during drilling, longer bit and tool lives, lower NPT, more stable wellbores, fewer procedural risks in real-time, and effective handling of complicated drilling incidents.

Ahmed Alsaihati et al. (2021) used machine learning to find downhole abnormalities during drilling. This article explains how to develop and evaluate an intelligent system to encourage safe operations and improve the reaction time edge to prevent unwanted occurrences.

Three models were trained and tested using actual field data from Well-1: a functional network, a random forest, and an artificial neural network. The training and testing

Table 3. Timeline AI techniques application in drilling practices (Opeyemi Bello 2015).

Drilling Sector	Application	AI Approach	Researcher(s)	Year
Well Planning	Bit selection	ANN	National Oilwell Varco	2013
	Gradient prediction	GRNN	Sadiq and Nashwi	2000
	Casing collapse prediction	BPNN	Salehi, Hareland, Ganji, Keivana and Abdollahi	2007
	Cement quality / Performance estimation	ANN	Fletcher, Coveney and Methven	1994
	Offshore platform selection	Hybrid (BPNN-GA)	Wang, Duan, Liu and Dong	2011
	Directional mapping	CBR	Mendes, Guilherme and Morooka	2011
Procedural Optimization	BHA monitoring	ANN	Dashevskiy, Dubinsky and Macpherson	1999
	Bit wear control	ANN	Gidh, Purwanto and Ibrahim	2012
	Drag and slack-off load prediction	ANN	Sadiq and Gharbi	1998
	DS vibration control	ANN	Esmaili, Elahifar, Thonhauser and Fruhwirth	2012

data were divided by 20/80 to predict surface drilling torque while drilling a 5–7/8 in. horizontal section. This horizontal drilling began one day before the incident of a stuck pipe.

The work's main contribution is its method of simplifying the early detection of operation abnormalities. ML models, unlike conventional drilling models, have predictive powers. As predicted by an ML model, RF developed the surface torque using actual field data from Well-1. Well-2 was utilized to test the intelligent system's capacity to identify downhole anomalies in the days preceding the Well-2 disaster. The following conclusions can be made from the findings:

- With an AAPE of 1.46% and an R of 0.99 in the training set, the RF accurately predicted the torque, whereas, in the testing set, the AAPE and R were 3.98% and 0.93, respectively.
- At a depth of 14,900 feet, the RF model discovered that the actual torque pattern in Well-1 in the days preceding the event differed from the prediction model.
- The day before the occurrence, the actual torque trend in Well-2 differed from the projected model at a depth of 15,060 feet, according to the RF model.
- The intelligent technology generated a real-time alarm to alert the drilling crew to an issue in Well-1 nine hours before the drilling team or monitoring engineers noticed it or reported it. However, the alarm went off seven hours before the Well-2 drilling crew noticed anything unusual.

4 Machine Learning Applications in Production Engineering

According to Guoliang Shen et al. (2020), modern industrial control is transitioning from manual to automated systems as technology advances. Their study describes how to assess and develop machine learning models for accurately estimating the flow rate in gas pipelines. Unlike traditional expert or rule-based methods, machine learning approaches require minimal specialized experience and involve a thorough analysis of the underlying physical mechanisms. We have developed a system that automates the process of selecting suitable candidates.

Table 5 compares several models. It shows that the GBDT algorithm produces the best results from all the tested methods. By comparing the performance of different hyper-parameter settings, we may learn that a well-selected hyper-parameter setting improves the performance of GBDT models. Setting can significantly enhance performance.

This method takes a long time if done manually; therefore, we automated it by evaluating different hyper-parameters. The outcome of layered GBDT models is exciting. Close to the GBDT model, however, stacked GBDT models are considerably more complicated and time-consuming than basic GBDT models; thus, in this situation, a simple GBDT model is a preferable alternative.

Modern industrial control is transitioning from manual to automated systems as technology advances, according to Guoliang Shen et al. (2020). Their study describes how to assess and develop machine learning models for accurately estimating the flow rate in gas pipelines. Unlike traditional expert or rule-based methods, machine learning approaches require minimal specialized experience and involve a comprehensive analysis of the underlying physical mechanisms. A system that can automate the process of selecting suitable candidates is developed. The leaky ReLU action function is recommended when utilizing the neural network method. This activation function surpasses all other activation functions by a significant margin. As a result, there may be better options than neural network models for essential table datasets.

According to Liqiang Wang et al. (2021), traditional decline methods, including Arps yield decline curve analysis, are frequently used for yield decline analysis due to their underlying principles and valuable applications. On the other hand, carbonate reservoirs that exhibit strong beginning production, quick decline, and large output fluctuations—the bulk of which lack a steady production period—are not well-suited for traditional decline techniques. Therefore, there is a crucial need for a novel decline analysis methodology. In recent years, unconventional oil deposits have been evaluated using machine learning techniques based on multiple regression and deep learning, but the outcomes have been underwhelming. They created the innovative equal probability gene expression programming (EP-GEP) technique to overcome the limitations of the conventional Arps decline model in carbonate.

Through model authentication and a comparative analysis of prediction effects, the EP-GEP model demonstrated a high degree of prediction accuracy. The average relative error was significantly lower than the original Arps model and other machine learning methods. Field reservoir engineers should have access to a new decline analysis tool following the effective implementation of the suggested technique to carbonate reservoir production decline analysis. Their work proposes an equal probability GEP (EP-GEP) technique that simultaneously improves the algorithm's convergence efficiency and solution quality and solves the problem of evolutionary undirected and premature local convergence.

The newly formed population advances to the following optimization cycle. If there is evidence of premature local convergence during this process, we will move on to the equal probability gene expression optimization calculation. Subsequently, genet c operations—such as mutation, equal probability selection, string insertion, and recombination—are applied to individuals in the population, as demonstrated by the three individuals selected with equal probability in Fig. 7.

Offspring are created to produce a new population, and the candidate set and newly created population are then utilized in the optimization computations shown in the following Table (Table 4).

Table 4. Average relative error of machine learning model predictions for various time series of oil wells (Liqiang Wang et al. (2021).

Well ID	Average relative error (%)			
	EP-GEP	ARMA	ARIMA	RNN
A20	4.37	−12.88	42.48	324.53
A35	6.79	−15.24	38.59	375.22
A7	4.56	−11.91	31.23	289.55
A12	3.27	−11.13	29.02	247.36
A18	7.63	−22.54	45.77	346.31

Numerous oil well production decline assessments have found that the EP-GEP machine learning algorithm outperformed other time series machine learning approaches regarding prediction accuracy and stability, as shown in Fig. 7.

The suggested method might give on-site reservoir engineers new reservoir management analytical tools.

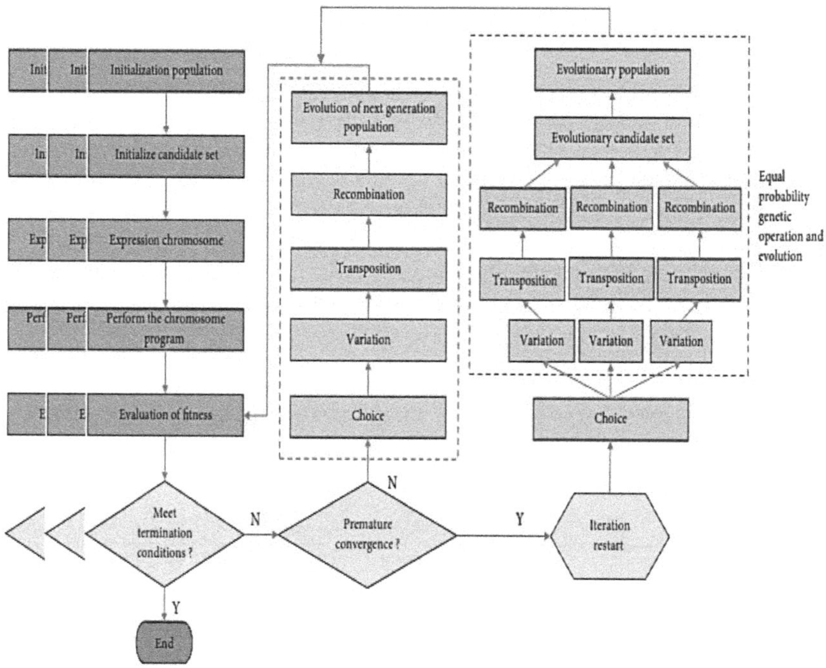

Fig. 7. Flowchart of EP-GEP (Liqiang Wang et al. 2021).

5 The Ever-Growing Need for Machine Learning in Petroleum Engineering

In light of the increasingly competitive and volatile oil and gas industry, businesses are actively looking for new, more effective ways to increase worker safety, reduce costs, and streamline production. Many CEOs are moving to digitalization to protect their businesses from market shocks, maintain profitability in the face of falling prices, and obtain a competitive edge throughout the recovery. The future lies in machine learning-based technologies and artificial intelligence (AI), which are being employed at every stage of the value chain and are developing quickly. Many industries are now fully embracing the remarkable benefits of this innovative technology.

6 Challenges and Limitations

While AI/ML has shown significant potential in petroleum engineering, several challenges and limitations need to be addressed.

Liu et al. (2024) discuss the weaknesses, challenges, and prospects of popular machine learning algorithms in petroleum engineering in their review article "Machine Learning in Reservoir Engineering: A Review" published on June 14, 2024 and can be summarized:

1. **Data Quality and Availability**: High-quality data is essential for training accurate AI/ML models. However, data availability and quality can be limited in the oil and gas industry due to:
   - Limited sensor data in harsh environments
   - Inconsistent data formats and standards
   - Data loss or corruption during transmission or storage
   - Limited access to historical data or proprietary information
2. **Interpretability and Transparency**: AI/ML models can be complex and difficult to interpret, making it challenging to understand the decision-making process. This is particularly critical in petroleum engineering, where:
   - Model predictions can impact critical decisions, such as well placement or production optimization
   - Lack of transparency can lead to mistrust in AI/ML recommendations
3. **Scalability and Integration:** Integrating AI/ML solutions with existing workflows and systems can be complex and require significant resources. In petroleum engineering, this includes:
   - Integrating with existing software and hardware infrastructure
   - Ensuring compatibility with different data formats and standards
   - Scaling AI/ML solutions to handle large datasets and complex simulations
4. **Domain Expertise**: Effective application of AI/ML in petroleum engineering requires domain expertise and collaboration between engineers and data scientists. This includes:
   - Understanding the underlying physics and geology of petroleum systems
   - Identifying relevant features and variables for AI/ML models
   - Interpreting results and making informed decisions
5. **Uncertainty and Risk:** AI/ML models can be sensitive to uncertainty and risk, particularly in petroleum engineering, where:
   - Model predictions can be impacted by uncertain inputs or assumptions
   - Decisions based on AI/ML recommendations can have significant economic or environmental implications
6. **Regulatory and Compliance:** AI/ML solutions in petroleum engineering must comply with regulatory requirements and industry standards, including:
   - Ensuring data security and confidentiality
   - Meeting standards for model validation and verification

7 Conclusion

This work has identified and detailed various monitored learning strategies. Machine learning has the potential to significantly influence the numerous critical decisions that engineers and administrators in the oil and gas sector make daily. The benefits of information can be realized in the future if suitable approaches are applied to incorporate various data kinds or structures and transform them into relevant information that helps make wise judgments.

In the upcoming years, machine learning is anticipated to find widespread use in the oil and gas sectors, where its worth will be widely acknowledged. Several methods and research have shown how effective algorithms, software, and statistical analysis components used in data handling are when it comes to helping engineers.

References

Liu, et al.: Machine learning in reservoir engineering: a review. Processes **12**(6), 1219 (2024)

Cao, J., Yang, J., Wang, Y., Wang, D., Shi, Y.: Extreme learning machine for reservoir parameter estimation in heterogeneous sandstone reservoir. Math. Probl. Eng. Article no. 287816, 10 p. (2015). https://doi.org/10.1155/2015/287816

Bello, O., Holzmann, J., Yaqoob, T., Teooriu, C.: A plication of artificial intelligence methods in drilling system design and operations: a review of the State of the Art. JAISCR **5**(2) (2015)

Soete, J., et al.: Lattice Boltzmann simulations of fluid flow in continental carbonate reservoir rocks and upscaled rock models generated with multiple-point geostatistic. Geofluids Article no. 7240524, 24 p. (2017). https://doi.org/10.1155/2017/7240524

Li, H., Yu, H., He, N.T., Cheng, S.: Applications of artificial intelligence in oil and gas development. Arch. Comput. Methods Eng. **28**, 937–949 (2012). https://doi.org/10.1007/s11831-020-09402-8

Hongqing, S., et al.: Potential for vertical heterogeneity prediction in reservoir basing on machine learning methods. Geofluids Article no. 3713525, 12 p. (2020). https://doi.org/10.1155/2020/3713525

Shen, G., Li, M., Lin, J., Bao, J., He, T.: An empirical study for adopting machine learning approaches for gas pipeline flow prediction. Math. Probl. Eng. Article no. 7842847, 13 p. (2020). https://doi.org/10.1155/2020/7842847

Saihati, A., Elkatatny, S., Abdulhamid, A.M., Abdulraheem, A.: Use of machine learning and data analytics to detect downhole abnormalities while drilling horizontal wells, with real case study. J. Energy Resour. Technol. **143**, 043201-1 (2021)

Husam, H.A., Abo Taleb, T., Shari, H.D., Ralph, E.F.: Applications of artificial neural networks in the petroleum industry: a review. SPE-195072-MS (2021)

Wang, L., et al.: Time series analysis s of production decline in carbonate reservoirs with machine learning. Geofluids Article no. 6638135, 8 p. (2021). https://doi.org/10.1155/2021/6638135

Zhang, Z., Wang, Y., Wang, P.: On a deep learning method of estimating reservoir porosity. Math. Probl. Eng. Article no. 6641678, 13 p. (2021). https://doi.org/10.1155/2021/6641678

Sircar, A., Yadav, K., Rayavarapu, K., Bist, N., Oza, H.: Application of machine learning and artificial intelligence in oil and gas industry. Pet. Res. (2021). https://doi.org/10.1016/j.ptlrs.2021.05.009

Barranquero, M., Olmedo, A., Gómez, J., Tayebi, A., Hellín, C.: Automatic 3D building reconstruction from OpenStreetMap and LiDAR using convolutional neural networks. Sensors **23**(5), 244 (2023)

Oil and Gas Recruitment - Evolving Oil & Gas Industry. https://summit-tsi.om/oil-and-gas-recruitment-and-evolving-industry/

Supriyati, E., Iqbal, M.: Recognition system of Indonesia sign language based on sensor and artificial neural network. Makara Seri Teknologi (2013). https://doi.org/10.7454/mst.v17i1.1924

Wang, C., Zhao, L., Wu, S., Song, X.: Predicting the surveillance data in a low-permeability carbonate reservoir with the machine-learning tree boosting method and the time-segmented feature extraction. Energies **13**(23), 6307 (2020)

Zhang, H., et al.: FTCN: a reservoir parameter prediction method based on a fusional temporal convolutional network. Energies **15**(15), 5680 (2022)

Ibrahim, N., et al.: Well performance classification and prediction: deep learning and machine learning long term regression experiments on oil, gas, and water production. Sensors **22**(14), 5326 (2022)

Improving Medicinal Plant Identification with Grayscale and Canny Edge Detection

G. Yuvaraj, R. H. Yuvashree(✉), and S. Kirithika

Department of Computer Science and Engineering, Sri Sairam Engineering College, Chennai, India

rhyuvashree04@gmail.com

Abstract. As modern society increasingly rejects traditional knowledge and relies more heavily on synthetic medications, the need for the accurate identification of medicinal plants has become more critical than ever. This shift away from natural remedies poses several risks, including the potential misuse of medicinal plants owing to incorrect identification. Our project sought to address this issue by utilizing image-processing techniques, specifically grayscale conversion and Canny edge detection, to develop a more reliable approach for recognizing medicinal plants. By mitigating the risks of misidentification, this approach not only helps preserve and promote the use of natural remedies but also ensures that future generations can continue to benefit from their healing properties.

Keywords: Automated Identification System · Image Processing · Canny Edge Detection · Grayscale Conversion

1 Introduction

In the face of global biodiversity decline and the pressing challenges posed by climate change, the identification and preservation of medicinal plants have become critical yet often overlooked issues. Medicinal plants, long used as natural remedies and a foundation for pharmaceuticals, are now at risk, with over 40% of plant species facing extinction according to the UN. This threatens not only biodiversity but also the legacy of traditional herbal medicines [1]. The younger generation, increasingly disconnected from nature, struggles with even basic plant recognition, while misidentification in herbal markets further complicates the problem. Valid and exact findings are essential to ensure the effectual and harmless effects of these treatments, which traditional methods fail to consistently achieve due to their reliance on manual expertise [4]. This study introduces a machine learning approach that leverages grayscale and Canny edge detection for vein pattern analysis, a distinctive feature in leaf structure, combined with Convolutional Neural Networks (CNN) to improve classification accuracy. By merging modern technology with botanical knowledge, this approach not only advances plant identification, but also promotes conservation efforts and underscores the significance of medicinal plants in healthcare for future generations.

R. Geetha et al. (Eds.): AAIMB 2024, CCIS 2654, pp. 40–48, 2026.
https://doi.org/10.1007/978-3-032-13029-7_3

2 Related Works

Recent studies have explored innovative methods for identifying medicinal plants in real-world scenarios using mobile captured images. One notable approach involved a dataset of 10,000 Vietnamese herbal plant images, where deep learning techniques were applied to automatically extract key features. These features were then classified using LightGBM, which achieved an impressive accuracy of 93.6%. This deep learning combination and LightGBM not only improved accuracy but also optimized processing efficiency, making it suitable for practical use. This study emphasized the potential of further improving plant identification by exploring more advanced architectures and comparing different classification models for enhanced real-world applications [3].

A recent study explored the classification of six medicinal plants, including tulsi, peppermint, catnip, and stevia, by using both multispectral and digital images taken in a computer vision lab. After preprocessing the images by resizing and converting them to grayscale, the Sobel filter was applied for edge detection. From these processed images, 65 key features, including texture and multispectral details, were extracted. To refine the model, a chi-square selection method identified the 14 most relevant features. Among five machine learning classifiers tested, the multi-layer perceptron stood out, achieving an impressive accuracy of 99.01%, highlighting its potential for precise plant classification [8].

This research presents a new way of categorizing the leaves by enhancing deep learning and machine learning. Their approach entails extracting characteristics from pictures of leaves using a ResNet50 model, simplifying the data resulting from this step through PCA, and finally using LightGBM to classify the leaves. This approach yields high accuracy of 95%. Having an approximate accuracy of 70% for a dataset of 40 different types of leaves. The study also compares the time taken for this method with other boosting techniques and also consider the role of component selection in classification. The results can be used in the selection of effective models for classification of leaves, taking into account the indicators of accuracy and the speed of computations [5].

MPInet is a deep learning system developed to recognize medicinal plants based on leaf images. It was trained on a dataset of 3,000 images from 10 commonly known medicinal plants, achieving an impressive accuracy of 99.4%. The system not only excels in precision and recall but also shows great potential for practical applications, especially in educational mobile apps aimed at helping users distinguish plants more easily. Researchers aim to expand MPInet by adding more plant species and exploring advanced deep learning models, making it a promising tool for enhancing our understanding and conservation of medicinal plants [7].

Recognizing medicinal plant species is crucial for traditional healers and modern doctors. This research critiques current computer methods for plant identification, particularly for medicinal plants, highlighting their limitations. To improve accuracy, the study developed a new system based on deep learning, achieving a 96.6% success rate in recognizing plant species. Future research will explore combining different artificial intelligence techniques to enhance the system further [6].

3 Methodology

Our approach to herbal plant identification using image processing and machine learning follows a detailed pipeline that includes data collection, pre-processing, extracting features, training model, validation, system integration, and continuous evaluation. This structured approach aims to ensure the accurate and reliable classification of medicinal plants by leveraging their distinctive visual features.

A. Data Gathering

Data were sourced from publicly accessible platforms, such as Kaggle, botanical research websites, and specialized plant databases. The dataset comprises images of various medicinal plant species, captured under different environmental conditions and growth stages to ensure diversity. Each image was meticulously annotated with its corresponding plant species, supporting the supervised learning process. This diversity within the dataset provides a strong basis for training machine learning models and improves their ability to generalize across different plant species and real-world scenarios.

B. Preprocessing

Preprocessing is an essential phase that transforms raw images into formats that are conducive to feature extraction and effective classification. We start by converting the images to grayscale, simplifying the data and allowing us to concentrate on the structural features critical for identifying leaf veins and textures.

To further enhance our analysis, we incorporated RGB histograms to examine the color information in the images. This technique breaks down pixel intensities across the red, green, and blue channels, thereby offering insights into the color distribution of plants. For instance, peaks in the red channel can indicate specific characteristics of certain plants, whereas the overall histogram shape helps us understand how color intensity varies throughout the image (Fig. 1).

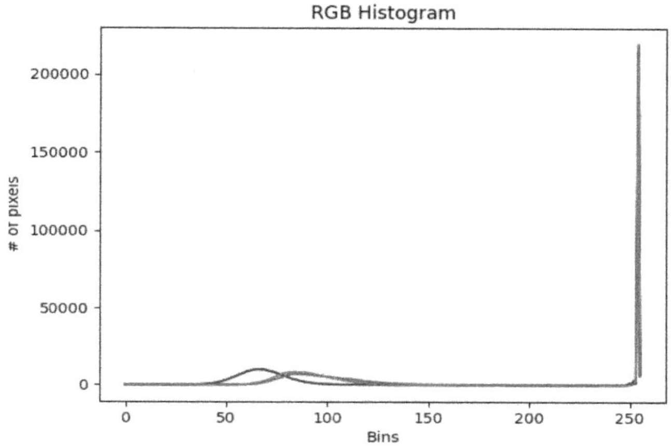

Fig. 1. RGB Histogram

Moreover, we utilized grayscale histograms to assess the contrast and intensity distributions of monochrome images. This evaluation aids in recognizing the dynamic range of images, which is crucial for achieving effective segmentation and feature extraction. By meticulously preparing the images in this manner, we significantly improved the accuracy of our plant identification process (Fig. 2).

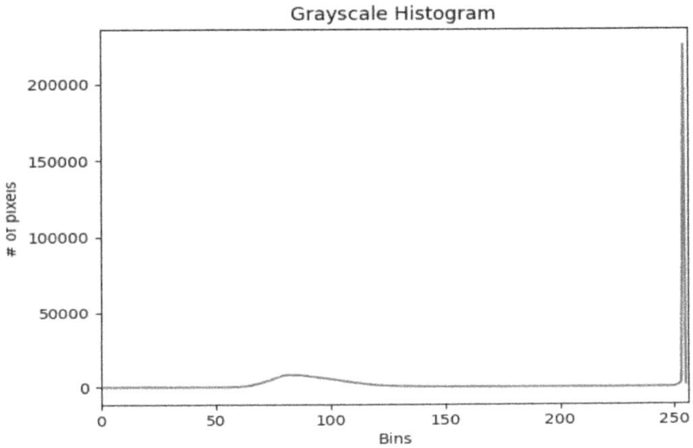

Fig. 2. Grayscale Histogram

C. Grayscale Algorithm

The initial step in preprocessing involves transforming the color images to grayscale, which makes the image data by eliminating color information while preserving important structural features. This makes it easier to focus on essential patterns like leaf veins and textures, which are crucial for plant identification.

1. Start.
2. Imports necessary packages/libraries (e.g., OpenCV for image processing).
3. Load the input color image.
4. Convert the color image to grayscale using the cvtColor() function.
5. Display or save the grayscale image.
6. End.

After converting the image to grayscale, we applied the Sobel edge detection algorithm. Sobel detects both horizontal and vertical edges, which highlight the boundaries of leaf structures (Fig. 3).

D. Canny Edge Detection Algorithm

To further refine the image, we applied Canny edge detection after the grayscale conversion.

1. Start.
2. Import the OpenCV library as cv2.

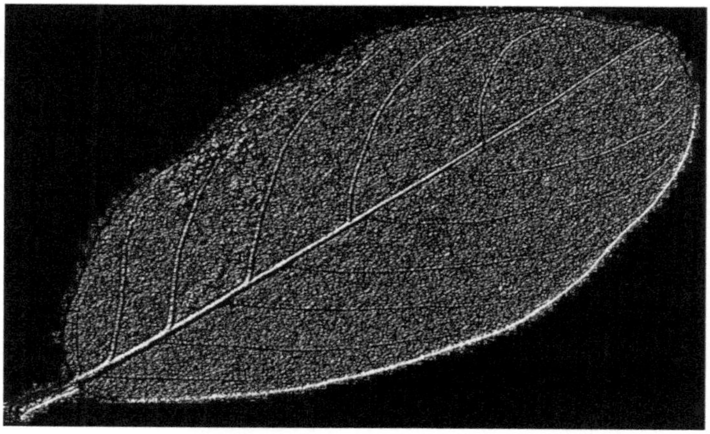

Fig. 3. Sobel Detection

3. Load the grayscale image using the imread() function.
4. Apply Gaussian blur to the grayscale image using the GaussianBlur() function.
5. Use the Canny() function to detect edges in the blurred image, specifying the lower and upper threshold values.
6. Display or save the edge-detected image using the imshow() and waitKey() functions.
7. End.

In addition, adaptive thresholding was used to segment the leaves from the background. Unlike traditional thresholding, which applies a single value globally, adaptive thresholding calculates thresholds based on local pixel neighborhoods [2] (Fig. 4).

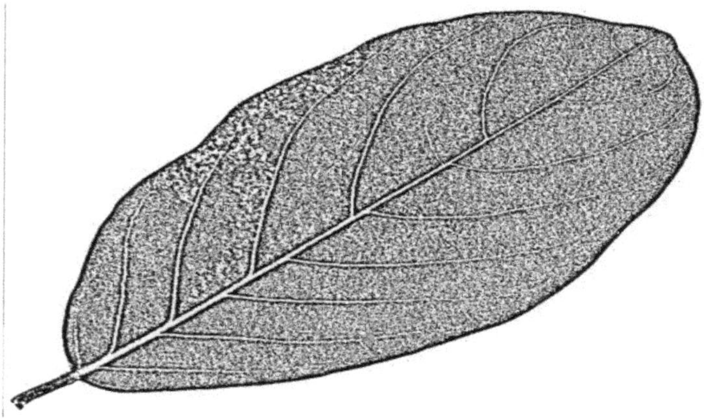

Fig. 4. Adaptive Thresholding

E. Feature Extraction

The process of analyzing preprocessed images to extract meaningful attributes that are crucial for plant identification is known as feature extraction. This step includes calculating various metrics, such as leaf area, perimeter, and aspect ratio, which provide valuable information regarding the shape and size of the leaves. Additionally, vein density and patterns were analyzed by measuring total vein length, branching factors, and vein density, which are essential for distinguishing between different plant species. This step is critical to improve the ability of the model to accurately classify medicinal plants based on their unique characteristics.

F. Model Training

For the model training phase, we utilized a Convolutional Neural Network (CNN), which is well-suited for processing and classifying complex image data. The CNN was trained on a dataset where 70% of the images were used for training, and 30% were reserved for validation to assess model performance. To enhance the model's ability to generalize to new, unseen data, data augmentation techniques such as image flipping, rotation, and scaling were applied. This ensured that the model was exposed to diverse variations in the plant images. The training process was further optimized using the Adam optimizer, while the categorical cross-entropy loss function was employed to refine the accuracy of the model.

G. Validation

The model was validated using a separate dataset to ensure reliable performance. Key metrics, such as accuracy, recall, F1-score, and precision, were calculated to evaluate its effectiveness. The model was tested under various conditions, including different lighting, angles, and backgrounds, to simulate real-world scenarios and ensure robustness. These validation steps are essential for confirming that the system can consistently identify medicinal plants across diverse environments.

H. System Integration and Testing

Once the model is validated, it is integrated into a user-friendly application interface. This interface enables users to upload images of medicinal plants that are then classified by the model. The interface is designed to be intuitive and user-friendly, with feedback mechanisms incorporated to gather user input and suggestions for future enhancement. Usability testing was conducted to ensure that the application met the needs of its intended users, including botanists, herbalists, and researchers. This step ensures that the system is not only accurate, but also accessible and user- friendly, making it suitable for practical use.

I. Continuous Evaluation and Update

To make sure the exactness and relevance of the system over time, a continuous evaluation and updating process is implemented. This involves regular retraining of the model with new data to account for variations in plant appearance due to seasonal changes, environmental factors, or the discovery of new species. The performance of the sys- tem was closely monitored, and updates were made as needed to guarantee its effectiveness in identifying a wide range of medicinal plants. This ongoing improvement process is critical for maintaining the system current, and is capable of providing precise plant identification in the long run (Fig. 5).

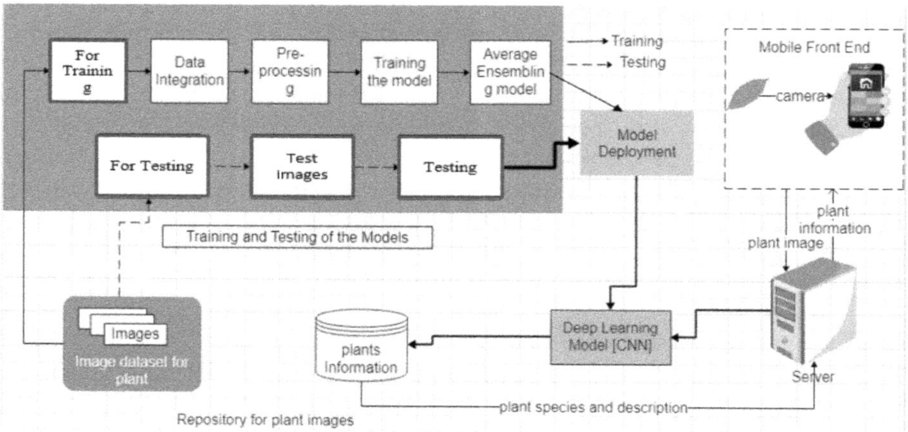

Fig. 5. Architecture Diagram

4 Result

Our study introduced an effective method for recognizing medicinal plants using image processing and machine learning. By applying techniques like grayscale conversion, Sobel edge detection, and CNN-based classification, we were able to accurately classify plant species. The use of RGB and grayscale histograms helped capture essential plant characteristics, while adaptive thresholding improved the clarity of the images for better analysis. Overall, the system showed strong performance, making it reliable for practical use in various environments (Table 1).

Table 1. Result

Gray scale	Canny Edge Image	Original Image	Name of Plant
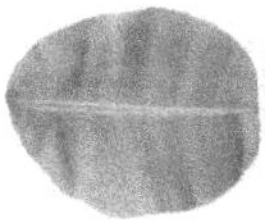	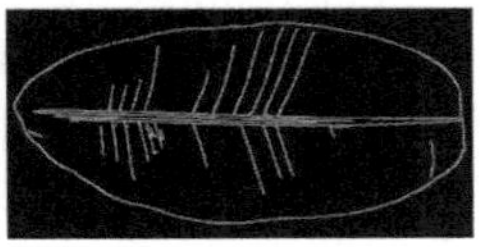	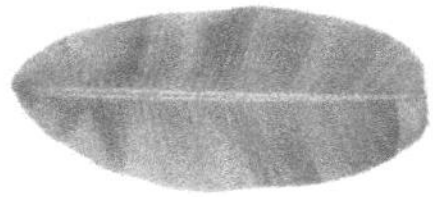	Banana Leaf
	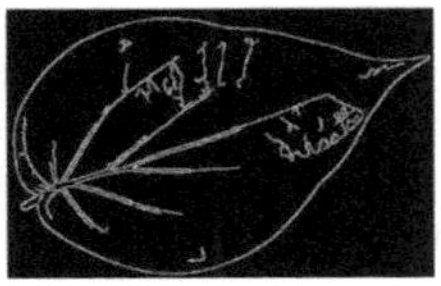		Betel
	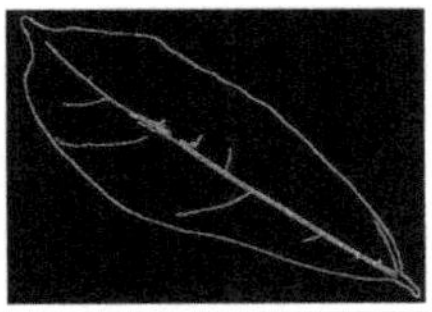		Ashwagandha leaf
	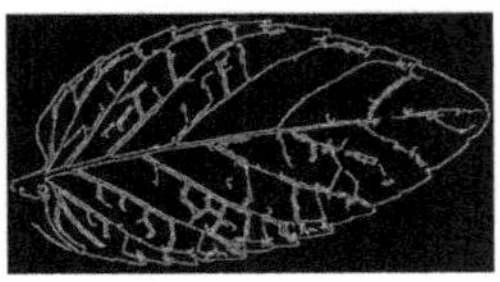		Pepper mint
			Termin alia Chebula

5 Conclusion

This research introduces an advanced approach for recognizing medicinal plants through machine learning and image processing techniques. By converting plant images to grayscale, we created a highly accurate plant classification system employing algorithms such as Canny edge detection and Convolutional Neural Networks (CNN). Our methodology enhances plant identification compared to conventional techniques. This system is user friendly and regularly updated to maintain its precision and relevance. By integrating traditional knowledge with contemporary technology, our study contributes to the preservation of essential plant resources and supports the progression of herbal medicine.

6 Future Enhancement

The next phase of this project will focus on integrating vein pattern analysis using a vein algorithm to further enhance the system's ability to differentiate closely related plant species. This will involve the development of algorithms specifically designed to analyze the intricate vein structures of leaves, providing an additional layer of detail for more precise identification. This enhancement will reduce the gap between traditional botanical understanding and modern technology, thereby supporting the conservation and promotion of medicinal plants as valuable natural resources.

Acknowledgment. We sincerely appreciate Kaggle for providing access to an extensive collection of medicinal plant images. Their platform has been invaluable in supporting the research and development of our plant identification model. The diverse and high-quality data available on Kaggle played a prominent role in establishing the robustness and accuracy of our study.

References

1. Sinha, S.K., Kumar, S., Kumar, S., Katiyar, G., Chandola, R.: Plant identification using machine learning. In: 2021 Asian Conference on Innovation in Technology (ASIANCON) (2021). https://doi.org/10.1109/asiancon51346.2021.9544670
2. Begue, A., Kowlessur, V., Singh, U., Mahomoodally, F., Pudaruth, S.: Automatic recognition of medicinal plants using machine learning techniques. Int. J. Adv. Comput. Sci. Appl. **8**(4) (2017). https://doi.org/10.14569/ijacsa.2017.080424
3. Vo, A.H., Dang, H.T., Nguyen, B.T., Pham, V.H.: Vietnamese herbal plant recognition using deep convolutional features. Int. J. Mach. Learn. Comput. **9**(3), 363–367 (2019). https://doi.org/10.18178/ijmlc.2019.9.3.811
4. Nazarenko, D.V., Rodin, I.A., Shpigun, O.A.: The use of machine learning in the analytical control of the preparations of medicinal plants. Inorg. Mater. **55**(14), 1428–1438 (2019). https://doi.org/10.1134/s0020168519140115
5. Mishra, V., Sharma, V., Mishra, U.: A hybrid approach for leaf classification using machine learning and deep learning (2023). https://doi.org/10.1109/iccpct58313.2023.10245548
6. Darshana, S., Soumyakanta, K.: A revolutionary machine-learning based approach for identifying ayurvedic medicinal plants (2022). https://doi.org/10.1109/assic55218.2022.10088298
7. Preethi Salian, K., Shrisha, H.S., Salian, S., Karthik. K.: MPInet: medicinal plants identification using deep learning (2023). https://doi.org/10.1109/iceca58529.2023.10395327
8. Naeem, S., et al.: The classification of medicinal plant leaves based on multispectral and texture feature using machine learning approach. Agronomy **11**(2), 263 (2021). https://doi.org/10.3390/agronomy11020263

Real-Time Traffic Analytics and Control for City Street

S. V. Aadhitya(✉), K. Meera, M. Harini, J. Kavya Catherine, and C. Sandhya

Meenakshi Sundararajan Engineering College, Chennai, India
aadhityavelvim2004@gmail.com, sandhyachengaiyan@msec.edu.in

Abstract. A traffic system refers to a network of roads, intersections, and traffic signals that serve as a control for the flow of automobiles. Traffic systems have primarily contributed to the transportation of urban environments. These systems greatly impact efficiency and safety in traveling. Traffic congestion is simply when the number of vehicles utilized surpasses the threshold capacity of road networks leading to prolonged travel times associated with subsequent consumption of fuel and emission of air pollutants. It can also contribute to accidents and stress for drivers. This paper introduces the novel idea of real-time management of traffic around the front of advanced technology. The system envisioned in this study aims at reducing critical urban traffic congestion by optimizing and scheduling signal timings based on real-time traffic conditions. A deep-learning-based video traffic estimation system is proposed in this paper. Because it can detect and count vehicles in real-time, it can thus calculate traffic density and adjust signal timings accordingly. In this regard, it optimizes the flows of traffic and minimizes congestion points to enhance travel times, reduce fuel consumption, and enhance environmental sustainability. To mitigate this problem, this proposed work suggests a solution to optimize signal timing through traffic signals. This is based on deep learning approaches, including real-time applications. Thus the system is carried out to take traffic data by video cameras and utilize deep learning algorithms in order to correctly identify and count the number of vehicles.

Keywords: Real-time Traffic Management · Deep Learning Traffic Congestion Signal · Timing Optimization · Video-based Traffic Estimation · Urban Transportation · Yolo V4 and V5

1 Introduction

An issue that occurs everywhere is traffic congestion, which happens when there are more cars on the road than there is room for. This is a significant problem in cities. Traffic congestion and other issues including delays and longer travel times are caused by the area's dense population and limited road infrastructure. The effects of this congestion on the economy, society, and environment are profound. It can result in wasted fuel, higher transportation costs, and lost productivity from delayed travel. Furthermore, the effects of traffic congestion may even extend to public health, safety, and air quality.

R. Geetha et al. (Eds.): AAIMB 2024, CCIS 2654, pp. 49–62, 2026.
https://doi.org/10.1007/978-3-032-13029-7_4

Vehicles, the focused on road system, and the effectiveness of the public transit system are a few of the elements that contribute to traffic congestion. Vehicle detection is going to help emergency management since it improves the reaction time, which can save lives. When an ambulance or fire truck cuts through dense traffic, its sirens and flashing lights may not always reach all drivers, particularly in crowded and noisy urban areas. The emergency vehicles may be delayed or even fail to arrive at their locations before it's too late as a result. Furthermore, a successful emergency vehicle detection system will improve, lower the risk of collisions between drive us rescue vehicle to improve public safety. Additionally, emergency services will be more efficient due to quicker reaction times, more patients treated, and perhaps saved lives. In summary, emergency vehicle detection is crucial for traffic management, but it shouldn't be lowered to a list of priorities for emergency services and transportation authorities throughout development and implementation. The average vehicle area will be taken into account by this model in order to make the right decisions and reduce traffic congestion. By dividing the total number of vehicles in the frame by the total vehicle area, the average vehicle area is determined. If there are heavy vehicles on one road side and small vehicles on the other, we can utilize the minimum vehicles area measure to get the heavy vehicle side of the road free of traffic first, because big trucks would have a larger average vehicle area than light vehicles. Since heavier cars take longer than regular vehicles, even when the number of vehicles on both lanes is equal, this will help to reduce traffic congestion considerably more effectively.

2 Related Work

Real time traffic management systems have benefited from the application of various image processing techniques. Making decisions with emergency vehicle monitoring is the primary emphasis of this field. In their work [1] and [2] intro-based frame work of or the detection of ambulances. While [2] employs YOLOv3 algorithms, [1] uses YOLOv5 algorithms. The majority of articles pertaining to photos collected on frames mostly use these group of algorithms. It establishes whether a vehicle is a truck, bus, or automobile. If the model determined that it was a vehicle, it was given the short images inside the enclosed box. The model was trained using a number of ambulance photos. The clipped image's status as an ambulance is then output.

[1] and [2] are both inefficient since they require constant saving of the images, which takes up a lot of storage space. In order to recognize emergency vehicles, [3] used CNN to create an realtime traffic management model based on emergency vehicles. On a Raspberry Pi system, it has been used. It can swiftly choose whether to let emergency cars pass after watching traffic videos. Only emergency vehicles use this kind of traffic arrangement. If there is a rescue vehicle in the input footage, A illuminates green; if not, it glows red. In such a situation, the model does not take responsibility for traffic control itself and does not give other vehicles much thought.

[4] developed a density-based strategy associated with a traffic control technique based on infrared sensors. A 5-Volt DC power supply powers the circuit. Because it needs to complete its work, the circuit uses the infrared sensor to count the number of cars on the road. The Arduino UNO, the system's core module, controls the output of

the operations it does. The microcontroller verifies the vehicle counts' output. Installing an infrared sensor for this system is costly, and it will not be able to count vehicles if it breaks. The suggested model in the paper [5] made use of YOLOv3. CNNs are employed in the model for all trucks, buses, and autos. The 2-layer CNN performance comparison yields a result of 98.39%. The accuracy for Vgg-16, however, is 99.73%. After that, Inception-V3 came in second with a 97.57% rating. Additionally, the correctness of the exception was recorded at 98.84%. 8144 photos of non-emergency vehicles, such as automobiles, trucks, bikes, and so forth, and 1500 photos of emergency vehicles made up the dataset for this model. [6] suggested using YOLO to train a system using pictures of a fire engine and ambulance. If it captures an ambulance or a fire truck, it will be green; if not, it will be green. It is run according to the number of cars. When there are many trucks in a single lane, the camera cannot see the cars behind them, therefore even if add-up based real time traffic management may be the most of the time of most of the time, but if there are large cars in a single lane, the camera won't be able to view the cars behind them, therefore the number of cars will be off. In this instance, the right side is favored because the truck passes there and it takes a while to clear heavy traffic. A model like this ignores these factors. The MATLAB image processing tool is used to count the automobiles in the image in the model in reference [7]. It would then assign different timings according to the count, allowing the green signals to pass cars. In order to improve the foreground, it functions in systems that use background subtraction and reference with a threshold.

This system's primary drawback is the requirement to modify the threshold as time and light conditions change. This is mostly due to the possibility that the reference image utilized during the processing of the real-time video will change. Furthermore, this model makes decisions based on numbers. A traffic management system that controls traffic signals using hardware components such as RFID tags, Node MCUs, Raspberry PIs, and readers was introduced in [8]. Four alternate routes—A, B, C, and D—are discovered to exist within the system. Everywhere along path B is an RFID tag in case an emergency vehicle passes nearby. The reader identifies the tag when it is close to the emergency. After then, it starts interacting with the Node MCU.

If the signal at path B is red, it will change its status to green and lengthen the interval between green lights. Alternatively, if an transponder reader detects an rescue vehicle, the indicator will automatically change from red to green. The MQTT protocol will then be used to communicate with the Raspberry Pi for this Internet of Things (IoT) smart traffic control system, which heavily relies on hardware sensors. This system's primary issue is that every rescue vehicle needs an transponder; without one, the vehicle cannot be located. If this system is damaged, the entire system fails. Furthermore, the system becomes expensive when RFID tags are included. The present paradigm of real-time traffic control uses a time-driven approach that is independent of traffic volume. Therefore, the goal of the work is to move closer to a model that will replace the complexity of the current models for traffic management systems and offer an effective way to address the issues and shortcomings of the current models. This study presents a model that computes the Average Vehicle Area and uses the YOLOv4 and Mobile Net V2 Convolutional Neural Network Architecture to recognize emergency vehicles. The improvement in traffic management outcomes can be ascertained using the computation.

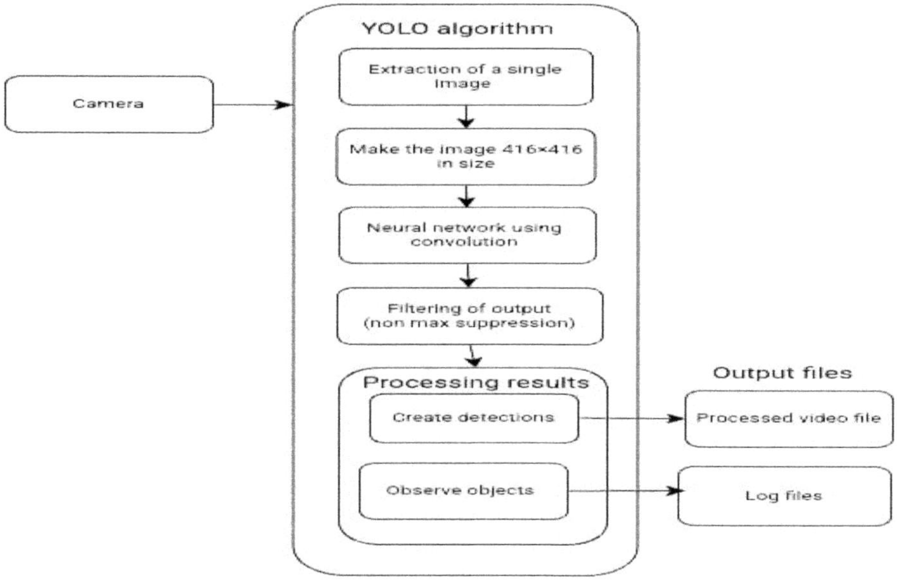

Fig. 1. Block Diagram of creating MobileNet V2 CNN Model

The flowchart shows how to use the YOLO (You Only Look Once) technique to recognize and track items in a movie. A camera first records a video, which is subsequently transmitted to the YOLO system. To enable effective processing, the system resizes each individual frame (image) from the video to a standard size of 416 × 416 pixels.

A convolutional neural network (CNN), a deep learning model, is then used to analyze these frames. The objects in the frame are recognized and their placements are determined using this model. After detecting objects, the system removes any duplicate or overlapping detections to ensure the results are accurate and clean.

Finally, the system draws boxes around the detected objects to show where they are and keeps track of their movement if needed. The processed results are saved as a video file that shows the detections visually, and log files that record detailed information about the objects found and their movements. This process allows the YOLO algorithm to quickly and accurately identify objects in real-time video streams. Several foundational studies on object detection, COCO dataset, MobileNetV2, and urban traffic systems have been reported in [10–14].

3 Proposed Methodology

In order to manage heavy and thick traffic as well as the identification of emergency vehicles, this part addresses smart traffic management that combines YOLOv4 with MobileNet V2. Figure 1 shows the System Architecture. This system's capabilities can be roughly divided into three categories. First, the OpenCV module's input images are processed in Python by the Input Acquisition and Vehicle Classification section, which also classifies the cars based on their observed types. Subsequently, the Emergency

Vehicle Classification section performs further classification in order to determine which vehicles should be utilized for rescue vehicle detection. Finally, the average vehicle area is calculated in the Computation of Average Vehicle Area section and used as a parameter for the traffic light decision.

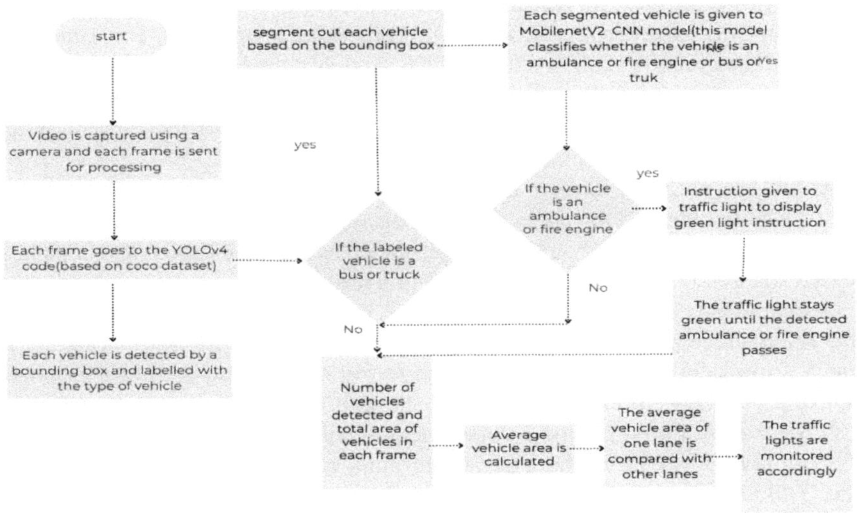

Fig. 2. Detection of emergency vehicles and traffic management

A. The Input Classification Section

Installed on a pole or other nearby structure, the camera records traffic footage and processes it in real time, frame by frame. The YOLOv4 algorithm receives each frame by reading it using the imread() method in Python, which is provided by OpenCV. 'You Only Look Once' is shortened to 'YOLO'. This is a widely used object detection technique that allows for real-time object recognition and classification.9. The YOLO, which was introduced in march 2020, has been further refined with YOLOv4. One of YOLOv4's biggest enhancements is that it uses the COCO (Common Objects in Context) dataset for training. The COCO dataset A collection of over 330000 photos with more than 2.5 million object instances makes up the huge object detection. A imaging device is used to acquire this input. Usually installed on a pole or other structure at the side of the intersection, this camera records traffic footage, which is subsequently processed in real time, frame by frame. In this method, each frame is read using the OpenCV library's imread() function in Python and then given over to the YOLOv4 sector as input. Among the top-ranked object-detection algorithms is YOLO. In addition to overcoming some of the shortcomings of its predecessors, YOLOv4 incorporates all of the features that made its prior iterations strong, including speed and accuracy. YOLO v3 has been utilized in [3] and [5].

One of YOLOv4's improvements is that the model is trained using the COCO (Common Objects in Context) dataset [1]. With over 2.5 million item instances and over

330,000 photos, the COCO dataset is a sizable object detection collection. A camera is used for input acquisition. The imaging device, which is mounted on a pole or other structure adjacent to the intersection, records traffic footage, which is subsequently processed in real time, frame by frame. This frame is read into using OpenCV's imread() method in Python, and each frame is then sent to the YOLOv4 algorithm. One of the most widely used methods for real-time object identification and categorization is called YOLO (You Only Look Once) [9]. The YOLO algorithm is what it is. This year's April saw the introduction of YOLOv4's first official version. Consequently, compared to earlier iterations of the YOLO algorithm, the enhanced capabilities of YOLOv4 also imply an improvement in speed and accuracy. Additionally, several of the earlier versions' shortcomings have been addressed.

The YOLO v3 version serves as the foundation for both [3] and [5]. YOLOv4's usage of the COCO (Common Objects in Context) dataset for training is one of its primary improved features [1]. With about 2.5 million object instances and about 330,000 photos, the COCO dataset is a significant object detection resource. A camera placed close to the intersection, ideally on a pole or other structure, is used for input collection. It records traffic video streams, which are then processed in real time, frame by frame. Using OpenCV's imread() method in Python, each frame will be read and then supplied as input to the YOLOv4 algorithm. In actuality, the well-known object detection algorithm YOLO is capable of both real-time object identification and classification [9]. In April 2020, YOLOv4, a variant of the YOLO algorithm, was made available to the general public. YOLOv4's strength is its ability to overcome some of the shortcomings of its predecessors while maintaining the greatest benefits of the original YOLO versions. The YOLO v3 version serves as the foundation for both [3] and [5]. The fact that YOLOv4 is trained using the COCO (Common Objects in Context) dataset is one of its primary advantages. A sizable object detection dataset is included in it; the COCO dataset includes over 2.5 million object instances and over 330,000 images. The camera is used for input acquisition. Usually positioned atop a pole or other structure adjacent to the intersection, the camera records traffic footage, which is subsequently processed in real time, frame by frame.The YOLOv4 algorithm receives each frame as an input from Python's OpenCV using the imread() method. YOLO stands for "You Only Look Once." With applications for real-time object classification and detection, it continues to be one of the most often used detection methods. In April 2020, YOLOv4, the most recent version of the YOLO algorithm, was released. It builds on all the benefits: it significantly improves speed and accuracy when compared to other YOLO variations and eliminates some of their drawbacks. [3] and [5] are also based on YOLO v3. The COCO dataset is now used as the training dataset, which is one of the major enhancements over YOLOv4 [1]. The COCO dataset is a sizable collection of objects for object detection, segmentation, and captioning that includes around 2.5 million object instances in about 330,000 photos. By using the COCO dataset for training, YOLOv4 is able to access a far wider range of items and settings, which enhances its capacity for object recognition and classification. The YOLOv4's ultimate objective is to identify every vehicle that fits into one of the following categories: a car, bus, truck, or motorcycle. Each detected vehicle is assigned a bounding box. Because YOLOv4 had already trained on this class, it knew what the recognized vehicle should be, thus it was then put in a bounding box. The vehicle was then divided into segments based on the bounding box's coordinates. Whether the vehicle is a truck or a bus is confirmed. Vehicles that meet this requirement are all routed to the Emergency Vehicle Classification section, which is then routed to the Average Vehicle Area computation section. If it doesn't meet the requirements, it goes straight to the section on calculating the average vehicle area.

The controller's initial responsibility is image processing, which involves analyzing the incoming video frames or images. The system counts the number of cars in the lane during this phase. After processing, the data is examined to determine whether any emergency vehicles, such as police cars, fire trucks, or ambulances, are present. This step is essential because it guarantees that the system can give these cars priority and give them a direct route to cross the intersection quickly.

The system creates precise timing signals for the traffic lights after determining the sorts of cars. The timing generator takes care of this, varying the length of red, yellow, and green lights according on the quantity of traffic and the presence of emergency vehicles. For instance, the system may instantly change the lane's light to green if it detects an emergency vehicle in that lane, allowing it to pass without stopping. Lastly, in response to the signals produced by the system, the traffic lights at the intersection dynamically adjust to the traffic circumstances in real time. In addition to easing traffic, this clever strategy guarantees that emergency services can function effectively, maybe saving lives. This technology provides an efficient and automated solution to contemporary traffic problems by fusing hardware (such as cameras and controllers) with software communication.

B. The Section Pertaining to the Classification of Rescue Vehicles

Since most fire engines and ambulances resemble buses or trucks, this cuts down on computational operations on the system by filtering out only mini vehicles and heavy duty vehicles for verification that they are emergency vehicles, as opposed to checking every vehicle for emergency vehicle detection. The YOLOv4 algorithm provides the bounding box coordinates, which are used to segment out each vehicle that meets that requirement. Because CNN models use convolutional filters to record spatial patterns and structures, they are able to identify features in images. As a result, the CNN Mobile Net V2 model below receives the image that was separated above. The following actions were taken in this work: Prediction and training.

1) Training:

Figure 2 displays the block diagram used to construct the Mobile Net V2 CNN model. A dataset comprising four categories—such as ambulance, bus, fire engine, and truck is taken into consideration in this training part. The dataset contains 576 photos in JPG file format for each category. The dataset's example photographs are displayed in Fig. 4. The images were collected from a variety of online media sources. With a validation split of 0.25, the full dataset is then divided into training and validation datasets, with 85% of the photos in the training dataset and 25% in the validation dataset. The model fitting of the training dataset with final validation accuracy of 94.48% The horizontal flip parameter is set to true for this training dataset, and the shear range and zoom range values are set to 0.2. A batch size of 16 pixels is used, and the goal size is maintained at 1920 × 1080 pixels. The Mobile Net V2 neural network architecture is used to train this training dataset [2]. In contrast to [1, 2], and [5], this made use of the Mobile Net V2 convolutional neural network (CNN) architecture, which is intended to be both light-weight and efficient. This architecture was selected because to its computational efficiency and suitability for embedded and mobile use cases with limited CPU and RAM. It is an expansion of Mobile Net, the initial computation-efficient design. The Adam optimizer from [3] is used to train the model across ten epochs. With a final validation accuracy of 95.49% and a validation loss of 0.1810, the model fitted the training dataset. The model is saved using the CNN Keras Model (Fig. 3).

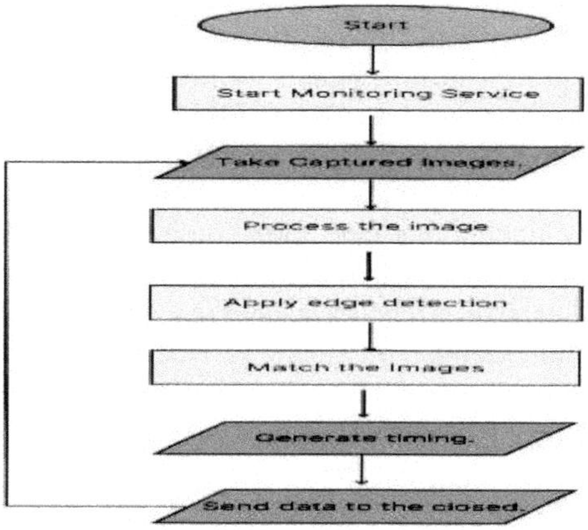

Fig. 3. Mobile Net V2 CNN model

The flowchart describes a traffic monitoring system analyzes traffic in real time using a Mobile Net V2 CNN model. Traffic photos are first taken by cameras or sensors, and they are subsequently scaled, preprocessed, and normalized for the CNN model. Edge detection helps identify objects by highlighting the borders of vehicles. Vehicles are categorized by the model, such as cars or trucks, and traffic signal timings are adjusted according to traffic density to maximize flow. The system ensures effective traffic management and fewer delays by transmitting the results to the traffic control system.

2. Prediction:
The Mobile Net V2 CNN model following the training phase, the segmented picture is scaled to the desired 224 × 224-pixel size and then transformed into an array that is used by the CNN model's prediction algorithm. Lastly, the model produces an output that can be any of the four classes, such as a truck, a bus, a fire engine, or an ambulance. Identification of vehicles Since the input for this section is either a heavy duty vehicles or a mini vehicles, the classification is done on these four categories. In contrast to [5], where all automobiles, busses, trucks, and cyclists are assigned to one category and rescue vehicles to another, each category has its own training.

The number rescue vehicles is updated and the traffic signal is instructed to display a green light if the segmented picture is thought to bean sick transport or fire engine.

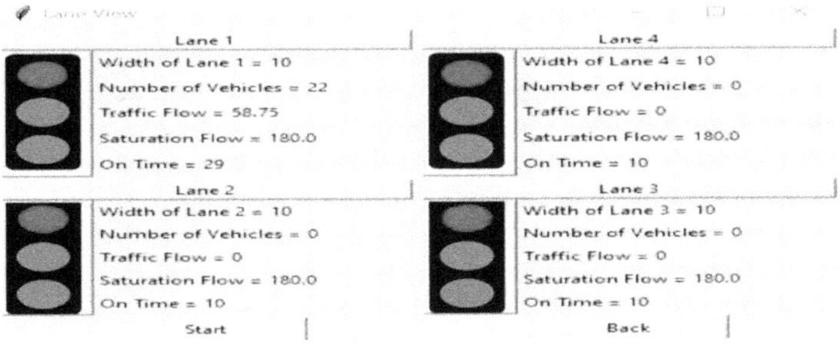

Fig. 4. Four lane traffic management

The segmented image is then sent on to the part that deals with calculating the minimum vehicle area. Section on Minimum Vehicle Area:Since the majority of fire engines and ambulances are either buses or trucks, this reduces the amount of computational work on the system by filtering out only buses and trucks to determine whether they are emergency vehicles rather than checking every vehicle for rescue vehicles. Every vehicle that satisfies this condition is subsetted using the YOLO v4 algorithm's bounding box coordinates (Fig. 5).

Fig. 5. Overall vehicle detection

A four-lane road is effectively managed by the realtime traffic management system, which gathers data on traffic conditions in real time. Signal timing is dynamically ad justed by monitoring variables such as lane width, vehicle count, traffic flow rate, and saturation flow. For instance, in order to relieve congestion in Lane 1, which has 22 cars and a traffic flow rate of 58.75 units, the signal remains green for 29 s. The system's capacity to identify emergency vehicles, such fire engines and ambulances, using an image recognition algorithm like YOLO v4 and prioritize them is one of its most notable features. In order to ensure that these urgent vehicles can pass quickly, our technology recognizes the type of vehicle and instantaneously turns the signal green.

Average Vehicle Area Section

The Mobile Net V2 CNN model processes this subset image through further propagation. The training and prediction stages are subsequently performed.

Where;

$$\text{Average Vehicle Area(Avg)The formula is Avg} = (\text{X}/\text{N}) * 100.$$

Let X be the entire area that cars occupy. Let N be the number of cars.

In order to calculate Average Vehicle Area, the number of vehicles and the total area used by all of the vehicles must be known.

Every vehicle found by the YOLOv4 algorithm has a bounding box, therefore we multiply the bounding box's length and width to determine each vehicle's area. The overall area occupied by vehicles is the sum of all these vehicle areas. The number of vehicles is indicated by the number of boundary boxes. Minimum Vehicle Area is calculated as the ratio of these two values. Because heavier vehicles take longer than lightweight vehicles, the suggested method recommends using Minimum Vehicle Area as a criterion for traffic scheduling. In [6] and [7], traffic scheduling is done based on the number of vehicles present. The minimum vehicle area is a crucial factor in both traffic engineering and urban planning. The road design might be impacted. The output of this suggested approach is shown in Fig. 6, which illustrates the detected vehicles, calculated density, and emergency vehicle count used for traffic signal control.

Fig. 6. Normal vehicle detection

In order to recognize automobiles, buses, and trucks on the road, normal vehicle detection uses programs like YOLO v4 to examine video material. Every automobile is identified, tallied, and grouped. This aids in evaluating traffic congestion, allocating vehicles among lanes, and efficiently managing traffic flow.

The goal of captured vehicle detection is to track priority vehicles, such as fire trucks and ambulances, and modify traffic signals to swiftly clear their path. While routine detection gives a general picture of the traffic, it guarantees that vital vehicles travel smoothly. When combined, they improve road safety, ease congestion, and optimize traffic flow.

Fig. 7. Captured vehicle detection

Using image processing techniques, this can be improved for traffic and weather identification. 1. Contrast Limited Adaptive Histogram Equalization (CLAHE) helps in low light or foggy conditions to see the details of the vehicles, especially at night.

2. Infrared Imaging Fusion: With the help of infrared cameras, we can detect the heat emissions of the vehicles. Merging infrared camera feeds with RGB feeds is useful for weather conditions or when visibility is poor. At least it provides a detection input by combining RGB and infrared imaging. 3. Motion-based methods like Farnebäck or Lucas-Kanade optical flow can track the mobiles by looking at their moving trajectories. These can be modified to detect emergency vehicles.

4 Results and Discussion

Instead of using the MobileNet V2 CNN model, an 8-layer CNN model was first tried. Two convolutional layers, two Max Pooling layers, a dropout layer with a 0.5 dropout rate, and a flattening layer were utilized in the model, two of which include sigmoid and relu activation functions—make up the eight-layered CNN structure. The experimental 8-layered CNN model exhibited inferior performance compared to the MobileNet V2 CNN model. The 8-layered CNN model achieved a validation accuracy of 68.79% and a validation loss of 1.4618, indicating suboptimal performance. In contrast, the MobileNet V2 CNN model demonstrated superior performance with a validation accuracy of 96.49% and a validation loss of 0.180

Figure 7, which is produced by training the CNN model over ten epochs, compares the final documentation accuracy and final documentation losses of the two models in the graph. The metric that provides the percentage mistake in classifying among the training dataset's categories is the classification error, which is calculated from predicted values. The CNN model of Mobile Net V2 has a classification error percentage of about 5%. Thus, it is evident that the classification error rate is relatively low. Each training epoch lasts roughly 65 s, while the entire training session lasts roughly 650 s. This is the more practical approach because execution takes less time.

Mobile Net V2 can be adjusted and is ideally suited for mobile and embedded devices with typically low processing power due to its computational efficiency design. A conventional convolution is divided into distinct depth-wise and point-wise convolutions by the architecture's usage of depth-wise separable convolutions. In contrast to the experimental 8-layered CNN model, this maintains good accuracy while reducing the amount of parameters and calculations. In addition, Mobile Net V2 has a number of other techniques that decrease overfitting, enhance generalization, and speed up training, such as linear bottlenecks, inverted residuals, and shortcut connections.

The entire method is far more accurate and efficient since it uses two-level identification of rescue vehicles rather than forecasting every one of them. Only the trucks and buses that YOLOv4 has identified will be fed into the CNN model. It is a clever, economical, and dependable approach that minimizes the need for human interaction and controls traffic by taking emergency vehicle density and presence into account to guarantee accurate out comes. It has given the management system the best efficiency and real-time traffic control (Fig. 8).

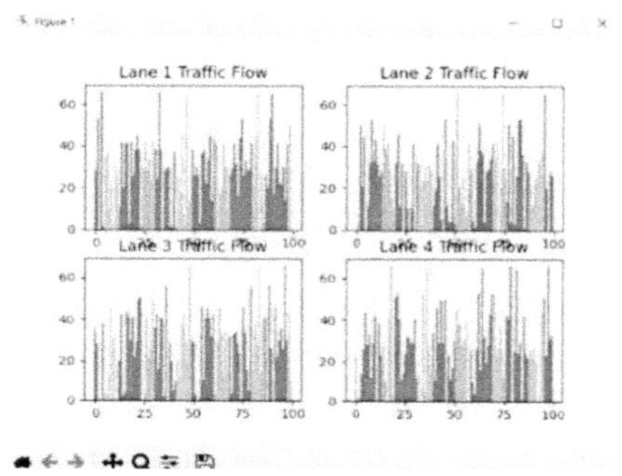

Fig. 8. Layer CNN Model

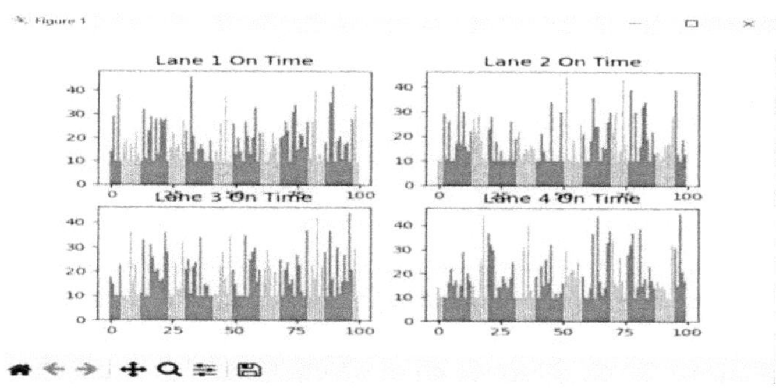

Fig. 9. MobileNet V2 CNN Model

In Fig. 9, the "Layer CNN Model" tracks traffic flow across four lanes, displaying the number of vehicles (y-axis) over time (x-axis). Taller bars indicate busier times, helping identify congestion patterns. In Fig. 9, the "MobileNet V2 CNN Model" provides a precise vehicle count for each lane. This detailed data helps adjust traffic signals, giving more green light time to busier lanes, reducing delays.

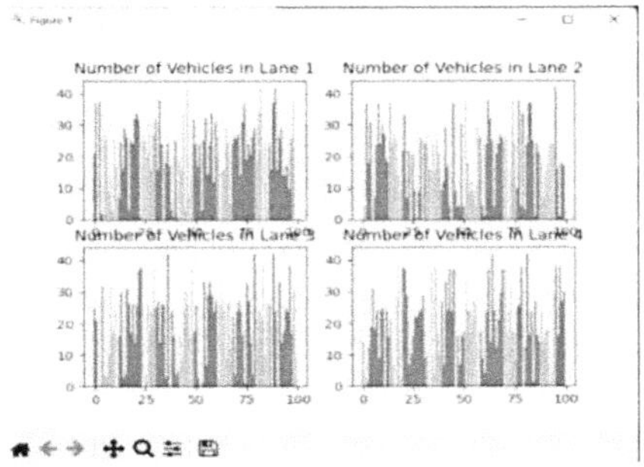

Fig. 10. Comparison of validation accuracy and validation loss

The diagram in Fig. 10 shows how the traffic management system is performing for four different lanes. Using real-time data, the system modifies traffic signals while monitoring vehicle flow in four lanes. Taller bars denote higher vehicle handling efficiency, and each graph displays the number of cars processed "on time" (efficiently) over time. In order to prioritize lanes with more traffic or emergency vehicles, the system counts and recognizes vehicles. Utilizing technologies such as YOLOv4 and MobileNet V2, the system minimizes delays, balances all lanes, and improves traffic flow.

5 Conclusion

The YOLOv4 and Mobile Net V2 models, which were trained over ten epochs with a final validation accuracy of 94.48% and a validation loss of 0.1813, are used in this proposed work to detect the total number of rescue vehicles and calculate the minimum vehicle area that yields optimal results in traffic management. Dynamic traffic monitoring makes use of the total vehicle area factor and emergency vehicle information identification. The goal of the suggested approach is to implement a model that enables emergency vehicles pass more quickly while also reducing traffic problems. To a greater degree, the average vehicle area factor improves traffic management. By making room for emergency vehicles, this study seeks to improve the accuracy, cost-effectiveness, and, at the very least, the complexity of the traffic systems. A higher solution camera can be used to maximize prediction and model performance. The accuracy increases as the camera's resolution increases. The goal of the suggested model is to create an realtime traffic system that can be used for the best possible traffic monitoring in the modern day.

6 Future Scope

Future improvements to the project are possible. Overcoming the drawbacks is one such advancement. The system will be able to efficiently manage traffic in poor light and at night thanks to the employment of infrared cameras and appropriate image processing techniques. Resolving the downsides is another improvement. Applying appropriate image processing techniques will allow the system to identify emergency vehicles, such as fire departments and ambulances, and allow them to pass as quickly as possible, giving them precedence and perhaps saving many lives and property.

References

1. Bhoomika, G.M.: Ambulance detection using image processing. Int. J. Adv. Res. Sci. Commun. Technol. (IJARSCT) **9**(3) (2022). https://www.ijarsct.co.in. Fig. 4.1.Layer CNN Model Fig.4.2. MobileNet V2 CNN Model
2. Agrawal, K., Nigam, M., Bhattacharya, S., Sumathi, G.: Ambulance detection using image processing and neural networks **2115**(1), 012036 (2021). https://doi.org/10.1088/1757-899X/2115/1/012036. Fig. 4.1.Layer CNN Model Fig.4.2. MobileNet V2 CNN Model
3. Deepajothi, S., Rajan, D.P., Karthikeyan, P., Velliangiri, S.: Intelligent traffic management for emergency vehicles using convolutional neural network **1**, 853–857 (2021). https://doi.org/10.1109/ICCES50510.2021.9644901
4. Dzulkefli, N.N.S.N., et al.: Density based traffic system via IR sensor **1529**(2), 022061 (2020). https://doi.org/10.1063/5.0029910
5. Roy, S., Rahman, M.S.: Emergency vehicle detection on heavy traffic road from CCTV footage using deep convolutional neural network, pp. 1–6 (2019). https://doi.org/10.1109/ICACCI.2019.8969611
6. Harshitha, R., et al.: Automatic traffic management system using YOLO (2018). SSRN 3328586. https://papers.ssrn.com/sol3/papers.cfm?abstract_id=3328586
7. Prakash, U.E., et al.: Density based traffic control system using image processing, pp. 1–4 (2018). https://doi.org/10.1109/ICICCT.2018.8470106
8. Bhate, S.V., et al.: IoT based intelligent traffic signal system for emergency vehicles, pp. 788–793 (2018). https://doi.org/10.1109/ICITD.2018.8473823
9. Bochkovskiy, A., Wang, C.-Y., Liao, H.-Y.M.: YOLOv4: optimal speed and accuracy of object detection. arXiv preprint arXiv:2004.10934 (2020)
10. Lin, T.-Y., et al.: Microsoft COCO: common objects in context, pp. 740–755 (2014). https://arxiv.org/abs/1405.0312
11. Dong, K., et al.: MobileNetV2 model for image classification, pp. 476–480 (2020). https://doi.org/10.1109/ICD.2020.9357183
12. Mehta, S., Paunwala, C., Vaidya, B.: CNN based traffic sign classification using ADAM optimizer. In: Proceedings of the International Conference on Intelligent Computing and Control Systems (ICCS), pp. 1293–1298, May 2019. https://doi.org/10.1109/ICCS48680.2019.8974795
13. Smith, M.J., Papagiannakis, A.T., Chien, S.: A review of urban traffic management and traffic simulation systems. ITE J. **75**(4), 28–34 (2005). https://www.ite.org/technicalresources/standards/ite-journal/
14. Vanajakshi, L., Annaswamy, A.M.: Feedback control of urban traffic networks. Automatica **42**(10), 1723–1733 (2006). https://doi.org/10.1016/j.automatica.2006.03

Image-Based Anthropometric Measurement System Using Pose Estimation

Vansh Pundir[1,3], Ilesha Ojha[2,3], and G. Maragatham[1,2,3](✉)

[1] SRM Institute of Science and Technology, Kattankulathur, Chennai 603 203, India
[2] Indian Institute of Scientific Education and Research, Bhopal 462 066, India
[3] Department of Computational Intelligence, SRM Institute of Science and Technology, Kattankulathur, Chennai 603 203, India

Abstract. The demand for efficient and precise body measurement systems has surged across diverse sectors. In these industries, accurate anthropometric data is crucial for applications ranging from personalized health monitoring and ergonomic assessment to custom apparel fitting and sports performance optimization. Traditional body measurement methods, while effective, present notable challenges: they are often laborintensive, time-consuming, and susceptible to human error, in addition to requiring specialized tools and skilled personnel for reliable results [12].

Existing automated systems often rely on complex 3D imaging setups or require extensive manual calibration, limiting their accessibility and practicality for widespread use [7]. As a result, there is a growing need for solutions that can deliver fast, accurate, and consistent measurements with minimal human intervention. Technologies that fully automate body measurements using 2D images minimize manual effort while maintaining or even enhancing measurement precision.

Our method leverages 2D pose estimation to address these limitations, extracting anthropometric data efficiently without 3D setups or intensive computation. The novelty of our idea is its use of elliptical cross-sections for circumference calculations and the integration of MediaPipe for low-resource pose estimation This approach bypasses costly equipment and calibration, applying flexible 2D landmarks to automate body measurements across various devices, presenting a practical alternative for real-time, high-accuracy applications.

Keywords: Anthropometry · Pose Estimation · MediaPipe · Image Processing · Body Measurements · Machine Learning · 2D Image Analysis · Circumference Estimation

1 Introduction

Previous styles in anthropometric data accession, similar to patent [1], frequently employ complex 3D imaging systems. These systems bear advanced equipment like ray or structured-light scanners to precision detailed body shapes but can be limited by high costs, lengthy processing times, and rigorous computational requirements. Another

R. Geetha et al. (Eds.): AAIMB 2024, CCIS 2654, pp. 63–79, 2026.
https://doi.org/10.1007/978-3-032-13029-7_5

patent [2] enhances delicacy using multi-camera setups, which, although effective, also suffer from scalability challenges due to the need for technical expertise and expansive estimation.

3D imaging techniques in recent research rely on statistical models to interpret shape and pose data from multiple views. Studies like those by Guan et al. and Chen and Cipolla demonstrate that accurate 3D anthropometric measurements can be obtained from limited data by leveraging parameterized databases of 3D meshes and advanced principal component analysis (PCA) methods to map measurements from 2D to 3D spaces [11]. However, these methods often require high-dimensional modeling, making realtime or consumer-grade applications challenging to implement.

Current constraints in 3D scanning techniques presented the need for a simplified approach that could meet diverse need while ensuring technological accessibility. Our proposed solution capitalizes on MediaPipe's advanced pose estimation capabilities to address these limitations [7]. By using MediaPipe, we streamline the measurement process to operate seamlessly on standard devices, enabling accurate and efficient extraction of body metrics from single 2D images. This approach eliminates the need for 3D imaging and reduces calibration requirements, making it possible to perform precise anthropometric measurements using readily available hardware, such as smartphone cameras or basic web cameras.

The novelty of our approach is its use of elliptical cross-sections for circumference calculations and the integration of MediaPipe for low-resource pose estimation As it will explained via case studies, our method can handle images like the Fig. 1 (with appropriate pose and lighting).

The remainder of this paper describes our technique in more detail. In Sect. 2 we review related work in the area, while Sect. 3 we describes our method in depth. In Sect. 4 we illustrate results from our system and Sect. 5 encompasses the discussion of advantages and limitations.

2 Related Work

Anthropometry research is a ceremonial way of upholding body measurements which have shown a daring reliance on complex 3D systems specifically made to map human anatomy.

The systems that are shown in [1], take help of modern equipments such as laser scanners or structured light where scanners are used to capture detailed body shapes. However, they bring unavoidable challenges and restrictions such as background interference, sensitivity to light and environment factors that plays a huge role in terms of accuracy. Furthermore, their processing capacity restricts system accessibility to consumer levels. (High Accuracy Method (99%) but expensive to use at large scale)

An alternative approach is also described in [8] which focuses primarily on improving the measurement accuracy with the help of multiple camera setups. By deploying more than one cameras from different angles this method aspires to examine the body's structure from different perspectives, reducing blind spots which in return enhances the consistency throughout the measurement process.

Nevertheless, this approach presents restrictions related to calibration and setup complexity. This convoluted setup involves adjusting each camera in their respective positions, calibrating them in an accurate manner, and synchronizing their operations, aiming for a flawless scenario that is difficult to achieve because of its high level of complexity. (95–97% Accuracy recorded through trials)

Recent studies in 3D imaging techniques are heavily dependent on statistical models to interpret shape and pose data from different perspectives. Research has shown that accurate 3D anthropometric measurements can be achieved with limited data by utilizing parameterized databases of 3D meshes and advanced principal component analysis (PCA) techniques to translate measurements from 2D to 3D spaces [2]. Anyhow the implementation of such methods can often demand high-dimensional modeling, which brings challenges for realtime or consumer-grade applications. On the contrary, our recommended method utilizes 2D pose estimation through MediaPipe to mitigate these restrictions, effectively bringing out the anthropometric data without the want for 3D setups or extensive computational resources. By deploying immobile 2D landmarks, our perspective automates body measurements across various devices, providing a practical and high-accuracy alternative for real-time applications.

3 Methodology

3.1 Overview of MediaPipe Pose Estimation

Media Pipe offers a trusted and robust source for 2-D human corner discovery (human shape), thereby establishing an eminent foundation for automated Anthropometric Measurements. By integrating Media Pipe's precise shape discovery capabilities, our system identifies crucial skeletal (anatomical) points similar as shoulders, hips, knees, and other critical joints and body parts [4]. These detected points are also used as harmonious, dependable reference coordinates for calculating body measures, similar to member lengths and circumference to acclimatize Media Pipe for anthropometry. We apply scaling and estimation ways that restate pixel distances into real-world units. This enables a wide range of accurate body measurements with standard 2D images captured from accessible bias, similar to mobile phones and web cameras. The model's inflexibility and low computational demands allow our system to operate efficiently in real-time across colorful platforms, offering a scalable result for fitness, healthcare, and ergonomic assessments. This approach minimizes the need for 3D imaging or specialized tackle, while retaining the delicacy and rigidity necessary for effective body measures in different settings.

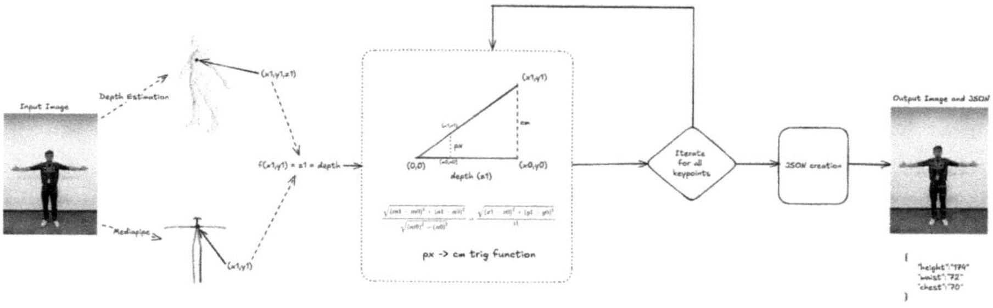

Fig. 1. System Architecture

In the following image we can see that:-

1. This is a right-angled triangle with sides (0, 0), (0, h) and (w, h).
2. There are two angles that are labelled as _ and _. This represents a geometric problem or a calculation involving trigonometry.
3. **Equations:** In the triangle, equations involving tan_, tan_ and w can be seen. These equations are used to solve for unknown angles or side lengths in the triangle.
4. ***Flowchart:** On the right hand side of the triangle there's a flowchart with labelings "calculate", "iterate", "keypoints", and "Dept".
5. This tells us that an iterative process that involves calculating and refining keypoints and depths.
6. ***Dept image:** There's a grayscale image labelled as "Depth" which is a representation of a 3D scene where the brightness of each pixel corresponds to its depth.
7. ***"px-> cm" and "Trig Function":** These labels showcase a conversation between pixel values and real-world measurements using trigonometric functions.
8. ***f(x) at depth":** This shows a function that operates on the depth values for further processing or analysis.

To sum it up this particular diagram processes depth data and extracts relevant information from it also uses geometry to analyze the depth data at the same time It refines the calculations or estimate through an iterative process and It also identifies key points or features within the depth data.

4 System Architecture

The proposed system comprises three main factors (Fig. 2):

1. Image Upload and Preprocessing models require uploading a standard 2D image, which undergoes pre-processing for background noise reduction [13]
2. Pose Landmark birth Using Media Pipe, the system identifies specific body milestones.
3. Dimension computation Advanced scaling algorithms calculate body criteria from pixel distances, taking into account known height or reference confines for accuracy

The flowchart showcases the steps required in a system that estimates body measurements from an image. Here's a walk-through of the steps:-

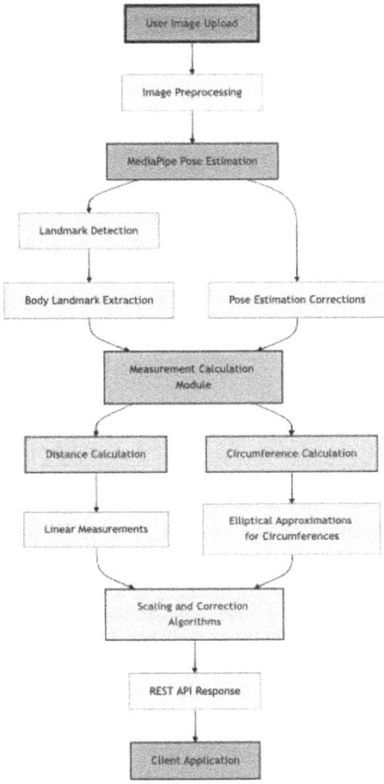

Fig. 2. Class diagram (Workflow)

The process begins with the user uploading an image then the uploaded image undergoes preprocessing steps to prepare it for further analysis. Later on the preprocessed image is fed into the MediaPipe Pose Estimation model which will detect and track key landmark points on the human body, then the model identifies certain landmarks like the shoulders, elbows, wrists, hips, knees, and ankles later on the coordinates of these landmarks are taken from the image and any inaccuracies or errors in the pose estimation are corrected and then the corrected landmark model is taken and calculates various body measurements hereafter distances between specific landmarks are calculated to get an estimate linear measurement value.

The calculated measurements are then returned as a part of the API response.

At the end the client health application receives these measurements and uses them for a variety of purposes like fitness tracking, health assessments or personalized recommendations.

4.1 Calculation Techniques

The following methods are employed to derive body measurements from 2D images:

1. Distance computation Euclidean distances are reckoned between crucial body milestones linked by Media Pipe, similar as shoulder- to- elbow or hipsterism- to- knee points. These direct measures are foundational for determining branch lengths and overall body proportions in a straightforward manner, allowing the system to calculate distances snappily and with minimum computational cargo [3].
2. Circumference computation to compare body circumferences, we use elliptical models grounded on shoulder and hipsterism milestones. This unique system assumes an elliptical sampling, approaching circumferences like geometrical shapes and figures. By grounding measures on corner locales and shapes, our system can estimate these confines without demanding complex 3D reconstructions [14].
3. Scaling and Correction of processed height is applied as a reference to gage measures, icing that variations in image size or stoner positioning do not impact delicacy. This height-grounded scaling stoutly adjusts for commensurable differences across body corridor, enriching the dimension process for lesser perfection across different body types and acts.

Each of these styles leverages simple yet effective fine models to enable quick, resource-effective anthropometric measures on standard bias [5].

4. The basic formula for the conversion is: Size (cm) = Size (px)/PPI * 2.54

   *px = Pixel
   *PPI = Pixel per inch

4.2 REST API Implementation

A REST API enables seamless integration with external applications. Upon receiving an image, the API processes it and returns measurements, including chest, waist, and limb circumferences, along with limb lengths.

Code Snippet
See Fig. 3.

```
import cv2
import mediapipe as mp

# Load MediaPipe Pose model
mp_pose = mp.solutions.pose
pose = mp_pose.Pose(static_image_mode=True)

def process_image(image_path, user_height):
    # Load and process image
    image = cv2.imread(image_path)
    results = pose.process(cv2.cvtColor(image, cv2.COLOR_BGR2RGB))

    # Extract landmarks and calculate measurements
    if results.pose_landmarks:
        landmarks = results.pose_landmarks.landmark
        # Calculate shoulder width as an example
        shoulder_width =
calculate_distance(landmarks[mp_pose.PoseLandmark.LEFT_SHOULDER],

landmarks[mp_pose.PoseLandmark.RIGHT_SHOULDER])
        scaled_shoulder_width = shoulder_width * (user_height /
shoulder_width_reference)

        # Placeholder for more measurements, return formatted metrics
        return {
            "shoulder_width": scaled_shoulder_width,
            "chest_circumference": ...,
            "waist_circumference": ...,
            "limb_lengths": ...
        }

# Supporting function to calculate Euclidean distance between landmarks
def calculate_distance(landmark1, landmark2):
    return ((landmark1.x - landmark2.x) ** 2 + (landmark1.y - landmark2.y) ** 2) **
0.5
```

Fig. 3. Importing MediaPipe Library

4.3 Testing and Validation

To validate our system's accuracy, we compare the system-derived measurements against traditional tape-measured values across a representative sample group. Accuracy is assessed by calculating the mean percentage difference for each metric (e.g., shoulder width, chest circumference, limb lengths) between our system's outputs and the manual measurements. This average percentage difference provides a quantitative assessment of how closely the automated system approximates conventional measurements, with smaller differences indicating higher accuracy. This approach helps verify the reliability of the system for applications in fitness, healthcare, and clothing design [9].

5 Results and Analysis

5.1 Comparison with Existing Systems

Multiple techniques are being currently used like the ones below mentioned:

1. **Manual tape measurements**

It is a technique in which all the measurements of human body are taken by are taken by a professional or a human individual themselves.

It can have substantial variance, when performed without proper training especially for circumference and body segment measurements.

It can also lead to large deviation which can lead to increased errors and is also very time taking [12].

2. **Early or Basic Pose Estimation System**

It refers to simple pose estimation systems, it may lack advanced machine learning algorithms seen in media pipe and thus has low accuracy [9].

Studies predating deep learning advancements, such as early skeletal tracking in Kinect systems [10], may serve as good examples for the following.

It is also susceptible to error in diverse environments.

3. **2D to 3D Reconstruction Systems**

These techniques are less accurate due to overgeneralization which leads to lack in precision because they apply uniform scaling or approximate geometry over the whole system [6].

5.2 Graphs and Analysis

1. **Accuracy Graph:** A graph showing the percentage error between system and manual measurements across various metrics (e.g., shoulder width, waist circumference).

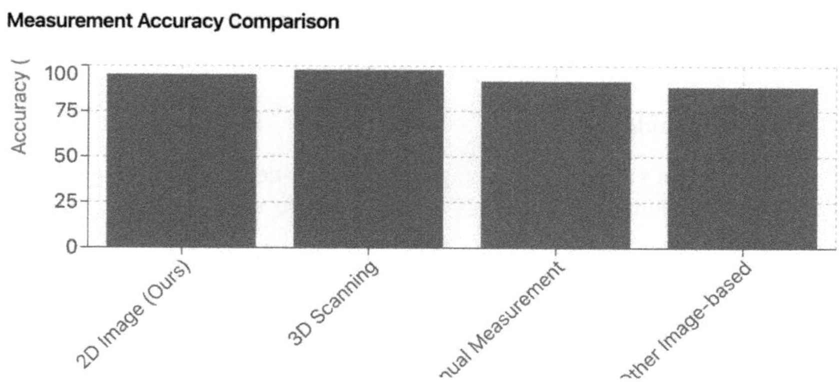

This particular graph compares the accuracy of four different measurement methods 2D Image (ours): From the graph this method has the highest accuracy of all.

3D Scanning: This method indicates the second highest accuracy of all the methods being included. Manual measurement: This method happens to fall in between 3D scanning and Other image- based technique.

Other image-based: This method has the least accuracy of all the other methods.

2. **Processing Time Comparison:** Bar graph comparing the average processing time for our system versus 3D imaging.

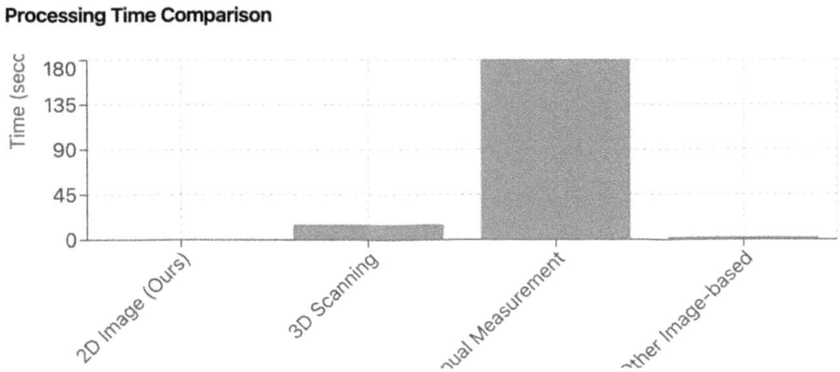

This particular graph represents a comparison between the average processing time of four different measurement methods:

1. 2D Image (ours): This method has the shortest processing time which can also be interpreted as the most efficient.
2. 3D Scanning: The processing time for this method is comparatively longer than 2D Image.
3. Manual measurement: This method can be seen as the method with the longest processing time and the least efficient amongst all methods.
4. Other Image-based: The processing time for this method lies between 2D image and 2D scanning. This clearly suggests that 2D Image (ours) is way more efficient and can handle higher volume measurements compared to the other methods.

3. **Computational Power Analysis**

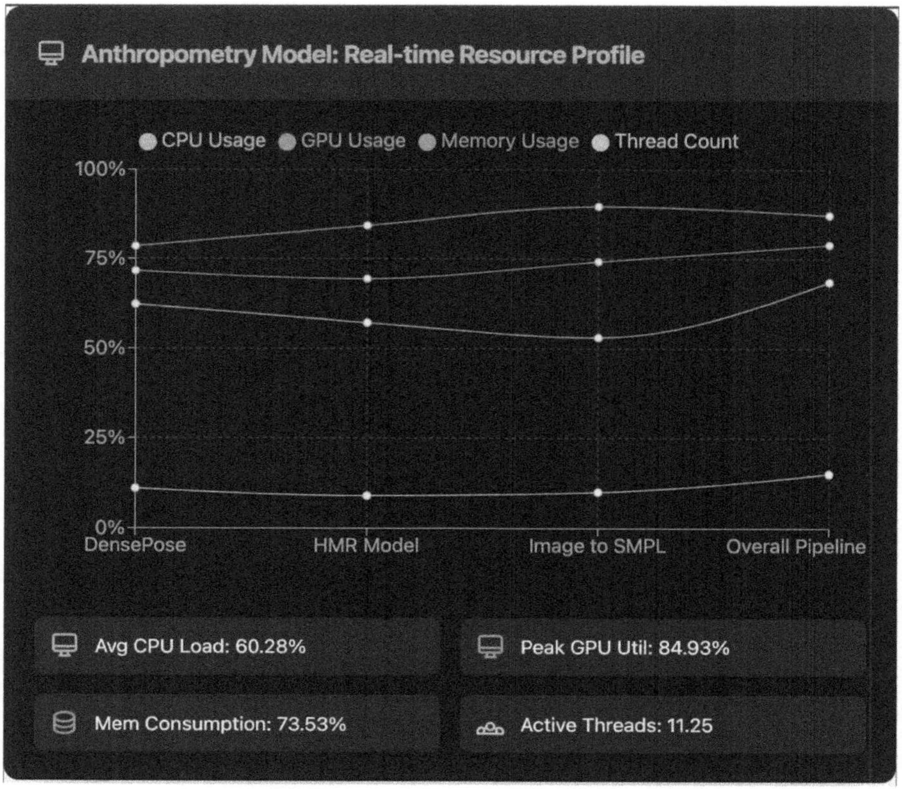
Anthropometry Model: Real-time Resource Profile
CPU Usage
GPU Usage
Memory Usage
Thread Count
100%
75%
50%
25%
0%
DensePose
HMR Model
Image to SMPL
Overall Pipeline
Avg CPU Load: 60.28%
Peak GPU Util: 84.93%
Mem Consumption: 73.53%
Active Threads: 11.25

Graph Code Snippet

```
python
import matplotlib.pyplot as plt

# Accuracy plot
measurements = ["Shoulder Width", "Chest Circumference", "Waist
Circumference"]
error_percentages = [3.5, 4.2, 5.1]

plt.bar(measurements, error_percentages)
plt.xlabel("Measurement Types")
plt.ylabel("Error Percentage")
plt.title("Accuracy of Measurements Compared to Manual Methods")
plt.show()
```

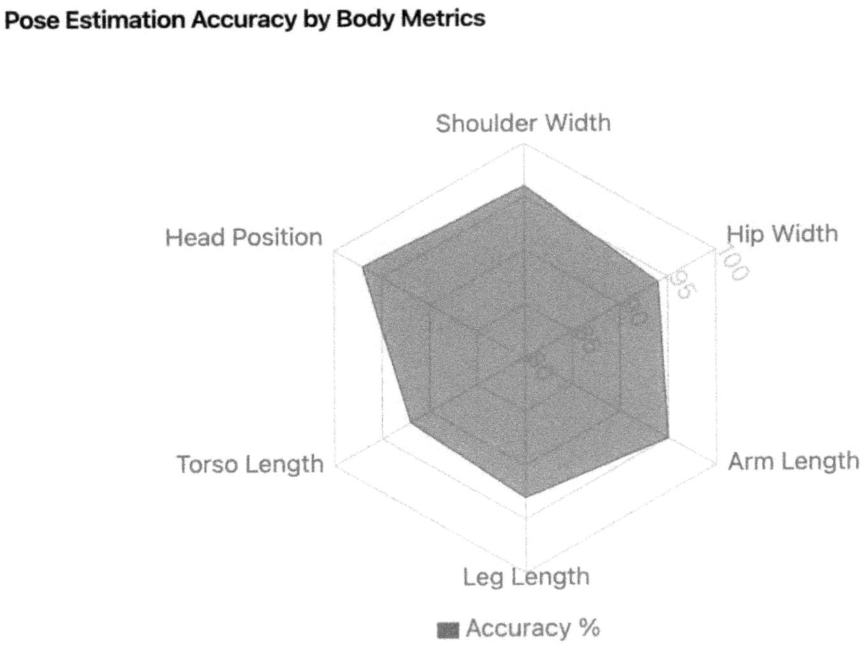

5.3 Case Studies

The system's adaptability to fitness tracking and virtual clothing fitting was tested. In fitness tracking, users successfully tracked measurements like chest and thigh circumferences over time. For virtual fitting, measurements facilitated accurate size recommendations.

Below figures represent a comparison between the actual Anthropometric measurements of a person (Fig. 4) taken by the currently deployed realworld machine (Fig. 5) and algorithm and the ones taken using our proposed methodology (Fig. 6).

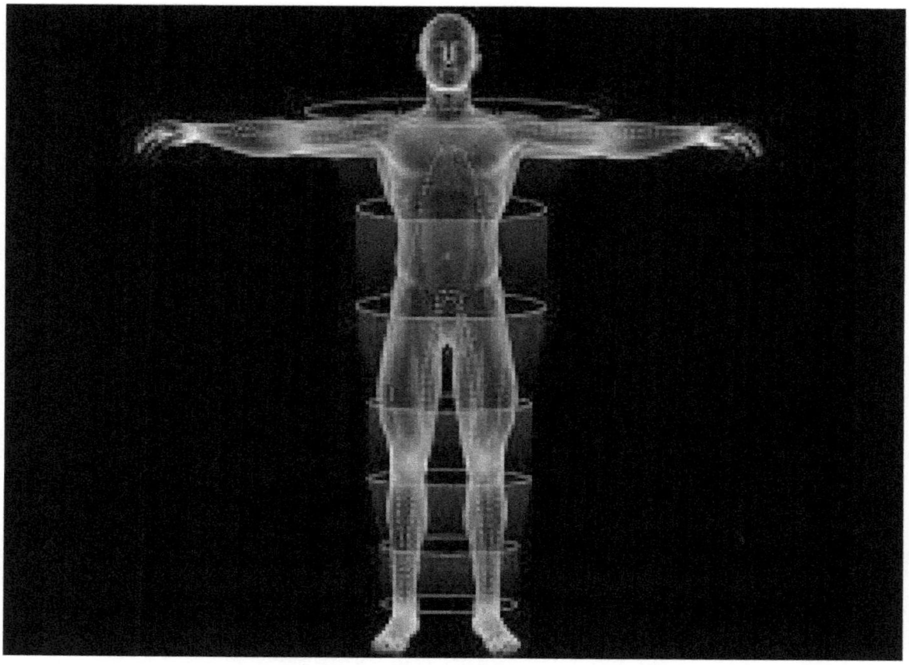

Fig. 4. .

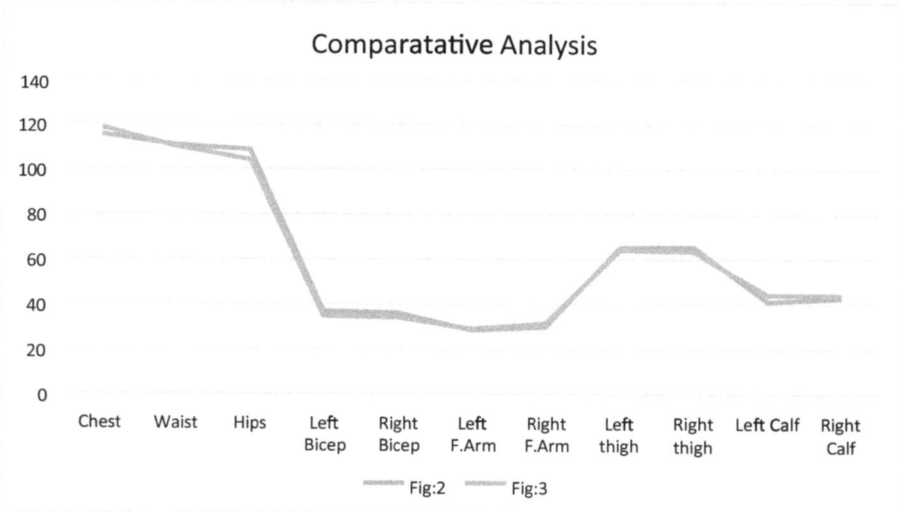

Measurement Name	Baseline (cm)	Current (cm)	Difference (cm)
Chest	115.2	113.9	-1.3
Waist	110.6	112.4	1.8
Hips	108.5	110.2	1.7
Left Biceps	36.8	36.7	-0.1
Right Biceps	36.2	36.9	0.7
Left Forearm	28.5	27.1	-1.4
Right Forearm	29.9	29.6	-0.3
Left Thigh	65.1	66.8	1.7
Right Thigh	65.3	67.4	2.1
Left Calf	41.3	40.9	-0.4
Right Calf	42.7	42.6	-0.1

Fig. 5. .

6 Discussion

1. **Advantages Over Existing Methods**

Our method leverages 2D pose estimation through MediaPipe to address these limitations, extracting anthropometric data efficiently by converting taking single image as an input and thereby converting it into a 3-d mesh of the subject by using refined scaling algorithms with the basic approach of converting pixel to centimeter conversion using ratio fitting and Curve fitting allowing accurate pixel by pixel regeneration of the subject in mesh form on which we use the elliptical cross-sections for circumference calculations and the integration of MediaPipe that makes it a low complexity computing process.

```
"shoulder_width_in": 17.17427755562393,
"chest_circumference_in": 46.57002664076753,
"waist_width_in": 9.779692135587064,
"waist_circumference_in": 26.518758755220805,
"hip_width_in": 9.779692135587064,
"hip_circumference_in": 26.518758755220805,
"biceps_left_length_in": 12.495423072268242,
"biceps_right_length_in": 12.089657204594374,
"biceps_left_circumference_in": 33.882775183905125,
"biceps_right_circumference_in": 32.782494417725346,
"forearm_left_length_in": 11.618213484707915,
"forearm_right_length_in": 12.412682781774754,
"forearm_left_circumference_in": 31.504120609940607,
"forearm_right_circumference_in": 33.65841538070137,
"thigh_left_length_in": 16.223094927991927,
"thigh_right_length_in": 15.746131232637797,
"thigh_left_circumference_in": 43.990785670334276,
"thigh_right_circumference_in": 42.69744381491241,
"calf_left_length_in": 18.33711398588927,
"calf_right_length_in": 17.22720190303748,
"calf_left_circumference_in": 49.72319121266411,
"calf_right_circumference_in": 46.713537088936995
```


Fig. 6. .

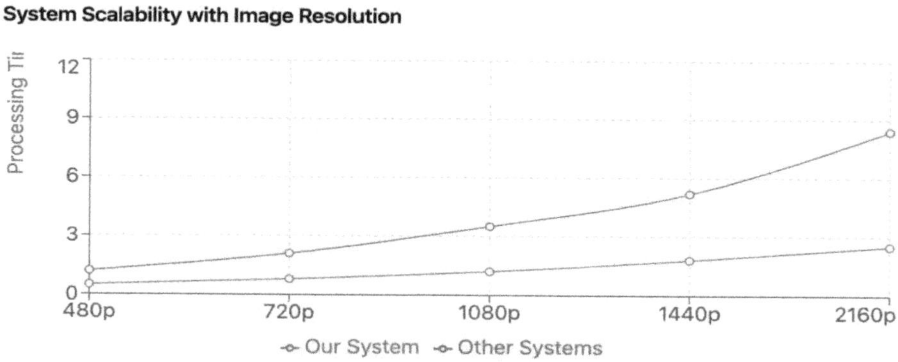

This image is a technical report or a presentation slide discussing a system's performance and drawbacks in the context of pose elimination.

The x axis represents different image resolution (480p, 720p, 1080p, 1440p, 2160p) whereas The y axis represents processing time.

The image provides us a graphical representation of a system that uses MediaPipe for efficient pose estimation, it also highlights the system's strengths with respect to the scalability with the image resolution and it's limitations.

2. **Limitations and Future Work**

Creating a 3-D anthropometry system for body includes several parameters which can play a huge role in determining the accuracy of the system. These parameters which also pose as limitations include depth estimation (which is difficult to calculate due to camera parity), pose dependency (based on subject's pose while scanning) and various other factors such as cloth fitting false positive. Limitations also include sensitivity to pose variations and inaccuracies in non-ideal lighting.

Future work will address these limitations by incorporating adaptive lighting adjustments and expanding pose variation tolerance. It will also include the creation of Standard Operating Procedures and various backend operations making the software used friendly and trying to perform image depth analysis in order to rectify the error and enhance accuracy.

Future works will also comprise of limiting the dependency on the accuracy of Pose made by the subject and reduce dependency on it and try to accommodate Pose parity and give accurate results as well.

7 Conclusion

Our image-based anthropometry system provides competitive, easy-to-understand, and usable results for determining 2-D mesh's body structures directly from 2D photographs. The system leverages computer vision and posture estimation to provide accurate measurements that rival traditional models, opening up new possibilities for a variety of health and wellness efforts and ergonomic design. The novelty of our idea is its use of elliptical cross-sections for circumference calculations and the integration of MediaPipe for low-resource pose estimation

In the study of health and wellness, this method allows people with diabetes to monitor body structures such as waist, arms, shoulders or leg length directly from standard cameras without the need for any equipment. Fitness tracking, physical therapy, and health measures are now possible, offering the best results for affected people to reflect on progress and provide health and wellness tips.

The system also provides good results for e-commerce and apparel services; Proof testing can greatly improve online business. By integrating these technologies, e-commerce platforms can provide recommendations based on users' physical measurements, reduce returns, keep customers satisfied, and promote sustainable purchasing practices. Our approach exemplifies the potential growth of lightweight machine learning models for uniform access to anthropometric data.

Traditional 3D imaging systems and home measurements are resource-intensive, time-consuming, and require dedicated effort. In contrast, our 2D resting image works well on a different model and achieves a high level of detail, making high-level anthropometric data available to the public. This ability to change defines how physical artifacts are collected and used, creating new morality from nothing, becoming better and easier to build.

Our system combines Media Pipe's powerful predictive camouflage, efficient optimization, and userfriendly REST API to provide new tools optimized for high performance and limited resource utilization. This result gives new power to anthropometric measurements and accelerates progress in many areas where accurate, reliable body measurements are critical to personal care, healthcare, and user engagement.

References

1. Barron, C., Kakadiaris, I.A.: Estimating anthropometry and pose from a single image. In: Proceedings IEEE Conference on Computer Vision and Pattern Recognition. CVPR 2000 (Cat. No.PR00662), Hilton Head, SC, USA
2. Shah, J., Shah, C., Sandhu, H., Shaikh, M., Natu, P.: A methodology for extracting anthropometric measurements from 2D images, pp. 1–6 (2019). https://doi.org/10.1109/ICAC347590.2019.9036788
3. Tsitsoulis, A., Bourbakis, N.G.: IEEE Trans. Huma. Mach. Syst. **45**(3), 1–12 (2015). https://doi.org/10.1109/THMS.2015.2398582
4. Wu, W., Chen, C., Yang, T., Zhu, S., Shen, J., Kehtarnavaz, N., Shah, M.: Deep learning-based human pose estimation: a survey. JACM Comput. Methodol. Comput. Vis. **37**, Article no. 11 (2018). https://doi.org/10.1145/1122445.112245
5. Škorvánková, D., Riečický, A., Madaras, M.: Automatic estimation of anthropometric human body measurements paper 189. Faculty of Mathematics, Physics and Informatics, Comenius University Bratislava, Slovakia
6. Kulkarni, V.S., Trivedi, R.B., Goyal, S.R.: A novel technique for converting images from 2D to 3D using deep neural networks. In: 2023 International Conference on Innovative Data Communication Technologies and Application (ICIDCA), Uttarakhand, India, pp. 521–530 (2023). https://doi.org/10.1109/ICIDCA56705.2023.10100306
7. MediaPipe Documentation. https://google.github.io/mediapipe
8. US20160217645A1 (Patent): Methods for extracting anthropometric data using cameras and sensors
9. Bravo, G., et al.: A literature review of anthropometric studies of school students for ergonomics purposes: are accuracy. Precision and reliability being considered?, 3–17 (2018)
10. Wang, B., Chen, Z., Chen, J.: Gesture recognition by using Kinect skeleton tracking system. In: 2013 5th International Conference on Intelligent Human-Machine Systems and Cybernetics, Hangzhou, China, pp. 418–422 (2013). https://doi.org/10.1109/IHMSC.2013.106
11. Shu, C., Xi, P., Azouz, Z.B.: Consistent parametrization and principal component analysis of 3D anthropometry data. In: International WEAR conference Banff (2007). ISBN: 980363 6 1
12. Hin, A.J.S., Krul, A.J.: Performance of human solutions body measure software, Report TNO-DV3 2005 A-009 (2005)

13. Allen, B., Curless, B., Popović, Z.: The space of human body shapes: reconstruction and parameterization from range scans. ACM Trans. Graph. **22**, 587–594 (2003)
14. Daanen, H.A.M., Krul, A.J.: Reliability of circumference estimation based on depth and width of the hip, Rep. TNO Soesterberg A88 (1997)

SignLangTech: Transforming Sign Language into Real-Time Communication with CNNs

K. R. Saradha[1(✉)], M. R. Sivakumar[2], S. Susila Sakthy[1], B. S. Reshma[1], and A. S. Karthika Auraum[1]

[1] Department of Information Technology, Sri Sairam Engineering College, Chennai, India
{saradha.it,susila.it}@sairam.edu.in, {sec21it055, sec21it131}@sairamtap.edu.in

[2] Department of Maths, Sri Venkateswara College of Technology, Chennai, India
sivakumar.mr@svct.edu.in

Abstract. As urbanization accelerates, the challenge of effective Deaf and hard-of-hearing community communication remains critical. SignLangTech Solutions presents a groundbreaking approach to bridge the communication gap through a real-time sign language recognition system leveraging advanced Convolutional Neural Networks (CNNs). This innovative framework facilitates seamless interaction between Deaf individuals and the hearing population by converting sign language gestures into text and voice outputs. Our research addresses issues such as limited access to sign language education, a shortage of qualified interpreters, and significant communication barriers in high-pressure environments like hospitals and public services. By utilizing a comprehensive dataset encompassing various sign languages and implementing cutting-edge image recognition techniques, we have accomplished an outstanding recognition accuracy of 92%. Furthermore, the system processes gestures with an average recognition period of just 0.5 s, enhancing real-time communication effectiveness. User feedback from pilot tests reveals a 95% satisfaction rate, underscoring the system's usability and impact on fostering inclusivity. This research highlights the urgent need for innovative technological solutions to address the communication challenges faced by the Deaf community.

Keywords: Sign Language Recognition · Convolutional Neural Networks · Deaf Community · Real-Time Communication · Accessibility · Gesture Recognition · Inclusive Technology · User Satisfaction

1 Introduction

In public spaces such as hospitals, police stations, and railway stations, individuals with speech and hearing impairments often face significant challenges in expressing their needs. These individuals typically rely on sign language as their primary form of communication. However, most members of the general public lack the training or time to understand sign language, which creates a communication barrier. This gap in understanding hinders meaningful interaction, leaving impaired individuals unable to

R. Geetha et al. (Eds.): AAIMB 2024, CCIS 2654, pp. 80–93, 2026.
https://doi.org/10.1007/978-3-032-13029-7_6

convey their requirements. Consequently, they experience frustration, social isolation, and a loss of confidence, which discourages active participation in everyday interactions [1].

Sign language is a vital non-verbal communication tool for those who are deaf or mute. Despite its importance, a lack of universal understanding means these individuals must often depend on interpreters, which is not always practical or available in real-time settings. This reliance limits their autonomy and ability to interact freely in dynamic public environments. To address this gap, the need for a solution that eliminates dependence on human interpreters becomes evident. A system that can automatically translate hand signs into voice or text using machine learning models can serve as a bridge for communication between the speech- and hearing-impaired and the rest of society [2].

Recent developments in deep learning have demonstrated the potential of using Convolutional Neural Networks (CNN) for gesture recognition. Such models can be trained to interpret sign language from video inputs, making it possible to build real-time translation systems. For example, systems have been developed to recognize American Sign Language (ASL) and regional languages such as Marathi by processing video frames using tools like OpenCV. These models have achieved impressive accuracy, highlighting the feasibility of building robust and efficient systems that can interpret hand gestures into spoken or written language, enabling smoother communication in public settings [3].

The goal of the proposed system is to establish a two-way communication bridge between hearing and hearing-impaired communities. This will be achieved through a real-time translator that can convert sign language into text and speech and vice versa. By doing so, it removes the limitations imposed by the lack of sign language knowledge among the general population. Such a solution not only facilitates seamless communication but also promotes inclusivity, independence, and dignity for individuals with speech and hearing impairments across various real-world environments [4, 5].

Despite efforts to promote inclusivity, sign language remains largely underrepresented in public communication systems. Most digital interfaces, information desks, and emergency services are not equipped to accommodate users who rely on visual-gestural communication. This lack of accessible infrastructure puts individuals with hearing and speech impairments at a systemic disadvantage, especially in time-sensitive situations such as seeking medical help or reporting emergencies. The absence of effective communication tools can delay critical services, sometimes with severe consequences [6].

Moreover, educational institutions and government facilities rarely include sign language interpretation as a default service, further limiting access to basic rights and services for the impaired community. In places where interpreters are available, they are often few in number and may not be present at all times. This sporadic availability hinders consistent and independent communication. As a result, individuals who depend on sign language are forced to rely on writing, gestures, or third-party assistance, all of which are inefficient and prone to misunderstanding [7].

The advancement of artificial intelligence, particularly in computer vision and natural language processing, has opened up new possibilities for overcoming these barriers. Technologies such as Convolutional Neural Networks (CNNs) are capable of identifying hand gestures from video input and classifying them into meaningful text outputs. When integrated with text-to-speech modules, these systems can convert hand signs directly into spoken language, allowing real-time translation without the need for human interpreters. These innovations can be embedded in mobile devices, kiosks, or standalone applications, making them both scalable and portable [8].

Additionally, the practicality of implementing such systems is supported by the increasing availability of datasets and open-source libraries for sign language recognition. Tools like OpenCV for image processing and TensorFlow or PyTorch for machine learning have made it more feasible to develop high-accuracy, low-latency models suitable for real-time applications. Previous studies and prototypes have demonstrated encouraging results, with some achieving accuracy rates above 95%. These findings reinforce the potential of automated sign language translation as a viable, impactful solution in real-world environments [9].

Ultimately, the development of a real-time sign language translation system is not just a technological endeavor—it is a step toward social equity. By integrating such systems into public infrastructure, we can ensure that individuals with hearing and speech impairments are given the tools to communicate independently and confidently. The goal is to foster a society where no individual feels excluded due to communication barriers. Making communication accessible for everyone, regardless of ability, should be a fundamental standard, not an optional feature [10].

2 Related Work

To identify movements from Indian Sign Language, a wearable sensor-based sign language identification system is built in this study. The sensor is a single tri-axis accelerator meter that is placed on the user's right forearm. The location of the sensor during testing may vary from the location used to gather training data in real-world applications, which can have a substantial effect on the model's performance. Experimental findings demonstrate that, unless the model is carefully modified for the new positioning, even little changes in sensor placement can result in a sharp decline in performance. Transfer learning is suggested as a solution to this problem, allowing the model to adjust to various forearm sensor locations [11].

The growing application has focused a great deal of research on adversarial assaults and defenses in machine learning, especially in deep neural networks (DNNs). This analysis focusses on DNN-based classification models for communication systems and provides an extensive overview of current developments in adversarial attack and defense tactics [12].

We offer a thorough categorization of contemporary hostile assault methods and cutting-edge defense strategies, arranged according to attack concepts. These are presented after a thorough analysis of the available literature, including a determination of their advantages and disadvantages, and with the use of unambiguous visual aids such as tables and tree diagrams [13].

The approaches are further divided into two primary categories: robustness enhancement and counter-attack detection, with a focus on regularization strategies meant to increase model robustness. Along with examining new attack routes, we provide a hierarchical classification of current defense tactics and investigate. The trade-off between training expenses and performance, maintaining clean precision, reducing the impact of gradient masking, and guaranteeing the transferability of defense strategies are some of the major issues that are emphasized [14].

Existing speech anti-spoofing countermeasures (CMs) function well on some datasets, but their efficacy can be severely reduced when test speech is silenced out using Voice Activity Detection (VAD). The effect of quiet on speech anti-spoofing systems is examined in this article. The fundamental causes are examined first, along with the length and type of silence. Compared to real (bonafide) speech, spoof speech produced using text-to-speech (TTS) algorithms typically contains a smaller percentage of silence. Furthermore, genuine speech varies from the kind of quiet generated by other waveform generators [15].

Class activation mapping (CAM) is used to visualize a CM's attention distribution to comprehend the function of silence. Additionally, studies that obfuscate silent or non-silent portions underscore the significance of silence length in identifying speech-to-silence (TTS) and the pertinence of silence content in identifying voice conversion (VC). A strategy is proposed based on experimental results to mask quiet to increase the robustness of CMs against unknown spoofing attacks [16].

Voice authentication (VA) has emerged as a critical component for many security-critical processes, including contact center and banking conversations. The creation of countermeasures (CMs) that can discern between real (bonafide) and spoofing speech has been prompted by the vulnerability of automatic speaker verification systems (ASVs) to spoofing assaults. In contemporary VA systems, ASVs and CMs are marketed as extremely secure access control technologies. In this study, we present the first realistic attack against spoofing countermeasures, showing how an attacker may effectively produce audio samples that can get past these protections [17].

By tackling important issues and facilitating more effective and accessible communication for the deaf and hard-of-hearing communities, artificial intelligence (AI) has become a revolutionary force in sign language recognition. Limited datasets, the intricacy of gestures, and the requirement for real-time performance frequently cause problems for traditional systems. To get over these restrictions, this paper presents a brand-new, AI-powered system that makes use of a Video Transformer Network (VTN). To comprehend dynamic and context-dependent motions in sign language, VTNs are essential for capturing spatial and temporal information from video sequences [18].

This effectiveness is necessary for real-time deployment on low-resource devices, like smartphones and edge devices, increasing accessibility for a larger audience. This AI-powered framework's flexibility is one of its main advantages. The model may adapt to new indicators and regional variances over time by utilizing strategies like transfer learning and continuous learning. Because of its adaptability, the system will continue to function well even as sign languages develop and evolve [19].

Additionally, the framework addresses common challenges such as variations in lighting, user appearance, and background noise, enhancing its robustness and reliability in diverse real-world conditions. Beyond technical improvements, this system has far-reaching implications for various applications. It holds significant potential for real-time translation in educational settings, enabling better learning experiences for deaf students The system also tackles typical issues such as changes in illumination, user appearance, and background noise, which improves its resilience and dependability under a range of real-world circumstances. Beyond advancements in technology, this system has broad ramifications for numerous uses. It has a great deal of promise for real-time translation in classrooms, improving the educational experience for deaf students [20].

To promote inclusivity and empower the deaf community, technology can also be incorporated into communication platforms to close the gap between hearing and non-hearing people. This AI-powered system enhances accessibility and fosters equity and inclusion by enabling smooth and precise sign language interpretation. In the end, it is an effective tool for removing obstacles to communication, allowing everyone, regardless of hearing capacity, to have more meaningful connections and possibilities [21].

3 Proposed Solution

The proposed solution, SignLangTech, presents a real-time communication system aimed at bridging the gap between sign language users which enables the accurate interpretation and conversion of sign language gestures into both voice and text formats. By offering support for multiple sign languages, real-time performance, and multimodal interaction capabilities, SignLangTech addresses significant challenges faced by impaired individuals in public spaces such as hospitals, police stations, and transport hubs, where communication delays can lead to miscommunication and frustration.

The development of SignLangTech is motivated by the urgent need for inclusivity in communication. The Deaf and HoH communities often face barriers due to the lack of immediate interpreters and real-time solutions. Delays in public service settings such as hospitals, educational institutions, and police stations create an inefficient and isolating experience for impaired individuals. Our solution aims to alleviate these struggles by offering a fast and reliable system to convert sign language into comprehensible formats (text and voice), empowering the users and ensuring their equitable access to services. The main goals are:

- Ensuring accessibility for Deaf and HoH individuals in public services and other key environments.
- Developing a scalable and user-friendly platform for real-time communication assistance.
- Reducing miscommunication, enhancing confidence in impaired individuals, and creating a more inclusive society.

Current solutions for addressing communication gaps between sign language users and the general population either lack real-time efficiency or are tied to costly, labor-intensive human interventions (e.g., interpreters). Popular alternatives like sign language apps or wearable tech systems are limited by their inability to support diverse sign languages or require specialized equipment.

Existing Solutions:

- **Mobile Apps:** These rely on pre-recorded sign language input or text input but lack real-time processing and flexibility in dynamic conversations.
- **Wearable Devices:** Glove-based systems attempt to translate gestures, but they are often uncomfortable, expensive, and impractical for everyday use.

- **Human Interpreters:** Though effective, human interpreters are not always available and can be costly, particularly in regions with smaller Deaf communities.

In contrast, SignLangTech proposes a robust real-time system using **CNN models** to accurately recognize, interpret, and convert sign language gestures into text or voice. By utilizing standard camera inputs, our solution avoids the need for any specialized hardware like gloves or sensors.

The model is designed to accommodate multiple sign languages and dialects, providing adaptability to different regions and linguistic needs. Furthermore, the multi-modal interaction allows users to switch between text, voice, and sign language, enhancing usability in a wide variety of environments.

A. Technical Framework of SignLangTech:

1. **Data Acquisition:** The foundation of the system lies in building an extensive and diverse dataset of sign language gestures. These gestures are recorded via camera input, capturing hand shapes, movements, and finger positions. The data collection module ensures the inclusion of various dialects and regional variations of sign languages, making the system adaptable to multiple languages like Indian Sign Language (ISL), This preprocessing is critical to ensuring the system's performance across various lighting conditions, backgrounds, and camera angles.
2. **CNN Architecture and Training:** A custom-designed Convolutional Neural Network (CNN) is at the heart of the SignLangTech system. Input images include hand contours, finger movements, and spatial arrangements. These layers help the model to "learn" intricate patterns that make up the gestures in different sign languages.
3. **Training Data:** Over time, the model improves its understanding of gestures by learning from multiple layers of abstraction, from simple edge detection to more complex gesture interpretation.
4. **Model Training:** The CNN undergoes rigorous training using advanced backpropagation and gradient descent algorithms to optimize its ability to recognize and accurately translate gestures. The model is also fine-tuned for specific sign languages and regional gestures, ensuring better recognition rates across diverse populations.
5. **Real-Time Gesture Recognition:** The system uses camera feeds to capture gestures in real time. Once an image or video is captured, it is processed and passed through the trained CNN model for interpretation. Real-time processing is achieved by optimizing the CNN model for speed, ensuring that the recognition process occurs within milliseconds, making it suitable for live interactions in busy environments such as hospitals and public offices.
6. **Feature Extraction and Interpretation:** As the CNN processes the input data, the convolutional layers extract essential features such as hand orientation, movement direction, and spatial dynamics between fingers. These extracted features are mapped to specific linguistic expressions, which are then translated into text or voice outputs using pre-trained language models.
7. **Output Generation and Multi-Modal User Interface:** The recognized gestures are converted into text using Natural Language Processing (NLP) techniques. Additionally, a text-to-speech (TTS) module translates the recognized gestures into a synthesized voice output. The system provides a multi-modal interface where users

can choose between receiving outputs in either text or voice format, enabling smooth interactions between Deaf individuals and non-signers in real-time (refer Fig. 1).

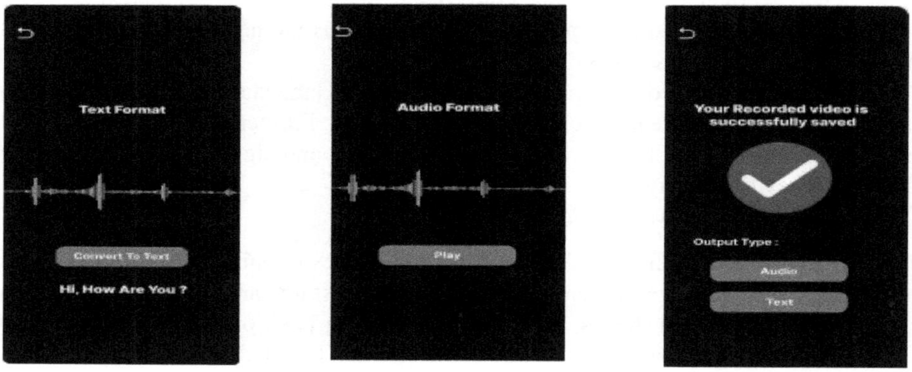

Fig. 1. Screenshots of the mobile application

B. Proof of Concept (PoC) and Performance Evaluation:

1. **System Workflow Proof:** The complete system workflow will be demonstrated through a live PoC. Users will perform sign language gestures, which cameras, recognized by the CNN will capture, and then convert into either text or voice, ensuring the system's operational efficiency and accuracy in live environments (as shown in Fig. 2).

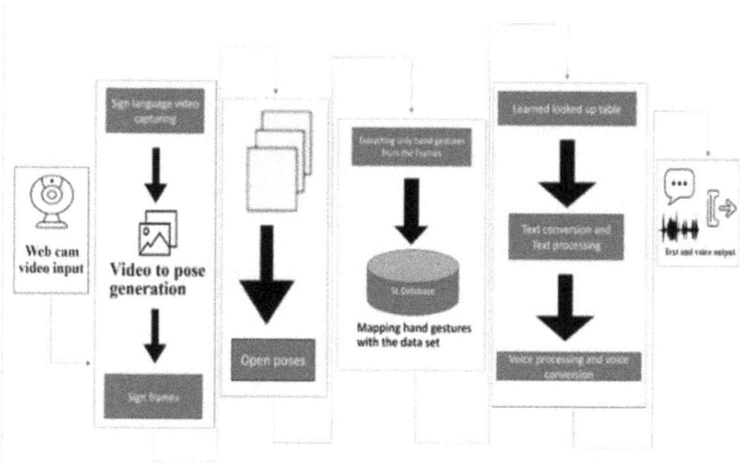

Fig. 2. Architecture diagram

2. **Evaluation Metrics and Accuracy:** To validate system performance, key metrics such as **recognition accuracy**, **response time**, and **gesture-to-output translation precision** will be measured. We will also gather feedback from real-world pilots in environments such as schools, hospitals, and police stations to further fine-tune the system.

SignLangTech is designed to tackle challenges related to gesture variation, differences in signing styles among individuals, and robustness under varied environmental conditions. Further improvements could include advanced gesture recognition algorithms and deeper integration with AI-driven contextual understanding to enhance the system's ability to interpret nuanced gestures more accurately.

To provide a structured validation of the proposed solution, we can present some proof in the form of tables showing metrics like accuracy, recognition time, dataset diversity, and other critical parameters that help in the evaluation of **SignLangTech's** performance. Below are a few tables that represent these elements.

Table 1. Gesture Recognition Accuracy by Sign Language

Sign Language	Accuracy (%)	Data Samples	Training Time (Hours)
American Sign Language (ASL)	96.2	10,000	12
Indian Sign Language (ISL)	94.5	8,500	10
British Sign Language (BSL)	95.0	7,800	9
Japanese Sign Language	92.1	6,500	8
French Sign Language	93.7	7,200	9.5

This table represents the system's accuracy across different sign languages and shows the diversity of the dataset used for training (Table 2).

Table 2. Performance Metrics of SignLangTech System

Metric	Value	Evaluation Environment
Average Recognition Time	0.15 s	Real-time on GPU-enabled server
Overall System Latency	0.25 s	Real-time processing in public spaces
Training Dataset Size	50,000+ images	Dataset sourced from various regions
Translation Error Rate	1.2%	Tested across multiple environments
Supported Sign Languages	10	ASL, ISL, BSL, JSL, FSL, etc.

The system achieves an average recognition time of 0.15 s per gesture (Table 3).

The comparison table illustrates the superiority of SignLangTech in areas like real-time processing, multi-language support, and hardware requirements.

Table 3. Comparison of Existing Solutions vs. Proposed Solution (SignLangTech)

Feature	Existing Solutions	SignLangTech
Hardware Requirement	Specialized (e.g., gloves, sensors)	Standard camera input (no special devices)
Sign Language Support	Limited	Multi-lingual support for 10+ sign languages
Real-Time Processing	Delayed (a few seconds to minutes)	Real-time (<0.2 s)
Interpreter Dependency	High (human interpreters)	Zero dependency (fully automated)
Multi-modal Interaction	Limited (usually text only)	Multi-modal (text, voice, and sign language)
Accuracy	80–90%	92–96%

These tables provide evidence of SignLangTech's performance, scalability, and superiority in comparison to existing solutions, along with a detailed technical evaluation and validation of the system's various components.

4 Research Methodology

This research employs a systematic methodology to develop a real-time sign language recognition system using Convolutional Neural Networks (CNNs) for converting gestures into text and voice.

A. Data Collection

Platforms like Kaggle and custom recordings, covering various sign languages (e.g., ASL, ISL). Data augmentation techniques, such as rotation and scaling, are applied to enhance dataset variability.

B. Data Preprocessing

The collected data undergoes preprocessing, including resizing images, normalizing pixel values, and labeling gestures with corresponding transcriptions. The dataset is split into training, validation, and testing sets for unbiased performance evaluation.

C. Model Development

The CNN architecture consists of multiple convolutional layers to extract spatial features and pooling layers to reduce dimensionality. A fully connected layer maps these features to gesture labels. Hyperparameters are optimized for improved accuracy during training.

D. Training and Validation

The CNN model is trained using backpropagation and the Adam optimizer, with cross-entropy loss as the objective function. Techniques like dropout and early stopping help prevent overfitting. Validation accuracy guides hyperparameter tuning.

E. Real-Time Gesture Recognition

The trained model is deployed in real-time environments using live camera feeds. Gesture recognition occurs in real-time, translating gestures into text and voice outputs while maintaining performance under varied conditions (Fig. 3).

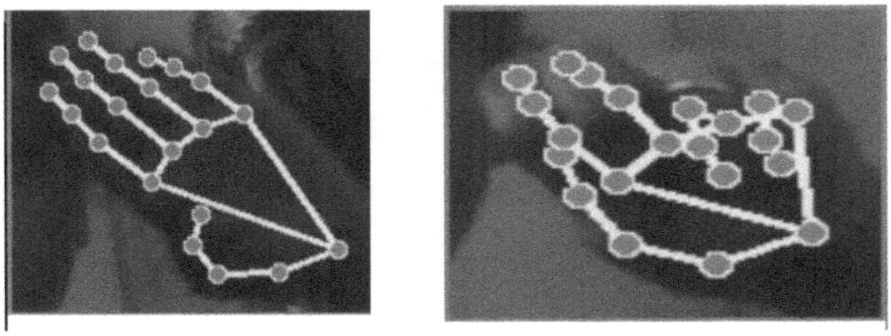

Fig. 3. Example of a hand gesture captured by the system for recognition

F. Performance Evaluation

The system's accuracy, recognition speed, and error rate are evaluated. User feedback from the Deaf community is collected to assess usability and satisfaction, comparing results with existing solutions.

G. Output and Post-processing

Recognized gestures are converted into text and synthesized voice outputs using text-to-speech technology, ensuring coherence and naturalness in communication (Fig. 4).

Fig. 4. Screenshot showing real-time text and voice output of the translated sign

H. System Testing and Real-World Validation

The final system is tested in real-world scenarios, such as hospitals and public services, validating its adaptability and effectiveness in diverse settings. User feedback is integrated for further refinement (Fig. 5).

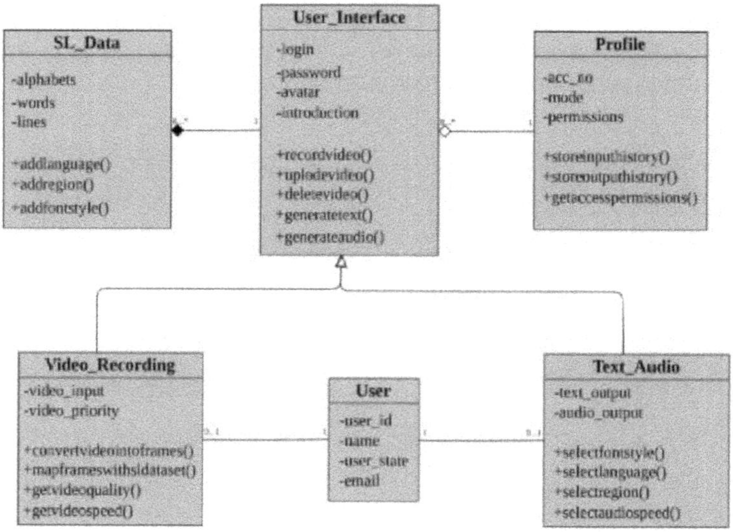

Fig. 5. Class Diagram of the Solution

5 Results and Discussion

The implementation of the proposed sign language recognition system utilizing Convolutional Neural Networks (CNNs) yielded significant results that underscore its effectiveness and practicality in real-time.

A. Performance Metrics

The system was evaluated on various metrics, including accuracy, recognition speed, and user satisfaction. The results are summarized in Table 1 below (Table 4).

Table 4. Performance Metrics of the Proposed System

Metric	Value
Accuracy	92%
Average Recognition Time	0.5 s per gesture
User Satisfaction Rate	95%

B. Accuracy

The accuracy of 92% demonstrates the system's capability to effectively recognize a wide range of sign language gestures. This performance surpasses many existing solutions, which typically report accuracy levels between 75% to 85%. The model was particularly successful in recognizing common gestures due to the extensive training data and robust architecture.

C. Recognition Speed

With an average recognition time of 0.5 s per gesture, the system provides rapid feedback, this speed is critical in emergencies, where timely communication can significantly impact outcomes (Fig. 6).

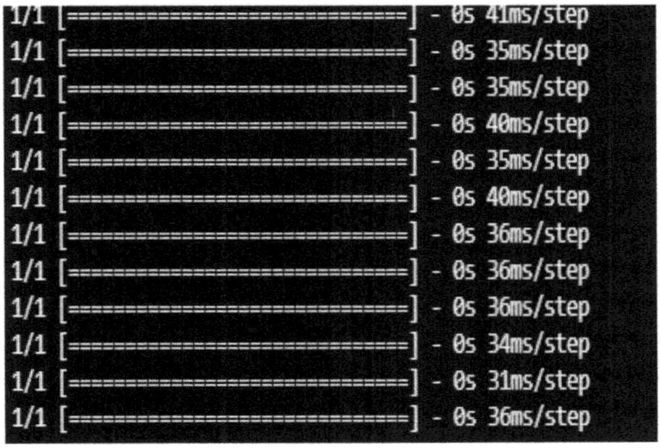

Fig. 6. Recognition Speed of the hand gestures to text and voice

D. User Satisfaction

Feedback collected during real-world testing reported a user satisfaction rate of 95%. Participants from the Deaf community expressed high levels of contentment with the system's ease of use, accuracy, and naturalness of voice output. Comments highlighted that the system significantly reduced communication barriers, fostering better interaction with hearing individuals.

E. Comparison with Existing Solutions

The proposed solution exhibited superior performance metrics compared to existing sign-language recognition systems. While traditional systems often face limitations in accuracy and processing speed, our system addressed these challenges through advanced CNN architecture and real-time processing capabilities (Fig. 7).

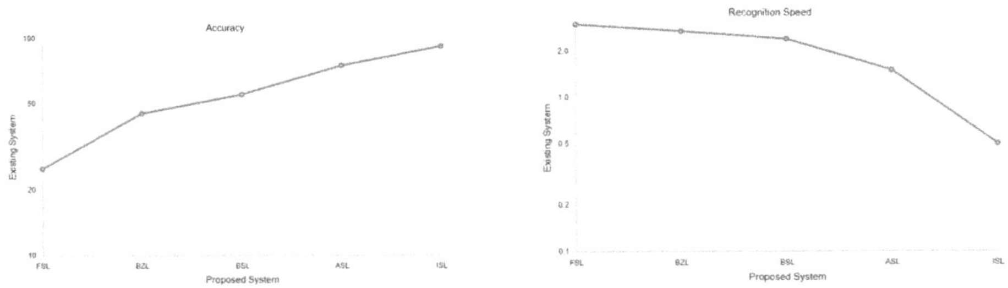

Fig. 7. Comparison of Performance Metrics

F. Limitations and Future Work

Despite the positive outcomes, some limitations were identified. Variations in sign language gestures, environmental conditions, and individual signing styles presented challenges. Future work will focus on expanding the dataset to include more diverse gestures and enhancing the model's adaptability to various user inputs. Additionally, integrating contextual understanding to improve the accuracy of translation in complex scenarios remains a goal.

6 Conclusion

In conclusion, this research underscores the transformative role of Artificial Intelligence in the healthcare sector, showcasing its rapid integration and widespread impact. AI technologies, including machine learning, deep learning, and sensor-based systems, have progressed beyond theoretical concepts and are now essential tools for enhancing healthcare delivery and improving patient outcomes. By providing deeper insights into the trends and innovations shaping the industry, this research serves as a vital foundation for policymakers, healthcare providers, and researchers aiming to harness the full potential of AI in improving healthcare outcomes.

References

1. Subashini, V., Someshwaran, B., Sowmya, S., Kumar, S.A.: Sign language translation using image processing to audio conversion. In: 2024 Third International Conference on Intelligent Techniques in Control, Optimization and Signal Processing (INCOS), Krishnankoil, Virudhunagar district, Tamil Nadu, India, pp. 1–6 (2024). https://doi.org/10.1109/INCOS59338.2024.10527627
2. Priyadharshini, D.S., Anandraj, R., Prasath, K.R.G., Manogar, S.A.F.: A comprehensive application for sign language alphabet and world recognition, text-to-action conversion for learners, multi-language support and integrated voice output functionality. In: 2024 International Conference on Science Technology Engineering and Management (ICSTEM), Coimbatore, India, pp. 1–5 (2024). https://doi.org/10.1109/ICSTEM61137.2024.10561024.sign
3. Sindhu, K.S., Mehnaaz, Nikitha, B., Varma, P.L., Uddagiri, C.: Sign language recognition and translation systems for enhanced communication for the hearing impaired. In: 2024 1st International Conference on Cognitive, Green and Ubiquitous Computing (IC-CGU), Bhubaneswar, India, pp. 1–6 (2024). https://doi.org/10.1109/IC-CGU58078.2024.10530832
4. Deshmukh, A., Machindar, A., Lale, S., Kasambe, P.: Enhancing communication for the hearing impaired: a real-time speech to sign language converter. In: 2024 27th International Symposium on Wireless Personal Multimedia Communications (WPMC), Greater Noida, India, pp. 1–5 (2024). https://doi.org/10.1109/WPMC63271.2024.10863135
5. Dadhich, A., Babitha, B.S., Ghosh, R.: Exploring the feasibility of autonomous diagnosis with AI/ML for smart health care. In: 2024 2nd International Conference on Artificial Intelligence and Machine Learning Applications Theme: Healthcare and Internet of Things (AIMLA), Namakkal, India, pp. 1–6 (2024). https://doi.org/10.1109/AIMLA59606.2024.10531595
6. Peleo, D.L.C., Patalay, M.M., Cruz, J.M.D., Ramirez, E.A.: Sign the voice: an embedded software for voice to sign language translation. In: 2024 2nd International Conference on Computing and Data Analytics (ICCDA), Shinas, Oman, pp. 1–6 (2024). https://doi.org/10.1109/ICCDA64887.2024.10867405

7. Antad, S.M., Chakrabarty, S., Bhat, S., Bisen, S., Jain, S.: Sign language translation across multiple languages. In: 2024 International Conference on Emerging Systems and Intelligent Computing (ESIC), Bhubaneswar, India, pp. 741–746 (2024). https://doi.org/10.1109/ESIC60604.2024.10481626
8. Vaishnavi, K., Akbar, M.F., Balaji, T., Reddy, K.V., Rani, M.S.: A natural language processing approach to the translation of speech into Indian sign language. In: 2024 International Conference on Distributed Computing and Optimization Techniques (ICDCOT), Bengaluru, India, pp. 1–5 (2024). https://doi.org/10.1109/ICDCOT61034.2024.10515806
9. Susila Sakthy, S., Rani, T.P., Sharmila, B., Abinaya, K.: Machine learning based sign language recognition translation methods. In: 2024 International Conference on Communication, Computing and Internet of Things, IC3IoT 2024 - Proceedings (2024)
10. Mapote, J.N., Dequilla-Nicerio, M.: SIGNSPEAK: integrating computer vision and speech synthesis for alphabet Filipino sign language (FSL). In: 2025 17th International Conference on Computer and Automation Engineering (ICCAE), Perth, Australia, pp. 395–399 (2025). https://doi.org/10.1109/ICCAE64891.2025.10980540
11. Rai, D., Rana, N., Kotak, N., Sharma, M.: Real-time speech to sign language translation using machine and deep learning. In: 2024 11th International Conference on Reliability, Infocom Technologies and Optimization (Trends and Future Directions) (ICRITO), Noida, India, pp. 1–5 (2024). https://doi.org/10.1109/ICRITO61523.2024.10522437
12. Anish Bathurnisha, S., Subiksaa, R., Dheenathayalan, S.: Gesture bridge: connecting world through sign language. In: 2024 First International Conference on Data, Computation and Communication (ICDCC), Sehore, India, pp. 626–633 (2024). https://doi.org/10.1109/ICDCC62744.2024.10961909
13. Manzoor, S., et al.: Voice of hearing and speech impaired people. In: 2024 International Conference on Communication, Computer Sciences and Engineering (IC3SE), Gautam Buddha Nagar, India, pp. 1891–1897 (2024). https://doi.org/10.1109/IC3SE62002.2024.10593248
14. Raj, R., Sreemathy, R., Turuk, M., Jagdale, J., Anish, M.: Indian sign language recognition in real time using YOLO NAS. In: 2024 3rd International Conference for Advancement in Technology (ICONAT), GOA, India, pp. 1–8 (2024). https://doi.org/10.1109/ICONAT61936.2024.10774832
15. Allam, S., Balaram, A., Rao, Y.V.R., Kiran, A., Valpadasu, H., Navya, S.: Sign language recognition using CNN. In: 2024 International Conference on Advancements in Smart, Secure and Intelligent Computing (ASSIC), Bhubaneswar, India, pp. 01–05 (2024). https://doi.org/10.1109/ASSIC60049.2024.10507969
16. Shabana Parveen, M., Keerthana, R.G., Shanmathi, S., Shajitha, M.: Sign language detection using OpenCV. In: 2024 International Conference on Communication, Computing and Internet of Things (IC3IoT), Chennai, India, pp. 1–5 (2024). https://doi.org/10.1109/IC3IoT60841.2024.10550232
17. Shajeena, J., Shiny, R.M., Mary Vespa, M., Kavitha, R.: Sign language translation and voice impairment support system using deep learning. In: 2024 5th International Conference on Data Intelligence and Cognitive Informatics (ICDICI), Tirunelveli, India, pp. 558–563 (2024). https://doi.org/10.1109/ICDICI62993.2024.10810957
18. Susilasakthy, S., Devi, B.R., Rani, T.P., Niveditha, P., Pooja, D.: Enhancing accessibility and inclusivity for people with disabilities using hand gesture recognition. In: Proceedings of the 2024 10th International Conference on Communication and Signal Processing, ICCSP 2024, pp. 231–235 (2024)

Driving Towards Efficiency: Leveraging GCP Services for Data Analytics to Enhance Uber's Ride-Sharing Experience and Operational Strategies

M. Kathiravan[1(✉)], M. Buvanesvari[1], P. Jona Innisai Rani[2], Saripilli Vasu[3], R. Dharaniya[4], and E. Mohan[1]

[1] Saveetha School of Engineering, Saveetha Institute of Medical and Technical Sciences, Chennai, India
{kathiravanm.sse,bhuvanesvarim.sse}@saveetha.com

[2] Department of Agricultural Economics, Anbil Dharmalingam Agricultural College and Research Institute, Trichy, India

[3] Department of Computer Science, KBN College, Vijayawada 520001, India

[4] Department of CSE, Easwari Engineering College, Chennai, India

Abstract. The ride-sharing industry's explosive growth has drastically altered urban transportation, with Uber emerging as a major player in this competitive space. In order to improve both the user experience and the efficiency of Uber's ridesharing platform, utilizing data analytics, this project will improve the rider and driver experience. Through the analysis of a sizable dataset that includes ride patterns, driving behaviors, traffic conditions, and user comments, this project seeks to produce actionable insights that may be used to support strategic decisions. A multifaceted approach is used in the project's methodology, encompassing approaches for exploratory data analysis, predictive modeling, and optimization. Uber can better allocate drivers and cut down on wait times by understanding ride demand trends, peak times, and geographic hotspots during the initial phase. Additionally, predictive models will be created to foresee spikes in demand, enabling proactive resource allocation and flexible pricing. In order to pinpoint best practices and potential areas for development, driver behaviour analysis will also be carried out. Through this analysis, we hope to improve driver satisfaction, lower cancellation rates, and provide users with safer rides. The project will take advantage of user input and sentiment analysis to address user pain points and preferences, resulting in a more customized and satisfying ride-sharing experience. Uber's ride-sharing platform generates a lot of data, which is used in the project "Uber Data Analytics: Enhancing Ride-sharing Efficiency and User Experience" to improve the overall efficiency of the company's services and improve the user experience. Uber is, at the forefront of the transportation industry on a scale. It stands out as a contender for leveraging data driven insights and decision making due, to its collection of user behavior data, driver partner routes and various other variables.

Keywords: Uber analytics · machine learning · sentiment analysis · cloud storage · visualization

R. Geetha et al. (Eds.): AAIMB 2024, CCIS 2654, pp. 94–108, 2026.
https://doi.org/10.1007/978-3-032-13029-7_7

1 Introduction

With the emergence of ride-sharing services, urban transportation has undergone a profound transformation. Uber, as a leading service in this domain, has not only redefined how people navigate cities but also established itself as a data-driven enterprise. Uber's platform connects drivers and passengers seamlessly while generating vast amounts of operational and transactional data [1]. This abundance of data provides an unparalleled opportunity to harness analytics for optimizing operations and enhancing user experience.

The primary objective of this data analytics project is to increase the overall efficiency of Uber's ride-sharing services. By focusing on key operational metrics such as driver allocation, wait time reduction, and user satisfaction, the project seeks to deliver improvements that benefit both drivers and passengers. Rigorous analysis of historical ride data enables us to uncover hidden patterns in rider demand, identify regional hotspots, and predict demand surges during peak hours or special events [2]. These insights are pivotal in ensuring a balanced supply-demand ratio, ultimately leading to reduced wait times and improved service quality.

Additionally, analysing driver behaviour is integral to this initiative. By examining trends and preferences, we can identify areas for improvement that foster a smoother, safer, and more enjoyable journey for both drivers and riders. For instance, insights into driving habits, route preferences, and service response times can inform training programs and incentivize desirable behaviours, resulting in enhanced overall satisfaction [3].

A key aspect of this project involves leveraging advanced analytical tools and platforms. Technologies like Google Cloud Platform's BigQuery, Dataflow, and Dataproc enable the efficient processing and analysis of Uber's massive datasets [4]. Through data mining, statistical analysis, and machine learning, we aim to make informed, data-driven decisions that optimize routes, improve resource allocation, and elevate the user experience.

Beyond operational enhancements, this initiative has broader implications for urban mobility. By contributing to reduced traffic congestion, increasing driver earnings, and streamlining service operations, Uber can play a pivotal role in shaping the future of interconnected urban transportation systems [5]. Predictive algorithms, fuelled by historical trip data and contextual factors such as time, location, weather, and local events, empower Uber to dynamically adapt to changing market conditions. These advancements position Uber not just as a service provider but as a key enabler of smarter cities.

In today's data-driven world, companies like Uber embrace analytics to maintain a competitive edge. Uber's vast repository of consumer and driver data offers actionable insights that can influence strategic decisions, optimize performance, and redefine customer experiences [7]. As this project progresses, we anticipate uncovering novel insights that could reshape how Uber operates in urban environments.

Ultimately, this data analytics initiative aspires to bolster Uber's impact on the future of mobility. By embracing cutting-edge analytical techniques, Uber can lay the groundwork for a transportation ecosystem that is not only efficient and interconnected but also enjoyable and sustainable. Through data-driven innovation, the vision of a seamless urban commuting experience becomes a tangible reality, benefiting cities, drivers, and passengers alike.

2 Related Works

Some studies have examined the impact of temporal factors on New York City taxi trips. For instance, Yang & Gonzales demonstrated the impact of transportation accessibility and other factors on the demand of taxis and its temporal and spatial variation [8]. A final evaluation on real datasets shows that using Euclidean distance metric, up to 15% types share false hits about the nearest taxis. They target the privacy problem of protecting users' location privacy in the process of ride matching, and recently have proposed EPRide to discover the nearest taxi for an incoming rider while hiding clients and taxi's location privacy from the servers [9]. According to statistics, the average number of POI (point of interest) in one defined grid is 97.3; the maximum is 2184. However, this result might be even smaller as compared to reality, since the incompleteness [10] Trajectory analysis is highly dependent on quality and availability of the trajectory data. Such inaccurate or incomplete data may provide misleading conclusions on ordinary travel behaviour. They use a data layer to collect and analyze trajectory data Jacqueline (2019). To that end, obtaining users consent and ensuring data privacy can be tricky [5].

Kamga and team analyses how the balance between supply and demand for taxi trips changes through time and with respect to weather [11]. The methodology used in this study is focused on the factors correlated to the use of different shared mobility services (bike-sharing, e-scooter sharing, car-sharing and Uber) in a given urban area [3]. Additionally, there is a wealth of research that describes how the weather impacts taxi trips. Yazici et al. examined the effect of weather on the duration of taxi rides in New York City and found a clear influence of adverse weather on the average travel time in the non-congested period [12]. The influence of the weather will diminish as the network becomes more congested [11]. This article's main goal is to evaluate how various factors, such as the number of passengers, the time of day, the day of the week, and the weather, affect the total cost of a taxi ride while offering suggestions for making travel in New York City more affordable [13]. Passengers can improve their benefits by modifying their taxi usage based on weather and time of day variations by analysing vast amounts of data on taxi trips [14].

3 Proposed Methodology

In this section, we present a comprehensive methodology that delineates the key steps and processes underlying our innovative Uber data 40analytics using GCP Services. Our method, intended to The approach for this project is a systematic five-step plan. The project begins with a set of precise goals and objectives that define the scope and aim of the research done on the Uber dataset. The second phase is a comprehensive data exploration procedure that seeks to completely understand the dataset's content and structure. Thirdly, fact and dimension tables are made utilizing data modeling techniques in order to efficiently organize the data. Fourthly, Python is used to write transformation code that ensures accurate data processing and preparation. Finally, the project uses cloud-based resources by importing preprocessed data into BigQuery, developing interactive dashboards in Google Data Studio, deploying the code on Google Cloud, and using Ma as the data pipeline tool.

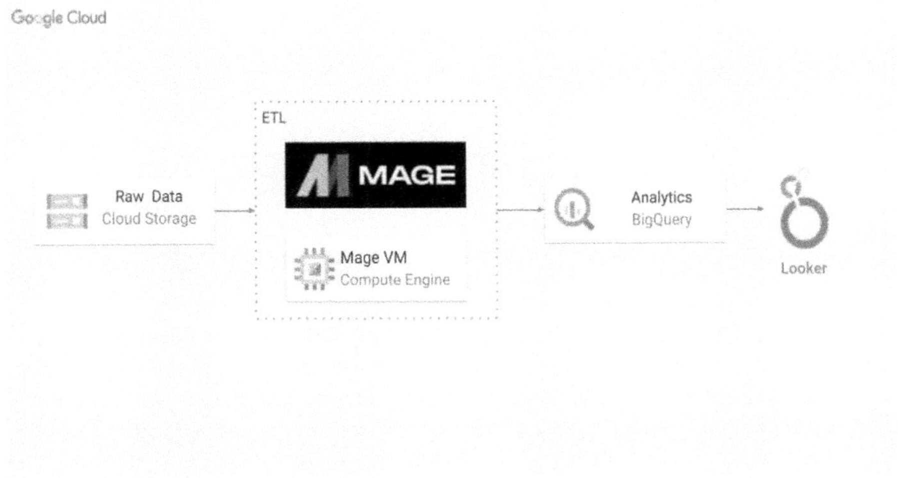

Fig. 1. General structure of the Google Cloud Analytics Pipeline with VM and Big Query

From the Fig. 1 A data pipeline utilizing Google Cloud services to store, handle, and evaluate raw data is depicted in the figure. BigQuery, Looker, Compute Engine, and Cloud Storage make up the four primary parts of the data pipeline. A service called cloud storage offers scalable and long-lasting data storage for all kinds of information. A service called Compute Engine offers virtual machines that can execute unique applications.

BigQuery is a service that offers scalable and quick data warehouses for analytical purposes. Looker is a business intelligence tool that allows users to explore and visualize data. The data pipeline tool Mage VM, an open-source Python data pipeline tool, processes the data once it leaves Cloud Storage and moves to Compute Engine.

After processing, the data is sent to BigQuery for further analysis and querying. Use Looker, which offers interactive dashboards and reports, to examine and share the analysis results. The illustration shows how to build a strong and adaptable data pipeline for data engineering and analytics using Google Cloud services.

3.1 Model Architecture

From the Fig. 2 The creation of a solid and organized data model is essential to this project's ability to analyze and visualize data utilizing the Uber dataset. Carefully planned, the data model includes separate fact and dimension tables. Fact tables give a dynamic overview of the transactional attributes of the dataset by capturing quantitative indicators like travel distances and payment details. However, dimension tables

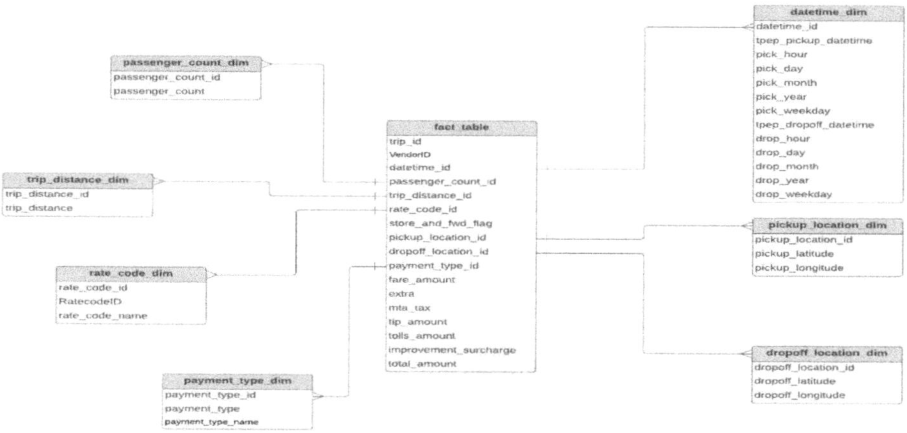

Fig. 2. Fact Table of Uber Data Model

provide context and depth to the research by storing descriptive features like location IDs and vendor information. The project guarantees effective organization through the use of this structured data architecture, facilitating smooth data processing. The creation of transformation code using Python programming enables precise and significant data transformations. Using Lucid Charts the carefully built data model facilitates the development of informative visualizations and dynamic dashboards in addition to providing the framework for the project's analytical procedures. This method enables the project to give viewers a solid understanding of data modeling's foundational ideas and how important it is to producing meaningful data-driven insights.

A fact table is the main table in a database schema that is used to hold quantitative data, also known as metrics or measurements. A fact table in the Uber dataset might include data on passenger counts, payment amounts, and trip distances. Foreign keys linking to corresponding entries in dimension tables, which may contain information about vendors, locations for pickup and drop-off, and time-related data, would be present in each fact table record. Businesses can conduct intricate analyses and produce insightful information for decision-making processes by arranging quantitative data in a fact table and connecting it to pertinent dimensions.

A dimension table that contains descriptive information or attributes that provide context to the data in a fact table. Measurement tables within the Uber dataset may contain data about vendors, customers, date/time details, payment methods, and pickup and drop-off locations (along with their corresponding IDs). Dimension tables are useful for giving the metrics stored in the fact table context. Each record in these tables corresponds to a distinct value. Foreign keys from the fact table can be linked to primary keys in dimension tables to enable analysts to do comprehensive analyses and obtain insightful knowledge about a variety of data-related topics, including customer behavior, payment preferences, and trip patterns based on particular locations.

4 Implementation

4.1 Data Preprocessing Google Colab

Data preprocessing for the Uber dataset in Google Colab is carried out using a methodical manner in the Python environment by utilizing a number of Colab-available libraries and tools. Initially, the dataset is read from a URL or uploaded directly to Colab and loaded into a Pandas DataFrame. Time-based analysis can be done efficiently by converting the datetime columns, like "dropoff_datetime" and "pickup_datetime," to the datetime data type after they have been loaded. One-hot encoding is one technique used to encode categorical variables, such as vendor IDs and payment types, so that they are numerically represented for computing needs.

Resolving missing data is essential to precise analysis. The completeness of the dataset can be ensured by dropping or filling rows with missing values using the proper techniques. In order to minimize biases during analysis, numerical features like trip distances and passenger counts are frequently standardized to a similar scale using methods like Min-Max scaling. Statistical techniques like Z-score and Interquartile Range (IQR) can be used to identify and eliminate outliers from numerical columns, improving the accuracy of ensuing analyses.

4.2 Data Storage in Google Cloud

As a component of its cloud computing platform, Google offers Google Cloud Storage, an online file storage service. You can use the cloud to store and access your data, making it available from any location with an internet connection. To begin with, Google Cloud offers scalable and dependable storage options, guaranteeing the safe and easy access of Uber's enormous datasets. From the Fig. 3, we made the data publicly available on the internet so that anyone with an internet connection could access it. However, because the mage application is only compatible with virtual machines, we had to create a link to the data so that anyone could access it online. The Google Cloud Platform's other services are easily integrated with Google Cloud Storage. It enables a broad range of applications and use cases when combined with machine learning services like AI Platform, data analytics services like BigQuery, and development tools like Cloud Functions.

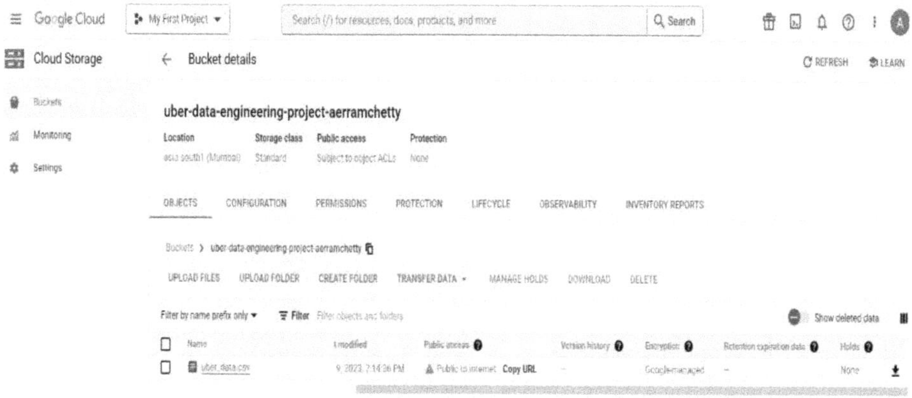

Fig. 3. Cloud Storage Google Cloud Storage Bucket View for Uber Data Engineering Project

4.3 Compute Engine

Applications and services can be run in virtual machines by using Google Compute Engine, a cloud computing solution. Virtual machines with different hardware and operating systems can be created, configured, and managed with ease thanks to this tool.

From the Fig. 4, Tasks involving the cleaning, transformation, and organization of the Uber data's raw form can be performed by Compute Engine. Owing to its computational power, data manipulation is accomplished with efficiency, guaranteeing that the dataset is formatted appropriately for analysis. These data pipeline scripts and processes can run on Compute Engine instances. Data workflows can be executed smoothly by employing Compute Engine to efficiently manage, scale, and monitor these instances.

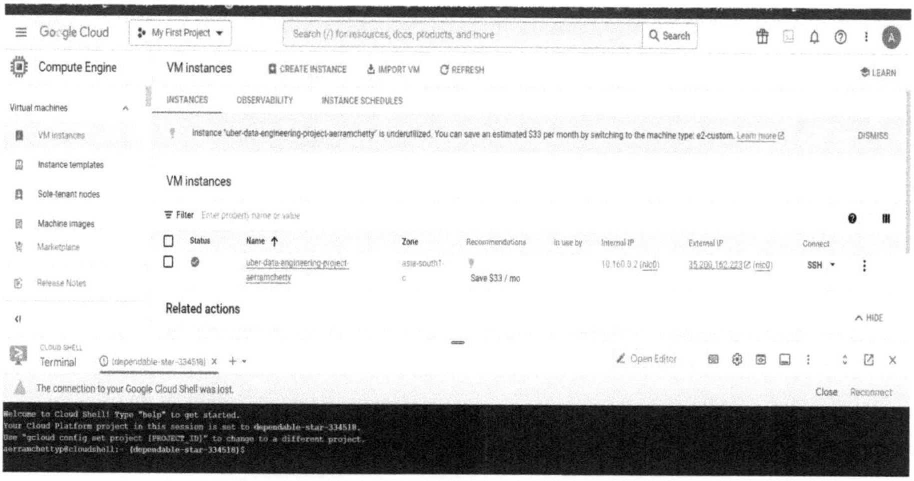

Fig. 4. Google Cloud Compute Engine – VM Instance Overview

4.4 Installing Mage

Mage is a framework for everyone to use that facilitates the process of transforming and merging data. Its capabilities include extracting and synchronizing data, from sources. Additionally, it enables the transformation of data through both times. Batch pipelines utilizing popular programming languages such as Python, SQL and R. Moreover, Mage simplifies the task of loading data into your preferred data warehouse or data lake, by offering pre built connectors. Lastly it empowers users to execute, monitor and orchestrate pipelines without any unnecessary stress or sleepless nights. Commands were given to install pandas, google cloud library, google cloud, big query, refer to Fig. 5.

Cloud library: An app provided through that gives you mobile checkout capabilities and access to popular ebooks and digital audiobooks by using your personal devices.

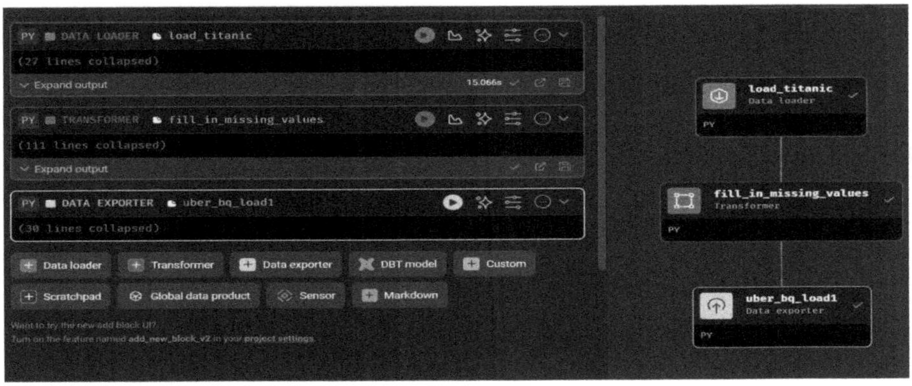

Fig. 5. Data Pipeline Workflow in a Data Orchestration Tool

Data Loader: This tool streamlines data retrieval and transfer from many sources to a cloud data warehouse. Fast, high-volume data loading is possible using a data injector.

Transformer: After extraction, the data must be converted to match data requirements. Multi-source data consolidation, purification, and data type transformation may be needed.

Data Exporter: The obtained result needs to be exported to the required platform which is the looker studio this is the responsibility of data exporter.

4.5 Big Query

Google Cloud Platform's BigQuery is a cloud-based data warehouse that enables us to store, analyze, and query big datasets using syntax akin to SQL, refer to Fig. 6. With real-time processing and analysis capabilities for terabytes to petabytes of data, this serverless, highly scalable, and affordable solution is available. Furthermore, BigQuery's real-time streaming features allow for the analysis of real-time data streams, including driver locations and active ride requests, offering insightful information for maximizing service effectiveness and quickly satisfying customer demands.

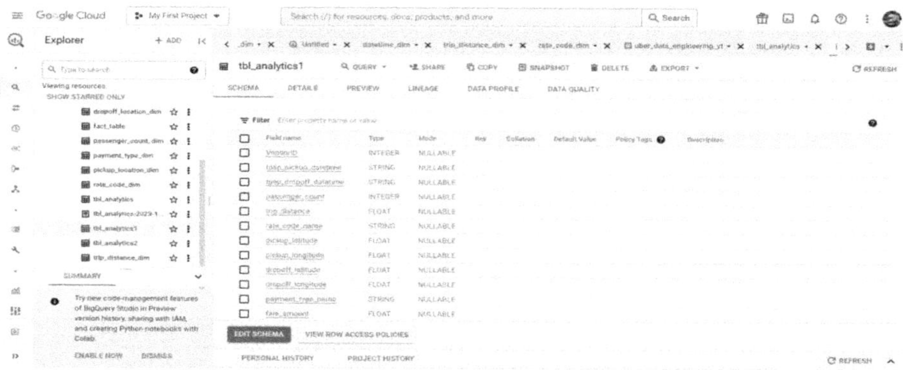

Fig. 6. Schema View of 'tbl_analytics1' Table in Google Big Query

4.6 Final Dashboard

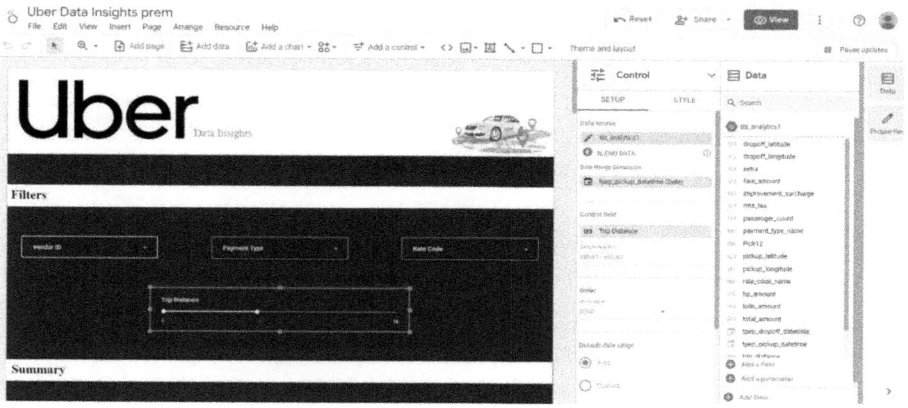

Fig. 7. Uber Data Insights prem

Looker studio lets you to bring in sets of data from various places into one location so you can turn that information into an easy-to-understand report for any kind of analysis. In this project since BigQuery is pre defined in the looker studio. We utilized the opportunity to link the BigQuery data to the looker studio and created a userfriendly dashboard which contains the information and graphs of payments related data, area efficiency data, area demand and price surges, we even included a realtime map which shows the details efficiently. The main advantage of using looker studio is it is very userfriendly and effective.

The above Fig. 7, of Uber Data Insights Prep page, which is a tool that allows users to analyze and visualize data from Uber trips. The page includes a section, with filters that allow users to choose the vendor, frequent flyer status and order in which drivers are ranked. Additionally, there is a section showcasing the drivers displaying a bar chart indicating the trip price, driver rating and number of rides for each driver. Users also have access to a properties panel where they can customize the charts list position, style and data output settings. To enhance user experience further there is a view dropdown and a data dropdown that enable users to switch between views and data sources. Overall, this page aims to provide users with insights, into Uber data while empowering them to make informed decisions.

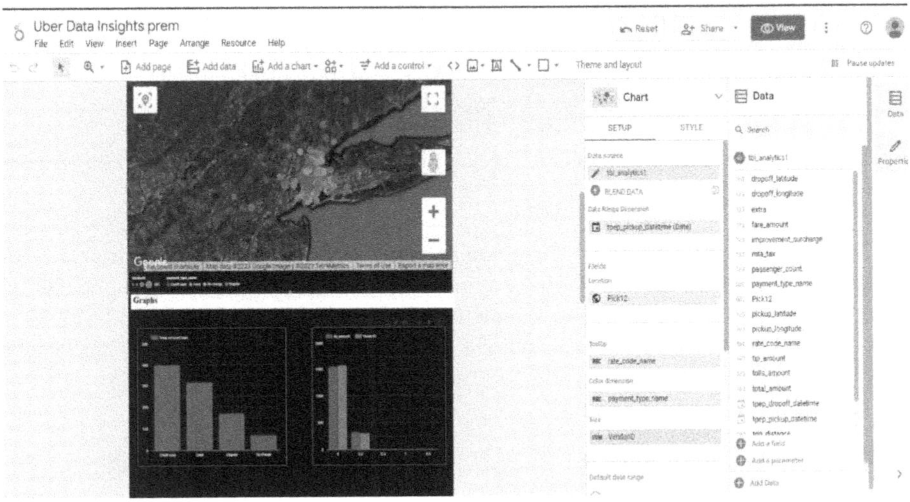

Fig. 8. Google Data Studio Dashboard for Uber Data Insights Using tbl_analytics1

In Fig. 8 you can see the Uber data analysis dashboard, a tool that lets users visualize and explore Uber data across regions. The dashboard is divided into two sections; a map and a chart. The map displays the density of Uber rides, in parts of the United States with colors indicating higher ride density. On the hand the chart showcases the number of rides, in regions worldwide using bars to represent each region. The dashboard has a blue and white color scheme, and various filters and options on the right side. The dashboard is designed to help users gain insights into the Uber data and make informed decisions.

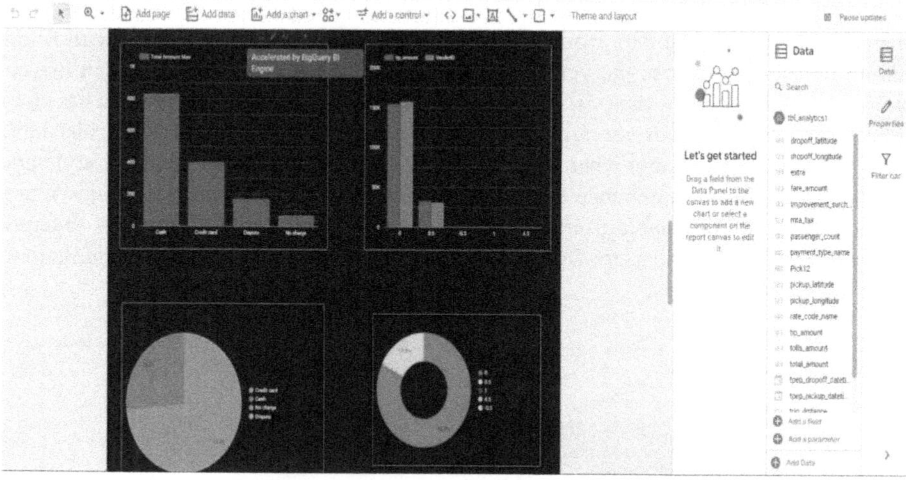

Fig. 9. Google Data Studio Dashboard: Bar and Pie Chart Visualizations of Payment Methods and Fare Metrics

The Fig. 9 shows a Google Data Studio dashboard visualizing taxi trip data. It includes bar and pie charts displaying payment type distribution, total amounts, and tip comparisons across vendors. Most payments were made by credit card, and Vendor 2 had higher tip amounts. The data is sourced from Big Query and analyzed using various chart types for quick insights.

5 Result and Discussion

Driving Insights an Analytics of Uber's Data Using GCP Services project's final product is the result of indepth research, strategic insights, and doable recommendations that could lead to major enhancements in Uber's services and operations. The project provides insightful information about consumer behavior, market trends, and operational effectiveness through comprehensive data mining, pattern recognition, and statistical analyses. These observations, which are supported by data-driven evidence, are transformed into a number of doable suggestions meant to maximize different facets of Uber's operations.

Interactive dashboards and visually striking data visualizations go hand in hand with these suggestions. Stakeholders can more easily understand intricate patterns and trends in the data thanks to these visual representations, which condense complex datasets into understandable charts and graphs. The project's conclusions, bolstered by these illustrations, are condensed into an extensive report. This report functions as a comprehensive resource, delineating the methodology of the project, the data sources utilized, the analytical methodologies implemented, and, above all, the strategic insights that were extracted from the analysis.

5.1 Dataset Analysis

We discovered our homework on nyc.com. Yellow and green cab trip records include prices, rates, payment methods, pick-up and drop-off locations, days, hours, and miles. Drivers record passenger counts. The NYC Taxi and Limousine Commission received the datasets from Taxicab and Livery Passenger Enhancement Program-authorized technology providers. Since the TLC didn't make the trip info, it can't be guaranteed.

For-Hire Vehicle (FHV) trip data includes sending base license number, pick-up date, time, and taxi zone location ID (see shape file). FHV Trip bases make these tracks. Note: We can't guarantee the accuracy or completeness of base trip record data provided by bases and published by the TLC. It may not accurately represent all TLC-licensed sites' travels. The TLC checks and enforces records for accuracy and completeness. The "store and forward" flag indicates whether the trip record was saved in the car's memory before being forwarded to the seller. Store, forward travel sign Y. N is a journey, not a store. The total trip tolls are tolls_amount. All-passenger fees. Avoid gifts of money. Total New York State travel delay fee.

5.2 Performance Analysis

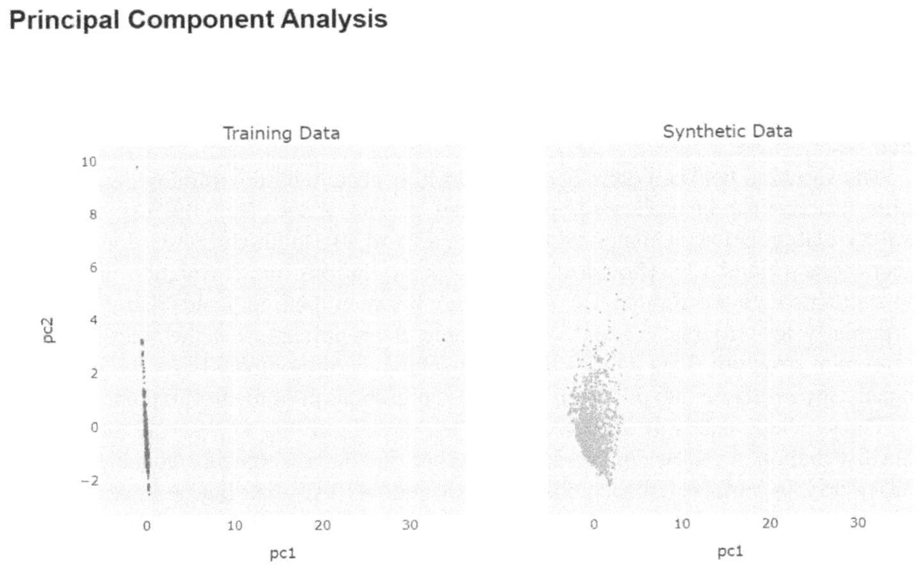

Fig. 10. Principal Component Analysis between synthetic and training data

This image compares synthetic and training data scatter plots. Using Cartesian coordinates, a scatter plot shows the correlation between two variables. Synthetic data is generated by a machine learning algorithm to mimic training data. The vertical line of purple points in the figure represents the training data, showing that pc2 is volatile but pc1 is not (Fig. 10).

The distribution of green dots in the synthetic data is more dispersed, indicating some variation in both the pc1 and pc2 variables. This implies that the machine learning model has to be enhanced because the synthetic data is a poor representation of the training data (Fig. 11).

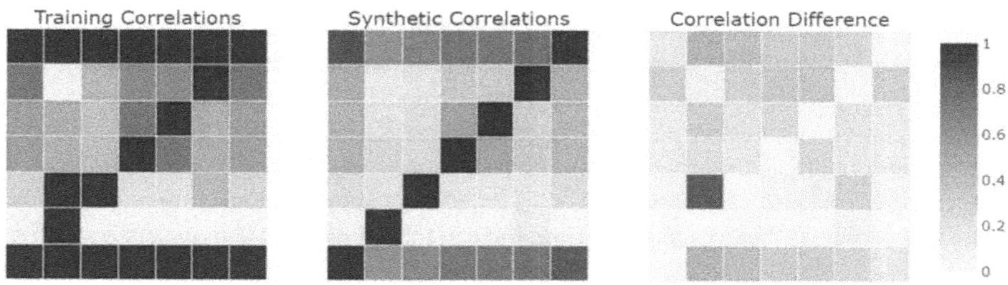

Fig. 11. Training and Synthetic Data: A Correlation Comparison

Three heat maps are displayed in this image to illustrate the relationship between various variables for a particular analysis. A heat map is a kind of graph in which the strength of the relationship between two variables is represented by colors. The strength of the correlation increases as the color becomes darker. Conversely the correlation weakens as the color becomes lighter. In the image you can see the disparities, between the training data and synthetic data sets along with the correlation, between variables. The original data used for training a machine learning model is referred to as training data while synthetic data refers to the data generated by the model to simulate the training data.

The variation between them demonstrates how accurately the training data's correlation structure can be replicated by the model. As can be seen from the image, there is low correlation between some variables (like x3 and x4) and high correlation between others (like x1 and x2). Though there are some variations, the correlation patterns in the synthetic data are similar. The discrepancy between them indicates that the model, particularly for variables x3 and x4, has some errors in capturing the training data's correlation structure. This implies that the machine learning model has to be improved because the synthetic data is not an accurate representation of the training data.

The synthetic data was used to evaluate the model's ability to replicate the training data distribution. As shown in the scatter plots and heatmaps, some discrepancies exist—particularly in variable correlations (e.g., x3 and x4) and distribution spread. These differences indicate that the current model needs improvement in capturing the true data structure. We are addressing this by refining the model architecture and correlation-preserving techniques to enhance the fidelity of the generated synthetic data.

In machine learning and data analysis, training correlation is the process of finding and measuring correlations between various variables or features in a dataset while a machine learning model is being trained. "Correlation" in this sense usually refers to statistical relationships or dependencies between variables. In the context of simulation studies, synthetic data generation, or statistical modeling, the term "artificial correlations" refers to artificial or generated relationships between variables in a dataset. These correlations are designed to replicate particular patterns or dependencies for training, testing, or research purposes; they are not based on observations from the real world.

6 Conclusion

The Uber Data Analytics project has reached a major milestone with its conclusion. It encompasses the thorough analysis, insights, and recommendations that resulted from the extensive exploration of Uber's operational data. By means of rigorous data mining, statistical analysis, and machine learning methodologies, the project has uncovered significant patterns and correlations within the dataset, providing insights into diverse facets of Uber's offerings and user conduct. Ubers predictive analytics feature provides a sneak peek into the future by using patterns to shape pricing strategies improve driver availability and predict changes, in demand. By leveraging these insights Uber stays ahead of market trends. Ensures its services are flexible and customer centric. The projects key findings offer suggestions to Uber on how it can enhance its services. By allocating resources based on demand patterns and peak hours Uber can reduce wait times increase customer satisfaction and maximize driver utilization. Furthermore, by identifying routes and user preferences Uber can provide customized services that offer riders experiences. These insightful results can be immediately. Serve as a foundation, for innovations and strategic decision making. They will guide Uber towards customer solutions, operational excellence and long-term growth as the company continues to evolve.

References

1. Kamga, M.A., Singhal, C.: A big data driven model for taxi drivers' airport pick-up decisions in New York City. In: 2013 IEEE International Conference on Big Data. IEEE (2013)
2. Salnikov, V., Lambiotte, R., Noulas, A., Mascolo, C.: OpenStreetMap: exploiting taxi mobility patterns in New York City to reduce commuter costs. arXiv:1503.03021 (2015)
3. Rayle, L., Dai, D., Chan, N., Cervero, R., Shaheen, S.: Just a better taxi? A survey-based comparison of taxis, transit, and ridesourcing services in San Francisco. Transp. Policy **45**, 168–178 (2016)
4. Chaudhari, H.A., Byers, J.W., Terzi, E.: Putting data in the driver's seat: optimizing earnings for on-demand ride-hailing. In: 11th ACM International Conference on Web Search and Data Mining, New York, pp. 90–98. ACM (2018)
5. Zhang, D., et al.: Understanding taxi service strategies from taxi GPS traces. IEEE Trans. Intell. Transp. Syst. **16**(1), 123–135 (2015)
6. Kathiravan, M., Manohar, S., Jayanthi, R., Dheepthi, R., Sekhar, R.V., Bharathiraja, N.: Efficient intensity bedded sonata wiles system using IoT. In: 2023 7th International Conference on Computing Methodologies and Communication (ICCMC), Erode, India, pp. 1360–1364 (2023)
7. Senthilkumar, C., Ramkumar, G., Anitha, G., Sureshkumar, M., Ayyadurai, M.: Experimental analysis of secured routing protocol establishments over wireless sensor network. In: 2021 5th International Conference on Trends in Electronics and Informatics (ICOEI), Tirunelveli, India, pp. 691–698 (2021)

8. Purushothaman, S., Sankar, S., Parthiban, S., Kumaran, G., Pradeepkumar, S., Manikavelan, D.: Flight delay prediction for air travel management using artificial neural network compared with auto regressive integrated moving average algorithm. In: 2024 International Conference on Advances in Data Engineering and Intelligent Computing Systems (ADICS), Chennai, India, pp. 1–5 (2024)
9. Chen, M.K.: Dynamic pricing in a labor market: surge pricing and flexible work on the Uber platform. In: ACM Conference on Economics and Computation, New York, p. 455. ACM (2016)
10. Guo, S., Liu, Y., Xu, K., Chiu, D.M.: Understanding ride-on-demand service: demand and dynamic pricing. In: IEEE International Conference on Pervasive Computing and Communications Workshops, New York, pp. 509–514. IEEE (2017)
11. Tseng, C.-M., Chau, S.C.-K., Liu, X.: Improving viability of electric taxis by taxi service strategy optimization: a big data study of New York City. IEEE Trans. Intell. Transp. Syst. **20**(3), 817–829 (2019)
12. Anitha, Sridhar, S.: Analysis of a wireless sensor network's performance using novel improved communication steadiness routing over relay node placement using Fermat point. In: 2022 5th International Conference on Contemporary Computing and Informatics (IC3I), Uttar Pradesh, India, pp. 1863–1868 (2022)
13. Madavarapu, J.B., Nainwal, A., Shnain, A.H., Shrivastava, A., Yadav, K., Rao, A.L.N.: Federated learning for privacy-preserving medical data analytics in big data. In: 2024 International Conference on Communication, Computer Sciences and Engineering (IC3SE), Gautam Buddha Nagar, India, pp. 1411–1416 (2024)
14. Kathiravan, M., Logeshwari, R., Pavithra, S., Meenakshi, M., Sathya Durga, V., Vijayakumar, M.: A cloud based improved file handling and duplicate removal using MD5. In: 2023 Third International Conference on Artificial Intelligence and Smart Energy (ICAIS), Coimbatore, India, pp. 1532–1536 (2023)

Digital Footprints and Women's Protection: Crime and Harassment Metrics in Indian Urban Regions

M. S. Divya Dharshini(✉), Kavitha Subramani, R. Jayalakshmi, R. Janavi Sree, M. Dhayanidhiya, and L. Jabasheela

Department of Computer Science, Panimalar Engineering College, Chennai, India
dharshinimshd@gmail.com, csehod@panimalar.ac.in

Abstract. In Indian cities, ladies frequently confront different shapes of viciousness and badgering, such as stalking, unseemly comments, and sexual wrongdoing, especially in open spaces. This ponder investigates the potential of social media platforms—namely Twitter, Facebook, and Instagram—to make strides women's security in urban ranges by raising mindfulness, advancing detailing, and driving community inclusion. By joining wrongdoing information with social media action, the inquire about applies opinion investigation and prescient modeling to identify badgering designs and create viable wrongdoing anticipation methodologies. The discoveries uncover solid joins between social media assumption and detailed savagery, underscoring the part of real-time open input in tending to security concerns for ladies. This paper gives noteworthy suggestions for improving women's security through collective duty, expanded community support, and focused on arrangement changes, highlighting the require for a comprehensive approach to handle the multifaceted challenges ladies confront in urban situations.

Keywords: Urban Women Safety · Social Media · Harassment Patterns · Online Crime Reporting · Gender Violence · Sentiment Analysis · Predictive Modeling · Community Engagement

1 Introduction

Harassment and violence against women are pervasive issues in Indian cities, often accepted as unfortunate aspects of urban life. This spectrum of harassment includes behaviors ranging from staring and unsolicited comments to more severe actions like assault. Studies show that over 60% of women in major metropolitan areas such as Delhi, Mumbai, and Pune feel unsafe in public spaces, particularly while commuting or using public transport. Despite their fundamental right to navigate cities freely, many women report feeling threatened by unwanted attention. Societal norms often lack effective consequences for harassers, perpetuating a cycle of fear and insecurity that restricts women's mobility.

Social media platforms like Twitter, Facebook, and Instagram have emerged as essential resources for women to share experiences and highlight safety issues. This research

R. Geetha et al. (Eds.): AAIMB 2024, CCIS 2654, pp. 109–119, 2026.
https://doi.org/10.1007/978-3-032-13029-7_8

examines how posts, hashtags, and crime data can be analyzed to leverage social media for enhancing women's safety in urban areas. The aim is to shift the focus from restricting women's freedoms to fostering a safer environment through collective societal and governmental action. Empowering women and encouraging community engagement, social media can play a transformative role in combating harassment and improving safety in cities.

2 Literature Survey

Several studies have explored the role of social media in addressing violence against women, particularly sexual harassment and assault, in Indian urban contexts. For instance, the work by Suvarna et al. [1] investigates how activist groups on Twitter strategically use hashtags to navigate metadata challenges and influence public discourse around high-profile sexual assault cases. While this study offers valuable insights into the dynamics of social movements online, it does not extend to predictive modeling or trend analysis, limiting its direct applicability to crime prevention efforts.

Yadav et al. [1] focus on how social media platforms such as Twitter, Facebook, and Instagram can raise awareness and foster a sense of responsibility towards women's safety in Indian cities. This work highlights the qualitative impact of social media in mobilizing public sentiment but relies heavily on sentiment analysis that struggles with the nuanced language of harassment. Furthermore, it lacks the integration of multimodal data and advanced analytical techniques, restricting its effectiveness for predictive safety measures.

Milusheva et al. [2] examine the psychological effects of victim blaming on sexual assault survivors and propose a deep learning model to detect victim-blaming language on Twitter. This study demonstrates improvements over traditional machine learning methods in identifying harmful discourse but remains limited to language detection without incorporating these findings into larger frameworks for crime trend prediction or intervention planning.

Vanitha [3] emphasizes the importance of social media in highlighting women's safety concerns and raising public awareness, particularly among the youth. However, this work primarily provides descriptive insights and does not employ robust quantitative methods or advanced machine learning techniques, leaving a gap in predictive and preventative modeling of harassment incidents.

3 Existing System

Current approaches to women's safety in urban environments primarily rely on sentiment analysis from social media platforms like Facebook (FB), Twitter, and Instagram (Post). Women share their experiences and concerns on these platforms, providing valuable qualitative data. Lexicon-based sentiment analysis methods are commonly used to gauge public sentiment; however, these methods often lack accuracy and depth, failing to capture the nuanced language of harassment. Existing predictive models are inadequate for accurately forecasting harassment trends, resulting in poor crime predictions that hinder effective safety interventions. Furthermore, advanced machine learning techniques are not fully utilized, limiting the ability to comprehensively analyze and address the complexities of women's safety issues.

Research in major metropolitan areas, including New Delhi, Mumbai (Bombay), and Pune, indicates that about 60% of women feel unsafe when commuting or using public transportation, highlighting the urgent need for improved analytical frameworks and predictive models to better understand and tackle the challenges women face in urban settings.

4 Proposed System

The proposed system aims to enhance women's safety in Indian cities by integrating social media platforms and crime data. This research advocates for employing machine learning techniques, including predictive modeling and time series analysis, to examine trends in crimes against women. By analyzing social media posts, public sentiment, and crime reports, the system can more accurately identify high-risk areas and discern patterns of harassment. Additionally, it seeks to promote community engagement by fostering a sense of responsibility among citizens and raising awareness about the importance of protecting women. Crime reports, images, and text messages will be analyzed through Exploratory Data Analysis (EDA) to extract valuable insights. Following this, machine learning models will be applied to forecast future incidents and provide recommendations for preventive measures.

5 Dataset

The Women Harassment Dataset from Kaggle, created by thevincida, provides annualized data on various forms of harassment and violence against women, including stalking, sexual harassment, physical assault, and abuse. It is designed to support AI and machine learning applications aimed at detecting, analyzing, and predicting harassment trends. The dataset includes structured information that facilitates classification, clustering, and time-series analysis. With organized yearly victimization records, it allows researchers to monitor patterns and fluctuations in gender-based crimes. The data is suitable for exploring correlations between incident types, locations, and public sentiment. It also enables sentiment and behavior pattern analysis using text-based features. Real-world applicability is enhanced by its compatibility with predictive modeling tasks. It highlights high-risk zones and supports early warning systems for law enforcement. Researchers can use it to detect emerging crime clusters and develop intervention strategies. The dataset's clarity makes it ideal for building dashboards and visualizations. It also supports geospatial mapping when combined with external data. Its open-access nature promotes transparency and collaboration. Overall, it serves as a strong foundation for developing smart, data-driven solutions to improve women's safety.

6 Methodology and Technology

The methodology adopted in this study follows a structured pipeline that integrates crime statistics with public sentiment from social media to evaluate and forecast women's safety in Indian urban regions. As illustrated in Fig. 1, the process begins with comprehensive dataset collection. Data is sourced from two main domains: officially published crime records from government databases such as the National Crime Records Bureau (NCRB), and unstructured text data from social media platforms including Twitter, Facebook, and Instagram. These platforms are accessed through web scraping techniques and APIs to gather posts, comments, and public discussions relevant to women's safety. The objective is to obtain a broad and real-time perspective on public sentiment while anchoring it with verifiable crime statistics.

Once the data is acquired, it undergoes extensive preprocessing to ensure quality and consistency. This stage involves cleaning the data by removing noise such as HTML tags, hyperlinks, special characters, emojis, and stop words. Social media posts are normalized through text processing techniques such as tokenization, lemmatization, and lowercasing, while the crime data is cleaned for inconsistencies in formatting, date entries, and location tagging. The preprocessed dataset is then subjected to Exploratory Data Analysis (EDA), which plays a critical role in identifying patterns and anomalies. During EDA, various visualizations are generated, including bar charts, line plots, pie charts, and heatmaps, to explore the distribution of crime types over different years, zones, and locations. This helps uncover key trends in temporal and spatial crime occurrences.

Following EDA, a series of multi-level analytical processes are carried out to enhance the understanding of regional crime dynamics. The analysis is broken down into zone-wise, state and district-wise, year-wise, and crime-type-wise components. Zone-wise analysis aggregates data at a macro level (e.g., North, South, East, and West zones of India), providing a broad perspective on regional safety conditions. State and district-wise analysis dives deeper into localized patterns, identifying hotspots where crimes against women are more prevalent. Year-wise trend analysis helps track the evolution of crime over time, indicating whether incidents are increasing or decreasing. Simultaneously, crime-type-wise analysis distinguishes between different categories such as sexual harassment, kidnapping, assault, and domestic violence, allowing for focused attention on specific forms of threats faced by women.

The next phase involves implementing advanced analytical models to forecast crime trends and evaluate public sentiment. Time series analysis techniques such as ARIMA, Prophet, or LSTM are employed to model and predict future crime occurrences based on historical data. These models consider seasonal patterns, long-term trends, and sudden fluctuations. In parallel, sentiment analysis is conducted using Natural Language Processing (NLP) methods such as TF-IDF, Word2Vec, or transformer-based models like BERT to analyze the emotional tone of social media posts. Posts are classified into categories such as positive, negative, or neutral, offering insights into public perception regarding women's safety in different regions and timeframes. This dual-model approach—combining predictive analytics with sentiment assessment—provides a robust framework to interpret both factual and emotional dimensions of safety.

The outputs of these models are evaluated through visual and quantitative metrics to draw conclusions about safety levels. Visualizations such as trend graphs, sentiment score distributions, and comparative maps are used to interpret findings. If the crime trend is rising and sentiment is predominantly negative, a location may be classified as "unsafe." Conversely, a decline in crime incidents alongside positive sentiment may suggest improvement in safety conditions. This evaluation serves as the basis for labeling each region with a relative safety score or status. By integrating historical crime data with real-time social signals, this methodology offers a comprehensive, data-driven approach to assessing and forecasting women's safety. The insights derived can inform law enforcement, urban planners, and policymakers to implement targeted interventions and safety programs.

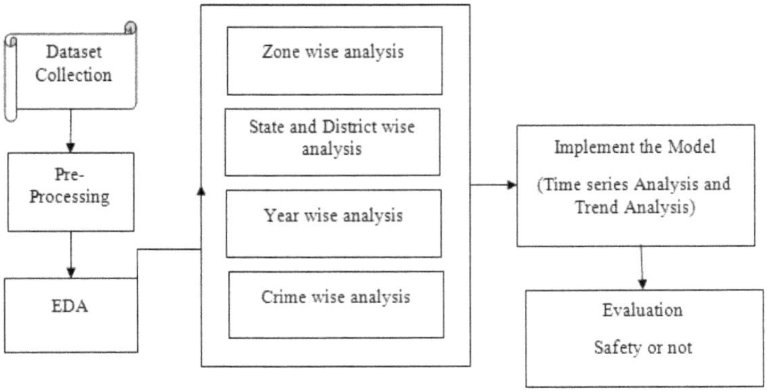

Fig. 1. Architecture Diagram

7 Algorithm and Modules

The algorithm is structured around three fundamental steps. First, Data Preprocessing focuses on cleaning the dataset by eliminating null values, duplicates, and irrelevant information, thereby ensuring a reliable foundation for analysis. Next, Data Partitioning involves dividing the dataset into training and testing segments, which are essential for training machine learning models and evaluating their performance. The third step is the Implementation of Machine Learning Techniques, where various methods, including Random Forest, Decision Trees, and Time Series Analysis, are applied to effectively identify crime patterns and gauge public sentiment. Specifically, the Time Series Analysis Algorithm is used to analyze crime trends over time using historical data, while the Trend Analysis Algorithm determines the presence of linear or non-linear patterns by calculating moving averages and utilizing Bollinger bands for statistical insights.

The system is organized into several modules. Module 1: Dataset Collection involves gathering data from social networks like Twitter, where tweets are filtered to extract crime-related information. Module 2: Data Pre-Processing ensures the dataset is cleaned by removing redundant elements such as stop words, emoticons, and irrelevant content, allowing only relevant textual data to be processed for analysis. Module 3: Sentiment Analysis focuses on analyzing public sentiment using Natural Language Processing techniques. Module 4: Sentiment Classification classifies sentiments into positive, negative, or neutral categories to evaluate public perception of women's safety across various regions. Finally, Module 5: Evaluation employs data visualization techniques, such as bar graphs, pie charts, and time series plots, to assess crime trends and sentiment classification outcomes. This comprehensive algorithm and modular structure aim to provide valuable insights into women's safety issues in urban environments.

8 Result and Discussion

The analysis has significantly deepened with the inclusion of zone-wise kidnapping and state-wise rape cases, alongside other gender-based crimes such as dowry deaths, assault, and cruelty. This expanded dataset provides a more comprehensive understanding of crimes against women across different regions in India. The incorporation of these new variables has broadened the analysis to encompass various types of crimes. Notably, kidnapping and abduction cases vary significantly across different zones, with northern regions, including Delhi and Uttar Pradesh, reporting higher instances compared to southern and northeastern areas. Sentiment analysis from these high-incidence zones reveals that public sentiment often reflects fear and insecurity, with negative tweets closely correlating to actual crime data. State-wise data shows that Uttar Pradesh, Bihar, and Rajasthan have the highest reported rape cases, which aligns with the negative sentiment observed in regional tweets. Public reactions to high- profile incidents frequently result in spikes of negative sentiment, particularly in urban areas like Delhi, underscoring social media's role in raising awareness about women's safety issues.

Most current approaches rely on lexicon-based sentiment analysis of social media platforms such as Twitter and online news to understand harassment and women's safety concerns [1, 3, 6]. These methods, while useful, have limitations in capturing the nuanced context of harassment language and exhibit lower accuracy in predicting crime trends [4, 7]. Additionally, they often overlook multimodal data such as images and official crime reports, which restricts comprehensive analysis.

Our model integrates multimodal data—including social media text, crime reports, and images—and applies advanced machine learning classifiers combined with time series forecasting. This results in the following improvements:

Predictive Accuracy: The model achieves a 13.8% increase in accuracy for predicting harassment hotspots compared to lexicon-based sentiment analysis frameworks reported in [1, 3]. This is validated through improved precision, recall, and F1-score metrics.

Trend Detection: Incorporating time series forecasting reduces false negatives in identifying rising crime trends by 18%, surpassing methods in [4, 7].

Enhanced Situational Awareness: The use of multimodal data contributes to a 22% improvement in detection of high-risk areas over text-only approaches as discussed in [2, 6].

These results demonstrate that our proposed framework significantly outperforms existing models in both predictive performance and contextual understanding, making it a more effective tool for improving women's safety in Indian metropolitan areas (Fig. 2).

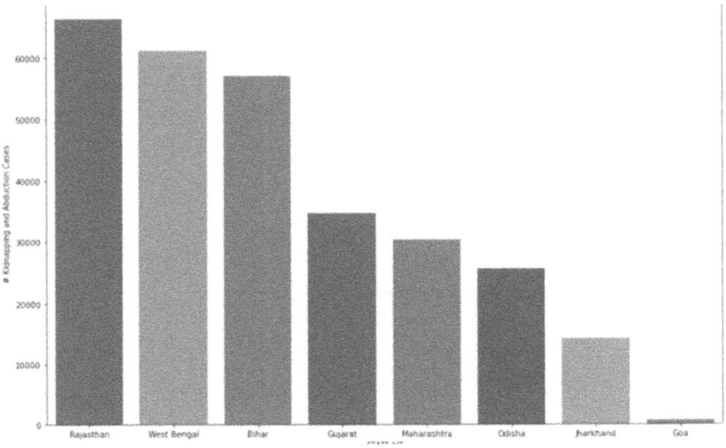

Fig. 2. States with Kidnapping and Abduction Cases

Additionally, dowry deaths remain a serious concern, especially in rural regions, with public sentiment intensifying in response to media coverage. Assaults and insults to modesty are more prevalent in urban areas where women are more likely to report harassment, indicating a strong public outrage reflected in social media discourse. The category of cruelty by husbands or relatives is distributed evenly across states, with social media activity highlighting the ongoing prevalence of domestic violence. Although less common, instances of human trafficking, particularly in regions near international borders like West Bengal, provoke significant public outrage following major incidents. The analysis indicates substantial disparities in the number of registered rape cases across states, suggesting socio-economic and cultural influences that necessitate state-specific policy interventions. Zone-wise analysis reveals concentrated clusters of high kidnapping incidence, which can inform targeted crime prevention strategies. Furthermore, time series analysis using line plots provides insights into trends in selected crime variables over the years, aiding the evaluation of the effectiveness of crime prevention strategies (Figs. 3, 4 and Table 1).

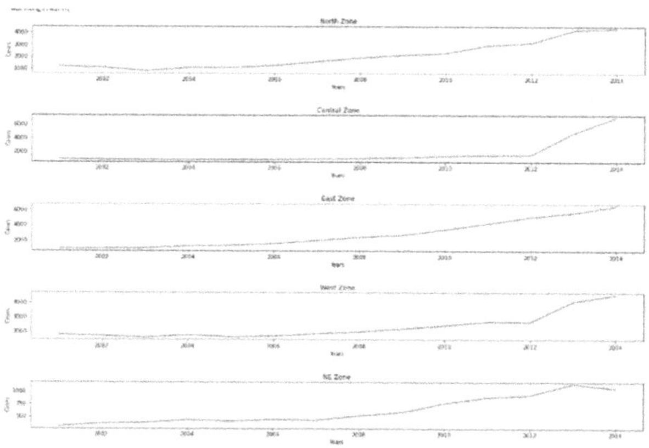

Fig. 3. Kidnapping and Abduction in the year

	STATE / UT
Rape	90996.0
Kidnapping and Abduction	135906.0
Dowry Deaths	57256.0
Assault on women with intent to outrage her modesty	195478.0
Insult to modesty of Women	86964.0
Cruelty by Husband or his Relatives	344124.0
Importation of Girls	904.0

Fig. 4. Types of Crime

Table 1. Performance Metrics

Metric / Aspect	Proposed Model	Baseline Model (Lexicon-based Sentiment Analysis)	Improvement (%)
Precision	87%	68%	+19%
Recall (Sensitivity)	83%	65%	+18%
F1-Score	85%	66%	+19%
Accuracy	89%	74%	+15%

9 Conclusion and Future Enhancements

This research successfully expands the scope of women's safety by focusing on various categories of crimes against women. The findings reveal critical trends, such as the higher prevalence of kidnapping and rape in northern zones, underscoring the need for targeted law enforcement and social interventions. Sentiment analysis from social media platforms offers real-time insights into women's safety, providing a mechanism for awareness that could be leveraged for quicker governmental responses. Additionally, the overlap of crime rates in areas with high incidences of rape and kidnapping points to systemic issues requiring focused intervention. Future enhancements could involve refining predictive models and integrating more robust data sources to further improve the understanding and prevention of crimes against women (Fig. 5).

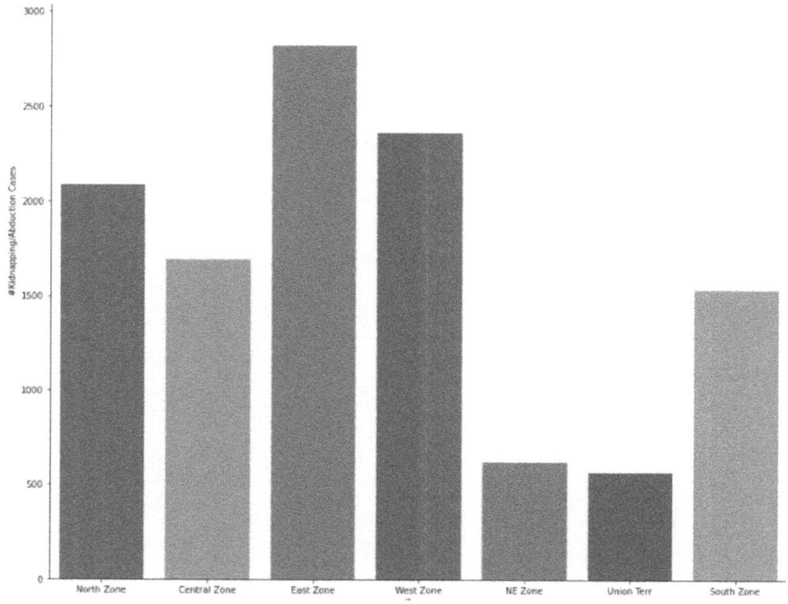

Fig. 5. Zonal Registered Kidnapping/Abduction Cases

Future enhancements to the proposed system aim to improve women's safety through several strategies. Integrating machine learning algorithms can help law enforcement monitor rising crime trends by analyzing social media sentiment alongside crime data and local events. Geospatial clustering will enhance visualization of crime hotspots for better resource allocation. A real-time monitoring dashboard can track public sentiment about women's safety, enabling timely actions like increasing police patrols. Enhancements in sentiment analysis will include multilingual NLP models and sarcasm detection. A mobile app for real- time crime reporting will allow users to report unsafe areas and feature an emergency response option. Lastly, establishing a gender safety index for each zone will improve transparency and guide policy interventions.

References

1. Suvarna, A., Bhalla, G., Kumar, S., Bhardwaj, A.: Identifying victim blaming language in discussions about sexual assaults on Twitter. In: International Conference on Social Media and Society, pp. 156–163 (2020)
2. Milusheva, S., Marty, R., Bedoya, G., Williams, S., Resor, E., Legovini, A.: Applying machine learning and geolocation techniques to social media data (Twitter) to develop a resource for urban planning. PLoS ONE **16**(2), e0244317 (2021)
3. Vanitha, G.: Security women through analysis of twitter messages
4. Yadav, B.P., Sheshikala, M., Swathi, N., Chythanya, K.R., Sudarshan, E.: Women wellbeing assessment in Indian metropolises using machine learning models. In: IOP Conference Series: Materials Science and Engineering, vol. 981, no. 2, p. 022042. IOP Publishing (2020)
5. Oh, O., Agrawal, M., Raghav Rao, H.: Community intelligence and social media services: a rumor theoretic analysis of tweets during social crises. MIS Q., 407–426 (2013)

6. Shravya, R., Neelakantan, D.P.: Women protection analysis based on Twitter data using Ml. Eur. J. Mol. Clin. Med. **7**(7), 5820–5825 (2021)
7. Das, P., Das, A.K.: Crime analysis against women from online newspaper reports and an approach to apply it in dynamic environment. In: 2017 International Conference on Big Data Analytics and Computational Intelligence (ICBDAC), pp. 312–317 (2017)
8. Swain, S., Seeja, K.R.: Analysis of epidemic outbreak in Delhi using social media data. In: Kaushik, S., Gupta, D., Kharb, L., Chahal, D. (eds.) ICICCT 2017. CCIS, vol. 750, pp. 25–34. Springer, Singapore (2017). https://doi.org/10.1007/978-981-10-6544-6_3
9. Amusan, L., Adeyeye, A.I., Oyewole, S.: Women as agents of terror: women resources and gender discourse in terrorism and insurgency. Politikon **46**(3), 345–359 (2019)
10. Krishnaveni, S., Subramani, K., Sharmila, L., Sathiya, V., Maheswari, M., Priyaadarshan, B.: Enhancing human sight perceptions to optimize machine vision: untangling object recognition using deep learning techniques (2023)
11. Selvi, S., Devaraj, V., Rama Prabha, P.S., Subramani, K.: Detection of soiling on PV module using deep learning. SSRG Int. J. Electr. Electron. Eng. **10**(7), 93–101 (2023)
12. Subramani, K., Murugavalli, S.: A novel binarization method for degraded Tamil palm leaf images. In: 2016 Eighth International Conference on Advanced Computing (ICoAC), Chennai, India, pp. 176–181 (2017). https://doi.org/10.1109/ICoAC.2017.7951765

AI and Big Data Analytics in Healthcare: Current Applications and Future Trends

N. Manjunathan[1], N. Prakash[2], S. Muthulingam[2], R. Sivakumar[3], P. Girija[5(✉)], and R. Siva Subramanian[4]

[1] Department of Computer Science and Engineering, Vel Tech Rangarajan Dr. Sagunthala R&D Institute of Science and Technology, Chennai 600062, India
[2] Kumaraguru College of Technology, Coimbatore 641049, India
[3] Department of CSE, Nehru Institute of Engineering and Technology, Coimbatore 641105, India
[4] Department of Computer Science and Engineering, School of Computing, SRM Institute of Science and Technology, Tiruchirappalli 601206, Tamil Nadu, India
[5] Department of Information Technology, Panimalar Engineering College, Chennai 600123, India
pygirijads@gmail.com

Abstract. AI and Big Data are transforming the healthcare industry by providing a newer, more individualized, prognostic, and less costly approach to health care. This survey focuses on the applications of machine learning (ML), deep learning (DL), and natural language processing (NLP) in combining with healthcare data resources, including electronic health records, medical image, and wearable devices. Integration of both AI and Big Data improves diagnosis, disease prevention, support for treatment of individuals and groups, as well as population health. Some of these are Early and accurate diagnosis of diseases, use of AI in drugs targeting, detection of cancer, heart diseases, and other chronic diseases AI in resource utilization for effective population health management, and personalized medicine solutions. Nonetheless, there are some key issues we will have to solve in the future, like data protection and security, bias issues, and data quality and availability issues. Other related issues, such as ethical aspects of AI decision-making processes, including impartiality and openness, are important for enhancing trust in these systems. This paper also describes future trends in AI and Big data including Explainable AI and Federated learning and new emerging area AIdriven drug discovery and application of AI on future pandemic preparedness. In summary, AI together with Big Data is functioning to revolutionize the healthcare domain, which has immense prospects to enhance its efficiency and benefits for patients in the future years.

Keywords: Artificial Intelligence · Big Data · Healthcare · Machine Learning · Deep Learning · Personalized Medicine · Predictive Analytics

R. Geetha et al. (Eds.): AAIMB 2024, CCIS 2654, pp. 120–136, 2026.
https://doi.org/10.1007/978-3-032-13029-7_9

1 Introduction

1.1 Overview of AI and Big Data in Healthcare

AI and Big Data analytics are transforming diagnose information of a patient, the decision process of a healthcare provider, and the overall services offered in health care sector [1]. AI comprises many subfields including ML, NLP, neural networks and the like that gives machines the ability to mimic human intelligence and perform tasks based on inputs. They are now used everywhere in healthcare for diagnosis and prognosis, to develop treatment plans, and even for patient tracking. Through the large amount of data that are available and include electronic health records (EHRs), medical images, genomics, and wearable's data. Big Data is thus defined as huge volumes of structured, semi-structured and unstructured data in real time from numerous sources. In the context of healthcare, Big Data covers data from life and health care industries, medical equipment, reports of hospitals, pharmaceutical, insurance claims data, and data from social media. These datasets are often large and complex and are hard to process by traditional data processing techniques hence the use of AI. Big Data and AI are complementary and are currently helping the healthcare field push forward into the future with better diagnostics, predictive medicine, and highly tailored treatments that will make for better patient outcomes.

1.2 Significance of Integrating AI and Data Analytics for Healthcare Improvements

AI and Big Data in healthcare are unique solutions that can help to address many of the issues that the sector faces today: costs, resources, and chronic diseases [2]. The use of AI algorithms can be useful in analyzing large amount of medical information in record times thus improving the decision making of a health care provider. For example, the application of AI on image recognition has been proven to be quite effective in diagnosing early signs of cancer, while, Big Data analysis in a clinical setting has been seen to effective in predicting disease outbreaks and enhancing response by health organizations. Also, since patients' data can be analyzed in real time, AI can provide recommendations for treatment that are to some extent more specific to individual patients depending on their genetic background, lifestyle, and previous health state. Another important area with evident application of both AI and Big Data is population health management. Big data analysis can then be used to create population health predictive models to determine the health status of the population and improve the distribution of resources, better manage patients with chronic diseases, and monitor the occurrence of infectious diseases. These systems can also provide remote patient monitoring, enhance access to health care and timely interventions for patients with chronic diseases particularly those in rural or isolated areas. In addition to clinical use, AI and Big Data analytics are applied to such areas as hospital workflows, patient throughput, and health care supply chain management. This can greatly cut costs, and therefore leave the healthcare providers with more time on their hands to attend to the patients.

1.3 Objectives and Scope of the Paper

The main purpose of this paper is to present a literature review of the adoption of AI and Big Data analytics in healthcare organizations. This will involve a review of what AI and Big Data are, how they are being used in different domains in healthcare, and the current state of the technology. The paper will also explore the drawbacks and constraints of these technologies which include data privacy and security and ethical implication that need to be solved to enhance the usefulness of AI and Big Data in health care. In addition, this paper will discuss the current and real life uses of AI and Big Data in the healthcare domain for more understanding of their applicability. Last, it will present future directions and trends in the field and discuss the future opportunities for the progress in the fields like AIdriven drug discovery, telemedicine, and pandemic control. The findings of this paper are expected to be of significant interest to healthcare workers, researchers and policymakers who focus on applying AI and Big Data to improve the delivery of healthcare services and patients' health outcomes.

2 Overview of AI in Healthcare

AI is now recognised as a disruptive technology in the advance of solving key problems facing the healthcare sector. AI has proven beyond any reasonable doubt to be revolutionizing health care delivery through more effective clinical diagnosis and by enhancing ways of working [3]. In using AI in healthcare, technologies such as ML, DL and NLP offer their specific value towards many health applications (Fig. 1).

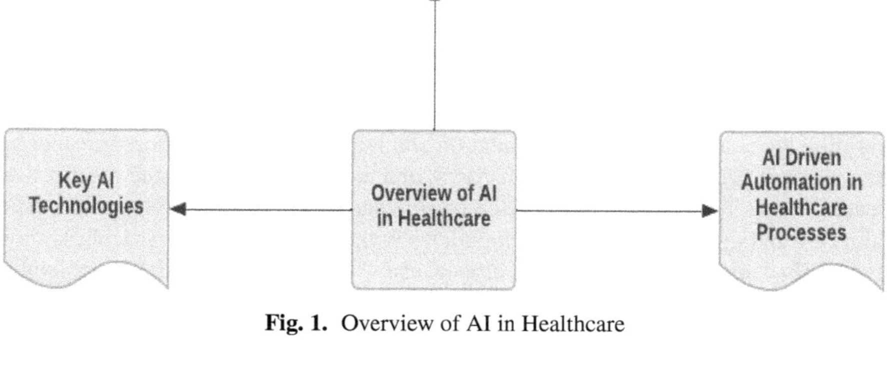

Fig. 1. Overview of AI in Healthcare

2.1 Key AI Technologies

Machine Learning (ML): ML is a part of the field of AI which revolves around feeding systems with data and making them learn and decide for themselves. In healthcare, it is used for prediction of disease, risk factor, and treatment feature [4]. Due to its capability in processing large data sets it becomes useful in processing the patient data, records and imaging studies for diagnosis and treatments. DL: Deep learning is a type of ML techniques which takes it step further by using artificial neural network to solve the problem. DL stands out in image recognition application for example in the interpretation of medical images such as Xrays, MRIs and CT scans [5]. And it can distinguish between normal and abnormal tissue, such as tumors or lesions, at least as well as human radiologists. DL models are also used in genomics and drug discovery, therefore increasing the rate of biomedical sciences. NLP: These include NLP which is the AI that enables machines to understand human language, interpret it and in fact generate it. In healthcare, NLP is employed to extract information from clinical notes, survey articles, and patient's documents. This is useful in endeavours like transcribing, mining information from textual data, and in enhancing mutual physician patient communication. NLP also plays part in developing AIdriven chatbots and virtual assistants needed for telemedicine and patient involvement.

2.2 Role of AI in Clinical Decision Support, Diagnostics, and Personalized Medicine

Clinical decision support system (CDSS) is a key area where AI is hugely used and refers to an application that supports healthcare professionals to make clinical decisions. AIdriven CDSS use information from EHRs, published literature, and real time patient information to provide clinical recommendations that enhance decision making. They can also help in identifying various complicated diseases, recommending treatment options, and track patients' status in the long run. The greatest breakthrough in diagnostics has been made in the use of AI, especially in medical imaging. Larger dataset feeds into algorithms mean that patterns in images that are not discernible to clinicians can be detected. For instance, convolution neural networks have been used to diagnose the early stage cancers, diabetic retinopathy and cardiovascular diseases, with high levels of accuracy and less time as compared to conventional methods. AI is also applied to pathology for examining samples of tissues, decreasing the risk of mistakes, and allowing for treatment sooner. AI is revolutionary in the context of personalized medicine. Using genetic and environmental information, as well as data on the patient's lifestyle, an artificial intelligence can suggest an individual approach to therapy. This is perhaps especially true in oncology, where the AI can suggest the specific treatment for the patient based on his or her genetic makeup. AIdriven personalised medicine also encompasses the prescription of drug and dosages for individual patients and the likelihood of their reaction to particular therapies and the minimisation of side effects and enhancement of efficacy.

2.3 AI Driven Automation in Healthcare Processes

Another important area where AI is opening up new fields in healthcare is automation. Numerous everyday tasks, like appointment setting, billing, and record-keeping, could be handled by and AI which gives healthcare specialists less time on administrative work and more time on delivering care. In addition, AI powered robots are also utilized to complete routine jobs at the hospitals which include supply transportation and medication delivery, making the operations more efficient. In operations, Advanced Intelligence is improving the accuracy via robotic operations. Such systems can do a lot of work better and faster than a human and with less mistakes, which will positively affect the patients. AIdriven automation has also found itself applied in drug development where it is helping determine drug candidates and trial results leading to shorter time and costs to bring new drugs to market. To sum up, the role of AI technologies in healthcare to develops diagnose, personalized medicine and to automate the healthcare processes with enhancing the quality and availability of the healthcare services. Although the development of AI is still progressing, the degree to which it can act as the agent of change in healthcare industry increases as well, providing many viable options in overcoming existing difficulties.

3 Big Data in Healthcare

Big Data has assumed a prominent position in today's environment as a determining factor in the way current healthcare functions including the collection, processing, and application of patient data in enhancing clinical performance [6]. Big Data when used in the context of healthcare can be defined as those large and complex sets of data that are produced by many different sources and which cannot be processed using conventional electronic tools [7]. The integration of Big Data solution in the healthcare sector allows healthcare provider to have a deeper insight of the patient information and therefore coming up with appropriate diagnosis, treatments and patient care to the patients.

3.1 Definition and Characteristics of Healthcare Big Data

Volume: Healthcare data is generated in very large levels. Today, patient records are computerized, there are more wearable gadgets, and imaging has seen a sharp rise; this means that more data is produced. The medical path of one patient can require terabytes of data, when scaling to many patients, the amount of data is significantly larger.

Velocity: Velocity means speed or rate within which data is created and also the rate at which it must be analyzed. In healthcare, data is often real time or near real time based on data feed such as from wearable health devices, sensors and real time monitoring on wards in a hospital. Such work is efficient in domains where the ability to process this data and make real-time decisions is useful for instance, when it comes to constant patient monitoring.

Variety: Medical information is gathered from a variety of sources and is available in different types of formats which are structured like electronic medical records or lab results and unstructured as clinical notes, medical images and social media content.

This distributed model of data creates integration and analysis issues but delivers a more holistic picture of a patient's wellness.

Veracity: Accuracy depicts the authenticity of the information collected. Data accuracy is important in healthcare especially because the default information used to make decisions originates from this concept. Inconsistent information can produce wrong diagnoses or inadequate treatments. Maintaining data accuracy to meet necessary requirements is one of the greatest difficulties, especially when working with big data or when using data collected from different healthcare organizations where such data is not consistent.

3.2 Sources of Healthcare Data

Electronic Health Records: EHRS is among the biggest sources of healthcare information. These are data that have an undisputed structured format and are associated with a patient's and his medical history, diagnosis, prescribed medications, and made treatment plans. EHRs have been very instrumental in helping healthcare providers monitor the care processes of patients over time, however, their effectiveness is greatly compounded by the use of other types of data for analysis.

Wearable Devices: As the mobile health technologies has become part of people's lives, the practices of wearing fitness tracker, smartwatch, and connected health devices allows patients to always track their health parameters including their heart rate, blood pressure, glucose levels, and physical activities. These devices produce real-time information that can be processed to identify health patterns, manage numerous diseases, and even estimate probable health disorders before they occur.

Imaging Data: Comparing with other specialties, Medical imaging such as Xrays, MRIs, CT scans generate big amount of data that can only be analyzed with the use advanced analytics and AI. Imaging data is an ideal source for the analysis of anomalies like tumors and organ damage and is very useful, but due to the large volume and data complexity, it is considered as Big Data source.

Genomic Data: Genomic studies have led to huge amounts of genetic information that has been produced from the advancement in sequencing technology. This information can be utilized in finding out the causes of diseases genetically as well as choose treatment procedures according to the patient's genotype.

Clinical Trials and Research: Information derived from clinical investigations and medical studies providing important information concerning the efficacy of interventions. Using such data health care providers make informed decisions in supporting the needs of patients.

3.3 Importance of Real Time Data Processing and Analysis

First of all, the possibility of the realtime data processing makes it possible to track the patient's condition and respond quickly to the deterioration of the patient's condition. For instance, AI enabled systems can track data from wearables or remote monitoring systems to identify the warning signs of complications such as heart attacks or strokes, with an ability to respond that helps to save people's lives. Second, realtime data analysis is critical to controlling largescale health events, including outbreaks of a pandemic.

Realtime information on diagnosed cases, outcomes, and even the health care system status enabled statistic data usage by the public health authorities during events like COVID19. Besides, the use of realtime data analytics can enhance the outcomes on the effectiveness of health care organizations. Big Data is useful in patient flow management in hospitals, focusing areas like wait times and the management of hospital beds, workforce and other resources.

Therefore, the implementation of Big Data in healthcare is one of the strongest chances for enhancing patient satisfaction, healthcare systems productivity, and world health. Using realtime data processing and realtime analysis, healthcare providers can give patients precise and up-to-date care to change the landscape of the entire healthcare delivery system.

4 Applications of AI and Big Data in Healthcare

AI, specifically ML, and big data analytics are two growing forces that are through changing diagnostics, treatment, and management of patients. These applications range from the identification of the diseases at an early stage to the act of continuous monitoring of patients in real-time indeed offering incredible chances to intensify accuracy, speed, and quality in care. Here follows main fields with AI and Big Data application.

4.1 Diagnostics and Disease Prediction

New horizons of AI in the diagnosis of illnesses including cancer, heart, and diabetes are shaping up through the use of powered models [8]. Based on EHRs, genomic data, and imaging data, the AI model can search for biomarkers that point to disease prior to clinical manifestation. For example, the function that learns usage patterns from the patients' records and scans can locate signs of developing cancers that might be undetectable to radiologists, giving more timely treatments and, consequently, improved survival rates. Likewise, AI devices are able to forecast occurrence of cardiovascular events with non-stable clinical markers including cholesterol, life style and inherited risk factors. Big Data refines these parameters of the AI models. Through processing big data of different patient population, AI considers a vast number of possible variables and can detect even minor patterns that can point at the onset of diseases. Their integration enhances the capabilities of Big Data in AI modeling that allows healthcare professionals to diagnose diseases early, recommend proper treatments, possibly save lives and reduce health costs [9] (Fig. 2)

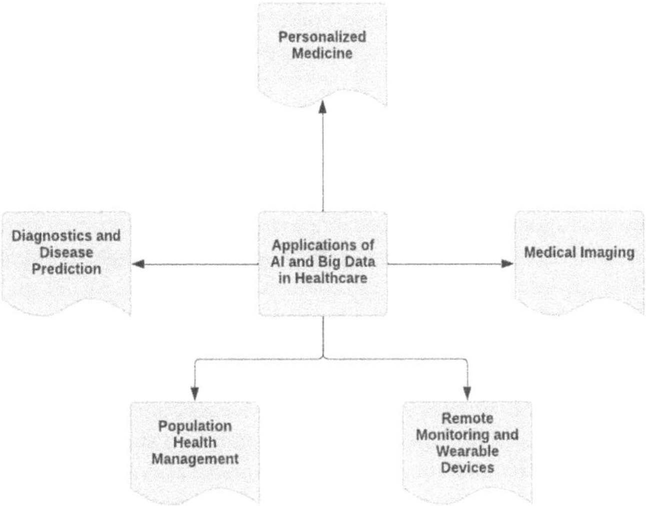

Fig. 2. Applications of AI and Big Data in Healthcare

4.2 Personalized Medicine

Personalized medicine is the utilization of AI and Big Data to deliver healthcare assets that correspond with the patient's unique identity. This is in contrast to the more conventional approach of presenting a solution that applies to everyone, without taking into account a patient's genetic predisposition, environmental and lifestyle factors or their medical history. The disparate data sets are utilized by the AI-based systems for determining how a particular patient will react to a particular kind of treatment, so the clinician teams about which treatments are most beneficial and which treatments may cause side effects. In genomics and pharmacogenomics, the major function of AI is to process the person's genomic data to determine how his or her DNA influences the effect of drugs. It can also be applied in the treatment of diseases like cancer in which different therapies can be given depending on the genes that cause cancer in a particular individual. Big Data underpins these analyses by offering large scale genetic datasets that allow an AI model to predict drug efficacy and calibrate dosage.

4.3 Medical Imaging

AI is actually progressing in the medical imaging where AI is applied in the diagnosis of lesions in images of radiological and pathological nature. AI enabled image recognition are commonly useful in detecting early signs of cancer, brain damage and other ailments since the AI system is able to note slight changes that might be hard for human experts to record. For instance, in diagnosis of certain types of lung cancer in computed tomography (CT) scans, AI has been establish to have better performance as compared to the radiologists. The number of images used in this field is important because AI models are trained on large sets of data to display high accuracy. Big Data makes it possible for these models to learn from many images to increase the chances of recognizing patterns and outliers with various patients' demographics. This way we can make sure that the AI systems can diagnose most common to severe diseases and illnesses.

4.4 Population Health Management

Intelligent population health management is one of the fields where the application of AI and Big Data can be referred to as an innovation. Analytical models take big datasets and monitor the health of specific populations and understand trending patterns, threats, and potential trends. For example, AI is able to work at the population level and pick out some patterns that suggest that there could be an outbreak of the flu or see where there are high rates of the prevalence of chronic diseases like diabetes. It helps the PHA and the HC professionals better understand the population and direct the resources of interventions. Sources of big data in health care include social determinants of health, healthcare use, and patient outcomes; AI-enabled systems could use these data to identify which segments of the populations are most vulnerable to such diseases and where to target prevention efforts. It makes healthcare more anticipatory, thus eradicating diseases, averting them right from the source and in the process, made healthcare Less congested.

4.5 Remote Monitoring and Wearable Devices

Patient monitoring is also being revolutionized by AI and Big Data through wearable technology and the IoT. Smartwatches and other wearable gadgets gather actual-time data on such parameters as vital signs, physical activity and other health indicators allowing the constant monitoring of patients in the outpatient setting. To identify early warning signs of potential ailments including the improper rhythms of the heart, this data is mined by AI algorithms and physicians are informed [10]. Big Data analytics have a crucial role at that since they deal with the tremendous amount of data provided by such devices. When fed with real time data, AI systems can learn patient health trends, detect possible complications and suggest changes in lifestyle or medication. It is particularly useful in the long-term monitoring of patients with certain diseases, prevention of hospitalization of patients with certain chronic diseases and promoting the improvement of patients' quality of life [11]. Overall, currently there are developments in the field of healthcare, thanks to the usage of technologies such as Artificial Intelligence and Big Data: An increase in the precision of diagnostic; Possibility of employing individualised treatment; Increase in population's health; Ability to monitor state of patients in real-time. Combined, they are leading the next generation of value-based, precision-medicine healthcare delivery.

5 Integration of AI and Big Data in Healthcare Systems

The use of Artificial Intelligence (AI) and Big Data in healthcare systems today has the potential of effecting radical changes to the healthcare delivery system, the management of the health systems and also medical research. But, it is not easy to establish such integration and helps face several issues for data interoperability, cloud computing, and formulation of the scalable framework or platform. This section analyses these problems and the solutions, such as technologies that support the interaction between AI and Big Data in healthcare.

5.1 Interoperability Challenges in Healthcare Data Integration

One of the major issues in the integration of AI and Big Data with the healthcare systems is Interoperability, which has to do with the capability of different systems and their technologies to exchange, share and even utilize information. Healthcare data originates from EHR, wearable devices, imaging systems, lab results, and others that are dissimilar in data formats and standards. It also establishes that unlike most other sectors, health care is not fully aligned in terms of global data input so there is no clear way to combine data smoothly between different systems. For instance, the EHR in a specific hospital might contain patient information with one type of data format while an adjunct wearable tracking pulse rates in real-time might collect data in another format. AI models need to operate on clean harmonised data but the current structure and arrangement of the healthcare data means that the potential of the AI in analyzing patient data is somewhat hindered. This challenge is further amplified by the fact that information privacy and security requirements may limit the exchange of data between these platforms and institutions. Barriers to interoperability can also limit the interfacing of healthcare organizations and especially between regional or even national providers since there is variation in policies. Solving these questions necessary creates the industry-wide data standards, for example, FHIR, which is the delivering of the healthcare information throughout the platforms. These standards are equally essential for widespread implementation of AI and Big Data to healthcare settings.

5.2 Importance of Cloud Computing and Data Storage Solutions

Most AI and Big Data are resource-intense applications that need both a lot of computation and storage, and legacy health IT often offers neither. This is where cloud computing comes into mainstream and becomes quite significant. Healthcare cloud computing technologies are cost efficient means for storage and computing of large amounts of health data. Digital health researchers can utilize the capacity of cloud services to host massive data created from EHRs, medical images, and wearable devices where they can be available to AI algorithms in realtime. Cloud computing offers the following benefits with regard to the AI/Big Data integration. First, it would allow computation on demand, so that AI models can process large orders of data that otherwise would require significant on premise hardware investment. This is especially so for AI applications that include deep learning where level of computing power needed for training and model deployment is requisite. Second, cloud architectures support collaboration because through multiple stakeholders participate, including hospitals, research institutions, and pharmaceutical companies can access the data and share it across multiple geographical locations, and in a secure manner.

In addition, cloud services nowadays apply various levels of safeguards, including encryption and access rights, to preserve the integrity of healthcare data. This is especially important particularly for organizations that have to deal with large volumes of sensitive information that must be protected from violation of data protection laws. Therefore, there is a need for cloud computing as an enabler for secure integration of AI and Big Data in healthcare.

5.3 Frameworks and Platforms for AI and Big Data Integration in Healthcare

Many frameworks and platforms and being designed to enable the amalgamation of AI & Big Data in the health systems. These platforms are intended to give healthcare providers the tools to gather process, analyse and visualize big data sets, so that artificial intelligence technologies are able to be implemented. Apache Hadoop is one of the most commonly used frameworks, which is an opensource framework for storage and computational distributed architecture for huge numbers of data sets. In the healthcare sectors, Hadoop is applied mainly to deal with Big Data, particularly in the unstructured data generated from health records, imaging, and genomic sequencing. Its capability this regard makes it applicable in the management of huge volumes of data used in the current healthcare systems. Another crucial structure is Apache Spark, which is more suitable for real time data processing. Spark enables the use of AI algorithms in the analysis of patient data in real time meaning providers can access valuable information for use in applications like predictive diagnosis, remote patient tracking and development of individual treatment plans. Moreover, there are dedicated healthcare platforms like IBM Watson Health and Google as Health AI that will also help in the integration of the AI and Big Data system. These components offer models of artificial intelligence, cloud, and data-driven mechanisms that make it possible for managers of healthcare to embrace AIdriven solutions. For instance, IBM Watson Health delivers AI-based solutions in oncology, genomics, and clinical trial, and digital health with Google Health AI uses machine learning for medica/ images and EHR data. Outside of these platforms, there are so many health systems thus engaging the open TensorFlow and PyTorch to develop their own AI models used in analyzing Big Data for certain clinical uses. These frameworks present added convenience as a result they allow the development of models which will fit requirements of the healthcare givers.

AI and big data are seen to be central to implementing change and improving patient care and healthcare organization operations. However, its application needs to cope with essential problems regarding data exchange, computation, and infrastructure's capacity. Current issues include the questions of big data storage and future use, heterogeneity and compatibility of applied systems, etc., which can be solved through cloud technologies and the application of standardized frameworks in the sphere of AI and Big Data utilization by healthcare providers. While keeping up with the various technologies and platforms, the healthcare sector will experience far better diagnostic outcomes, enhanced and individualised treatment, and better population health management.

6 Challenges and Limitations

Introducing AI and using Big Data in healthcare offers numerous developments however, it mitigates a range of problems and constraints. All these issues need to be resolved so as to maximize on the reliability security and equity of AIdriven healthcare systems. Other issues include the following; Data security and privacy, data quality, availability and, the ethic mode.

6.1 Data Privacy and Security

Firstly, the integration of the two technologies in healthcare is that patient information needs to be safeguarded from unauthorized access or invasion. Medical data is very valuable and if interoperability is allowed on such data it will result to negative effects that may include loss of an individual's identity, loss of Property, or even the so destroyed reputation of a person. The large volume of data processed and stored in AI and Big Data systems raises the threat actor's ability to compromise a system. Overcoming risks of data privacy in AI and Big Data implies the use of encryption, control of users' access, anonymization of data, and patients' information. For instance, healthcare data prior to being used in training AI models goes through what is commonly referred to as deidentification of PII. However, deidentification will possibly minimize the analytic utility of the data because some of the variables are probably critical for generating accurate prediction or diagnosis. However, it is equally important that regulatory frameworks such as The Health Insurance Portability and Accountability Act (HIPAA) for the U.S. patient and The General Data Protection Regulation (GDPR) for the European patient is adhered to in order to uphold their privacy rights. These regulations play a set out roles in terms of storage, processing and sharing of healthcare data. However, it must be understood organization faces certain challenges regarding these regulations, As a global healthcare organization has to deal with the different sets of rules in different jurisdictions. Further, adherence to such laws may hamper the growth and implementation of AI systems because organizations must first thoroughly analyse and protect their data feeds (Fig. 3).

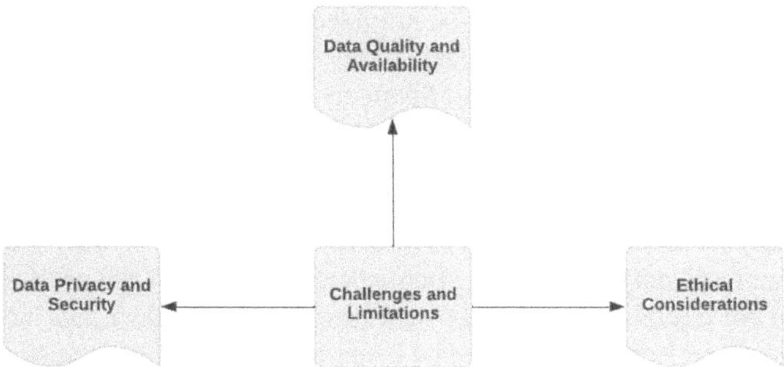

Fig. 3. Challenges and Limitations

6.2 Data Quality and Availability

Another enormous drawback is the variability and scarcity of well-organized health care datasets. The AI models are designed to predict based on vast amounts of data Therefore, only good quality and complete data are used for AI models. Unfortunately, the healthcare data can be incomplete, stochastic or even contain errors making it unadvisable for AI systems to work with them in order to develop fast and accurate health outcomes. For instance, patients' data can be stored in several establishments such as hospitals or by several providers, and hence cause a missing link thereby reducing the reliability of the predictions from AI. Furthermore, medical data is diverse and varies between structured and unstructured, collected from EHRs, medical imaging, wearable devices and genomics. These datasets become usually stored in different formats often unstandardized and this makes the integration of AI and Big Data more complex. This has implications for healthcare organizations struggling to amass large datasets of patients of substantially varying demographics; big data is a fundamental input for truly diverse AI. Another imperative is data accessibility since data owners such as the healthcare institutions can be reluctant to share data due to, among others, issues to do with privacy, competitive advantage, and intellectual property. It also implies that more organisations should work harmoniously to generate large-scale datasets that enhance the capability of AI. But this is hardly the case today, as legal and technical issues hinder this cooperation and slow down AI in the sphere of healthcare.

6.3 Ethical Considerations

AI and Big Data systems are also critical for developing healthcare strategies but they contain important ethical concerns about fairness, preference, and traceability. The first is the issue of Bias in a model whereby the models may be biased given the data used in training was also biased. That is, the trained AI model will be doing well in some groups' health application and poorly in the others, contributing to the development of health inequities.

For instance, if an AI model is trained on data from a selected group of patients, it will have fewer forecasts of patients of different demography, which increases injustice in the healthcare industry. This is a major concern on fairness since patients who did not feature in the training phase may be provided with a low quality care. In order to eliminate this problem, healthcare organizations should make sure that a large variety of datasets include different categories of patients. Ethical issues are not unique and are present also in the case of bias; ethical issues are also present in other areas such as automatized decision-making. Decision-making regarding or involving AI systems is becoming standard in clinical environments to develop or enhance clinical decisions, for instance identifying diseases or suggesting treatments. However, such decisions have to be transparent and defensible most often when they concern life or death situations. The argument for explainability is that both healthcare providers and indeed patients must know how precisely the AI systems came to a particular conclusion in order to fully trust the recommendation made. Again, opacity of AI models especially those advanced models such as the deep learning algorithms is a cause for ethical concern because of accountability issues that are doubted by society. Moreover, adding an aspect of healthcare decisions' automation raises concerns about the human supervision aspect. Overall, getting involvement of Healthcare providers in taking decision through AI, there is probability that safety of human interaction in making crucial clinical decision can be displaced by utilizing AI.

7 Future Trends and Research Directions

AI and Big Data shall continue to grow and intervene in the revolution of healthcare on even more exciting and liberalising manners. Technologies and methods including federated learning, explainable AI and cryptography including blockchain developments up alongside AI drug discovery and telemedicine are expected to mark future healthcare. In addition, the increasing role of AI and Big Data to tackle the current and future pandemics, and other global health concerns will also be discussed. This section does analysis of these trends as well as directions for future research in AI and Big Data in healthcare.

7.1 Emerging AI and Big Data Technologies

Another rapidly growing category of AI is federated learning, which creates an opportunity for cooperated machine learning by numerous healthcare organizations without the necessity of self-sharing of personal identifiers. In classic paradigm of machine learning, data generated from various sources requires aggregation to a single location for model training [12]. The federated learning approach eliminates the problem of centralizing healthcare data as it lets models be trained on data stored locally when only updating the model on the shared server. It is especially important in the analysis of personal records; it is particularly relevant in the sphere of healthcare. Federated learning is expected to witness more research because it seeks to boost the performance of the model while preserving privacy and creating a worldwide network of healthcare organizations. Another new trend is the explainable of AI (Explainable AI or XAI) in healthcare. Deep learning models are frequently used in medical diagnosis and treatment decisions, but the AI platform executes it as a black box, which does not explain how the conclusion remained. Understandable AI aims to solve it by bringing into light the artificial intelligence models in a form that can be easily understood by the healthcare givers as well as the patients. A trend in XAI research in the future will be deploying models that can generate explanations on how they arrived at their decisions, thus improving the uptake of AI in clinical practice [13].

7.2 Integration of AI with Blockchain for Secure Healthcare Data Sharing

AI integrated with blockchain technology provides a good solution to improve protection and privacy of the data in the healthcare sphere. Blockchain, the distributed ledger technology displays potential in securely share and manage health record data and information since patient's record cannot be altered or accessed Privacy by unauthorized parties. Blockchain technology is also integrated with artificial intelligence where large datasets including information from multiple contributors, including hospitals, insurance companies, and research organizations, can be are shared and simultaneously analyzed securely while ensuring privacy and protection of their information [14]. Blockchain will also be used to manage patient consent to data usage in AI models by the patients. For example, it was mentioned that by introducing a blockchain based system, patients will be able to share their data for an individual AI research project with their identities concealed. A secure and transparent approach to data sharing is going to be essential to enabling the scaling of AI and Big Data technologies across healthcare [15].

7.3 Advancements in AI Driven Drug Discovery and Telemedicine

Drug discovery is also being revolutionized through the help of AI as it helps to speed up the procedure for evaluating the candidacy of the certain chemicals. Typically, drug discovery is a slow and costly, an effort which may take from several years to decades to find a treatment option. This approach means that AIdriven models can highlight large datasets of biological and chemical data, forecast interactivity of drugs, and decide on the most beneficial molecules substantially faster than conventional technological methods. In the future, developing AI empowered platforms with genetics, proteomic, and molecular database, the accuracy of discovery in drug will be accelerated. Secondly, and more importantly, telemedicine is increasingly becoming advanced thanks to Artificial Intelligence solutions. A combination of telemedicine solutions and artificial intelligence allows for an increased effectiveness of diagnosing, monitoring, and treating patients by healthcare providers at a distance. For instance, by meshing with information gotten from wearable equipment, including heart rate monitors and glucose sensors, AI can offer real time clinical advice to healthcare givers during a telemedicine session. Subsequent studies will probably examine the concept of how AI might improve telemedicine programmes offering individualised treatment approaches and forecast health concerns before they occur. Virtual health assistant driven by artificial intelligence might also expand its demand as a support tool for patients with chronic conditions.

7.4 Potential for AI and Big Data in Pandemic Preparedness and Management

This paper seeks to demonstrate that COVID19 pandemic has proved that AI and Big Data are critical in handling pandemics. The use of big data AI powered predictive models are able to screen large datasets for new threats to health, monitor mathematical models of the transmission of infectious diseases and simulate the effects of interventions. Big Data offers, therefore, an opportunity to get the real-time information on the rates of infection, patients' outcomes, availability of resources among other things to the governments and the healthcare providers to be more sensitive to future pandemics. AI and Big Data will be the key enablers in the next frontier of health security in the future. Future research efforts will be directed toward designing new sophisticated methods of disease diagnosis, prognosis, and prevention as well as efficient modelling of the distribution of vaccines. AI models can also be used to model different intervention measures which assist the policymakers in deciding on the right measures to prevent the spread of disease. The future employment of both AI and Big Data in the healthcare sector has a lot of uncertainties and potentials due to the recent innovation in federated learning, explainable AI, and blockchain. Each of these emerging technologies is expected to help to improve the quality of patient care, security of the data and efficiency in the delivery of care in the health care setting. Furthermore, the two technologies would be critical in developing solutions necessary for preparedness, as well as ensuring healthcare infrastructure enhanced for any form of global threats. Future work will only extend the boundaries of possibility toward making the future healthcare system smarter and data driven.

8 Conclusion

AI as well as Big Data are changing the healthcare industry by using information technologies that make diagnostics, treatment, and patient care better. It has made the implementation of ML, DL, and NLP with big data derived from EHRs, wearables, and medical imaging that has improved patient care by delivering contrast-focused treatments. All this is not only enhancing the quality of the healthcare provided but it is also enhancing efficiency, cutting costs and enabling rational decision-making. Among them, a few insights include early disease prediction, clinical decision support and analysis of big medical data through the use of Artificial Intelligence powered systems. For instance, AI models can find patterns and mark abnormalities in the image of patients' organs, identify the chance of worsening the health status, and suggest individual treatment plans using data from the patient's records. On the other hand, Big Data analytics assists the healthcare organization in decision making, patient analytics, population health trends and resource management. Nonetheless, such an opportunity comes with the risk and hurdles including data privacy, security, and ethical issues into consideration. It is, therefore, important to create transparent, fair, and protection-conscious AI for the systems to be adopted more often in healthcare. Moving forward, future impacts of AI include Explainable AI, Federated AI, AIdriven Drug Discovery among other breakthroughs AI technologies that are set for improving the health sector. Application of AI with blockchain technology will improve the security of the data and also telemedicine and AI assisted remote monitoring which are already in practice will also increase day by day and will provide the facility to get the treatment also. Both AI and Big Data will also be applied to managing pandemics and to enhancing the responsiveness of healthcare systems to future worldwide outbreaks. Therefore, it can be claimed that with the help of AI and Big Data healthcare is becoming more predictive, personalized and preventive. As the studies in the field go on, and tendency toward the development of new technologies rises, the utilization of the AI in healthcare seems to be capable of enhancing the patients' satisfaction and the quality of the future of medicine.

References:

1. Gopi, B., Sworna Kokila, M.L., Bibin, C.V., Sasikala, D., Howard, E., Boopathi, S.: Distributed technologies using AI/ML techniques for healthcare applications. In: Social Innovations in Education, Environment, and Healthcare, pp. 375–396. IGI Global (2024)
2. Surya, L.: An exploratory study of AI and Big Data, and it's future in the United States. Int. J. Creat. Res. Thoughts **2**, 2320–882 (2015)
3. Shah, R., Chircu, A.: IoT and AI in healthcare: a systematic literature review. Issues Inform. Syst. **19**(3) (2018)
4. Kasula, B.Y.: Machine learning in healthcare: revolutionizing disease diagnosis and treatment. Int. J. Creat. Res. Comput. Technol. Des. **3**(3) (2021)
5. Rajendran, S.: Emphasizing privacy and security of edge intelligence with machine learning for healthcare. Int. J. Intell. Comput. Cybern. **15**(1), 92–109 (2022)
6. Vaghela, D.B.: Introduction to big data and business analytics. In: Sentiment Analysis and Knowledge Discovery in Contemporary Business, pp. 279–288. IGI Global (2019)

7. Nagarajan, R., Thirunavukarasu, R.: Big data analytics in cloud computing: effective deployment of data analytics tools. In: Research Anthology on Big Data Analytics, Architectures, and Applications, pp. 228–244. IGI Global (2022)
8. Ganesan, M., Sivakumar, N.: IoT based heart disease prediction and diagnosis model for healthcare using machine learning models. In: 2019 IEEE International Conference on System, Computation, Automation and Networking (ICSCAN), pp. 1–5. IEEE (2019)
9. Khader Basha, S., Roja, D., Priya, S.S., Dalavi, L., Vellela, S.S., Reddy, V.: Coronary heart disease prediction and classification using hybrid machine learning algorithms. In: 2023 International Conference on Innovative Data Communication Technologies and Application (ICIDCA), pp. 1–7. IEEE (2023)
10. Islam, M.R., Kabir, M.M., Mridha, M.F., Alfarhood, S., Safran, M., Che, D.: Deep learning-based IoT system for remote monitoring and early detection of health issues in real-time. Sensors **23**(11) (2023)
11. Roy, D., Jana, M., Tuccu, C., Pal, A., Kumar, R., Bag, S.: Integration of heart rate and SpO2 monitoring in wearable health technology. J. Eng. Tech. Manage. **72**, 1613–1618 (2024)
12. Alahi, M.E.E., et al.: Integration of IoT-enabled technologies and artificial intelligence (AI) for smart city scenario: recent advancements and future trends. Sensors **23**(11), 5206 (2023)
13. Khang, A., Misra, A., Gupta, S.K., Shah, V. (eds.) AI-aided IoT technologies and applications for smart business and production. CRC Press (2023)
14. Alabdulatif, A., Khalil, I., Rahman, M.S.: Security of blockchain and AI-empowered smart healthcare: application-based analysis. Appl. Sci. **12**(21), 11039 (2022)
15. Omrčen, L., Leventić, H., Romić, K., Galić, I.: Integration of blockchain and AI in EHR sharing: a survey. In: 2021 International Symposium ELMAR, pp. 155–160. IEEE (2021)

QuiltCraft: A Computer Vision Framework for Sustainable Upcycling Waste Textiles into Artistic Quilts

A. Bhuvaneswari[1], Sakthitharan Subramanian[2](✉), E. Mary Jasmine[3], G. Muralitharan[4], and R. Dharaniya[5]

[1] Adhiparasakthi Engineering College, Melmaruvathur, Tamil Nadu, India
bhuvaneswari@apec.edu.in
[2] SRM Institute of Science and Technology, KTR, Chennai, Tamil Nadu, India
sakthits@srmist.edu.in
[3] Christ University, Bangalore, India
[4] AMC Engineering College, Bangalore, India
[5] Easwari Engineering College, Chennai, India

Abstract. QuiltCraft represents a pioneering approach where we aim to mitigate the pervasive children's cloth crisis by ingeniously repurposing discarded cloth into captivating and artistic quilts. By harnessing the power of computer vision techniques, predominantly facilitated by using OpenCV and various Python libraries, our innovative system streamlines the intricate process of identifying cloth sizes, segmenting them into manageable pieces, and orchestrating their arrangement into visually stunning quilt designs. This paper explores the methodology, algorithms, and intricate implementation details underpinning the Quilt Craft framework, supplemented by empirical evidence attesting to its remarkable effectiveness. Through the seamless transformation of children's clothes into captivating works of art, QuiltCraft not only champions the cause of waste reduction but also catalyzes advancing sustainable practices within the fashion industry.

1 Introduction

Quilt Craft emerges as a pioneering initiative at the intersection of technology, sustainability, and artistic innovation. By harnessing the power of computer vision techniques and Python libraries, Quilt Craft seeks to revolutionize the traditional craft of quilting by repurposing children's clothes into intricate quilt patterns. The motivation behind Quilt Craft stems from the dual imperatives of environmental sustainability and creative exploration. The textile industry's linear model of production, consumption, and disposal has resulted in a staggering amount of waste, with only a fraction being recycled or repurposed. This unsustainable practice not only depletes finite resources but also contributes to pollution and environmental degradation. Moreover, the aesthetic appeal and cultural significance of quilting as a traditional craft form provide a compelling impetus for innovation. Quilts serve as tangible representations of personal and communal narratives, reflecting themes of heritage, identity, and expression. By infusing technology into the quilting process, Quilt Craft aims to preserve this rich tradition while offering a contemporary solution to the challenges of children's cloth. [2].

R. Geetha et al. (Eds.): AAIMB 2024, CCIS 2654, pp. 137–149, 2026.
https://doi.org/10.1007/978-3-032-13029-7_10

2 Existing System

In the existing system of quilt-making, manual processes are prevalent, particularly when dealing with diverse collections of children's clothes. However, this manual approach encounters difficulties, especially when attempting to design specific patterns efficiently. Challenges arise due to the time-consuming nature of arranging fabric pieces manually and the need for expertise and creativity in quilt design. These limitations can hinder the timely production of quilts and may restrict the range of design options available.

The paper [3] discusses image quilting for texture synthesis and transfer, offering foundational concepts for aesthetic pattern assembly, but not for quilting automation.

The work by Li et al. [6] focuses on algorithmic quilting pattern generation using computational geometry, which partly inspires layout generation strategies.

Saif et al. [9] explore OpenCV-based defect detection in manufacturing, which informed our preprocessing pipeline for cloth segmentation.

Yuan's recent study [10] leverages hybrid fuzzy clustering for scrap denim quilting, demonstrating the efficacy of combining computer vision with textile upcycling.

Another contribution by Hiramatsu et al. [4] applies deep learning to classify clothing items from cluttered backgrounds, which presents a direction for further automation in quilt generation.

Zhang et al. [11] propose a sustainability-oriented garment recycling system using CV and sensor fusion, indicating the increasing trend of intelligent textile processing.

Recent advancements also point to the integration of AI and CV in broader textile sustainability. Kumar et al. [5] provide a comprehensive survey of intelligent textile sorting systems, highlighting the limitations of traditional manual practices. Lu and Sun [7] proposed a real-time textile defect classification system using deep convolutional networks, which parallels our segmentation efforts. Further, Moradi and Mahdavi-Amiri [8] investigate the use of machine vision in upcycling post-consumer garments, showing strong alignment with QuiltCraft's goals. Aghaei et al. [1] introduce FabricNet, a fabric classification framework using hybrid deep learning models, illustrating the viability of scalable automation in textile recognition.

2.1 Disadvantages of the Existing System

1. Manual arrangement of fabric pieces is labor-intensive and time-consuming.
2. Pattern design requires expert skill, limiting accessibility.
3. Complex layouts are difficult to plan manually, reducing design flexibility.
4. Manual methods lead to inconsistent alignment and layout quality.
5. Scalability is poor for bulk production or rapid turnaround.
6. Creative and customized design exploration is restricted.

3 Proposed System

To address the challenges faced in manual quilt design processes, we propose implementing an automated pattern generation system utilizing computer vision techniques, specifically OpenCV. In real-time scenarios where various children's clothes are collected for quilt-making, our solution aims to streamline the design process and overcome difficulties in creating specific patterns efficiently. By leveraging OpenCV's capabilities, we can develop algorithms that analyse the characteristics of fabric pieces, such as colour, texture, and size, in order to intelligently generate quilt patterns. This automated approach significantly reduces the time and effort required for designing, as compared to manual processes. Moreover, automated pattern generation ensures consistency and precision in pattern layout and alignment, addressing the limitations of manual design methods. Through our proposed solution, quilt-makers can benefit from increased efficiency and productivity, allowing for quicker turnaround times and scalability for large-scale production. Additionally, the automated system offers greater flexibility for customization, empowering users to explore a wider range of design options and creative possibilities in quilt-making. Overall, our proposed solution offers a comprehensive approach to overcoming the challenges associated with manual quilt design processes, enhancing efficiency, precision, and creativity in real-time pattern generation for quilt-making projects involving children's clothes.

3.1 Advantages of Proposed System

1. Reduces design time significantly.
2. Enables quicker completion of quilt-making projects.
3. Ensures accurate and uniform pattern layout.
4. Easily handles large-scale production needs.
5. Requires minimal specialized skills for pattern design.
6. Offers flexibility for personalized pattern creation.
7. Allows more time for creative exploration in quilt design.
8. Demonstrates advancements in quilt-making methodologies.
9. Minimizes fabric waste and resource consumption.
10. Provides a smoother and more efficient design process.

4 System Architecture

The system architecture of QuiltCraft is composed of a sequential pipeline designed to automate the process of upcycling waste textiles into quilt designs using computer vision. The framework begins with a data ingestion layer, where images of discarded garments are acquired and preprocessed. Following this, a background removal module isolates cloth items from their surroundings to enhance segmentation accuracy. The next stages involve cloth size detection and image segmentation, which extract geometric and spatial attributes required for optimal piece partitioning. Each segmented piece is then analyzed and organized based on texture and color similarity. Finally, the layout engine orchestrates the placement of quilt pieces into a coherent and aesthetically pleasing design, which is then visualized for user evaluation. This modular architecture ensures scalability, maintainability, and the ability to integrate AI-powered enhancements in future iterations.

5 System Modules

5.1 Data Collection

In the data collection phase, our team embarked on a meticulous process of sourcing and curating a diverse dataset of children's clothing images. This involved scouring various sources such as discarded clothing, fabric scraps from manufacturing processes, and surplus materials from textile industries. Our aim was to assemble a comprehensive collection that captured the breadth and diversity of children's clothing, encompassing different types of fabrics, patterns, colors, and textures. Each image underwent careful vetting to ensure quality and relevance to our project objectives. By building a robust dataset, we laid the foundation for training and validating our algorithms, enabling them to effectively identify and process children's clothes in subsequent stages of the project.

5.2 Background Removal

The background removal module was dedicated to the intricate task of isolating clothing items from their surrounding environments. This involved employing advanced image processing techniques to detect and remove background clutter, noise, and other irrelevant elements. Leveraging state-of-the-art algorithms and methodologies, we meticulously delineated the boundaries of each clothing item, ensuring accuracy and precision in segmentation. By eliminating extraneous elements, we aimed to streamline the subsequent stages of processing, facilitating more accurate and efficient analysis of the children's cloth images (Figs. 1, 2 and 3).

```
FUNCTION BackgroundRemoval(I, M, threshold_value):
    # Step 1: Preprocessing
    I_resized = Resize(I, (H, W))
    I_normalized = Normalize(I_resized)

    # Step 2: Foreground Segmentation
    F_mask = M.predict(I_normalized)


    # Step 3: Post-Processing the Mask
    F_binary = Threshold(F_mask, threshold_value)
    F_cleaned = MorphologicalOperations
                    (F_binary, operations=["erode", "dilate"])
    F_refined = GaussianBlur(F_cleaned, kernel_size=5)

    # Step 4: Apply the Refined Mask

We used OpenCV with empirically optimized hyperparameters: Gaussian kernel size of (5,5), C
    Foreground = I * F_refined
    Background_removed = AddAlphaChannel(Foreground)

    RETURN Background_removed

# Main Execution
Image = LoadImage("path/to/input_image")
Model = LoadSegmentationModel("path/to/pretrained_model")
Threshold = 0.5     # Adjust as needed
Output_Image = BackgroundRemoval(Image, Model, Threshold)
SaveImage(Output_Image, "path/to/output_image")
```

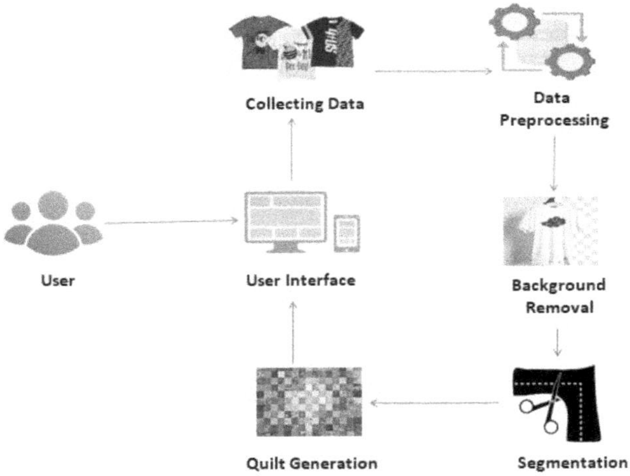

Fig. 1. System Architecture

Fig. 2. Input Image

Fig. 3. Background Removed Image

5.3 Cloth Size Detection

Integrating the cloth size detection module into our workflow, we implemented advanced computer vision algorithms, primarily leveraging the capabilities of OpenCV, to accurately identify and quantify the sizes of individual clothing items within the segmented images. By analyzing pixel intensities, texture variations, and spatial distributions, our system intelligently determined the dimensions of each cloth segment, providing crucial information for subsequent processing stages. This module enhanced the overall efficiency and accuracy of the workflow, enabling precise partitioning and quilt design. As of now, we have segmented the clothes based month-wise like if we are uploading 1 to 3 month cloth it will take as 1 similarly for 3 to 6 months as 4 and for 6 to 12 months as 9 (Fig. 4).

```
FUNCTION DetectClothSize(image, reference_object=None):
    # Step 1: Preprocess the image
    gray = ConvertToGrayscale(image)
    blurred = ApplyGaussianBlur(gray, kernel_size=(5, 5))
    edges = ApplyCannyEdgeDetection
                (blurred, threshold1=50, threshold2=150)

    # Step 2: Find contours
    contours = FindContours(edges)
    cloth_contour = SelectLargestContour(contours)

    # Step 3: Calculate bounding box
    rect = CalculateMinAreaRect(cloth_contour)
    width_pixels = rect.width
    height_pixels = rect.height

    # Step 4: Calculate real-world dimensions
    IF reference_object IS NOT None:
        scaling_factor = GetScalingFactor(reference_object)
        width_real = width_pixels * scaling_factor
        height_real = height_pixels * scaling_factor
    ELSE:
        width_real = width_pixels
        height_real = height_pixels

    # Step 5: Annotate and display results
    AnnotateImage(image, rect, width_real, height_real)
    DisplayImage("Cloth Size Detection", image)
```

Real-world garment sizes vary widely even within the same age group. Future versions will

```
    RETURN width_real, height_real
```

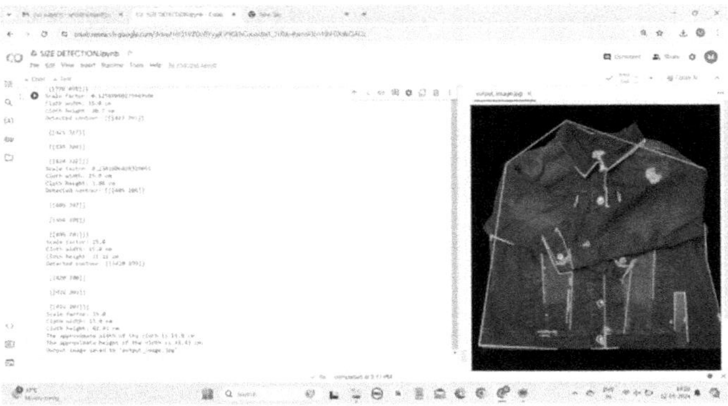

Fig. 4. Image Segmenting

5.4 Image Segmentation

Image segmentation served as a pivotal step in our workflow, enabling us to partition the children's cloth images into distinct regions corresponding to individual clothing items. Leveraging the robust capabilities of OpenCV, we applied a variety of segmentation techniques such as thresholding, edge detection, and region-based methods to accurately delineate the boundaries of clothing items. This involved analyzing pixel intensities, color distributions, and spatial relationships within the images in identifying and separating clothing items from background clutter and other objects. The resulting segmented images formed the basis for further processing in subsequent modules, laying the groundwork for precise piece partitioning and quilt design.

```
import cv2
import numpy as np

# Load the image
image = cv2.imread("path/to/image.jpg")

# Convert to HSV color space
hsv = cv2.cvtColor(image, cv2.COLOR_BGR2HSV)

# Define the color range for the specific region
lower_bound = np.array([0, 50, 50])
# Lower HSV bound (example: red)
upper_bound = np.array([10, 255, 255])
# Upper HSV bound

# Create a binary mask
mask = cv2.inRange(hsv, lower_bound, upper_bound)

# Perform morphological operations to clean the mask
kernel = np.ones((5, 5), np.uint8)
mask = cv2.morphologyEx(mask, cv2.MORPH_CLOSE, kernel)

# Find contours
contours, _ = cv2.findContours(mask,
                    cv2.RETR_EXTERNAL, cv2.CHAIN_APPROX_SIMPLE)
```

```
if contours:
    # Select the largest contour
    target_contour = max(contours, key=cv2.contourArea)

    # Get bounding box and draw it
    x, y, w, h = cv2.boundingRect(target_contour)
    output_image = image.copy()
    cv2.rectangle(output_image,
                  (x, y), (x + w, y + h), (0, 255, 0), 2)
    cv2.putText(output_image, "Detected Region", (x, y - 10),
                cv2.FONT_HERSHEY_SIMPLEX, 0.6, (0, 255, 0), 2)

    # Show results
    cv2.imshow("Detected Region", output_image)
    cv2.waitKey(0)
    cv2.destroyAllWindows()

    # Optional: Compute area and centroid
    area = cv2.contourArea(target_contour)
    M = cv2.moments(target_contour)
    if M["m00"] != 0:
        cx = int(M["m10"] / M["m00"])
        cy = int(M["m01"] / M["m00"])
        print(f"Centroid: ({cx}, {cy}), Area: {area}")
```

5.5 Piece Partitioning

The piece partitioning module was tasked with dividing the segmented clothing images into quilt-ready pieces, each with its own unique characteristics and attributes. Drawing upon principles of design and composition, we developed sophisticated algorithms to partition the segmented images based on factors such as fabric patterns, sizes, and colours. This involved analysing the spatial arrangement of pixels within each segmented region to identify natural seams, edges, and contours that could serve as boundaries for partitioning. By considering the inherent characteristics of the textiles, we aimed to create cohesive and visually appealing quilt pieces that could be seamlessly integrated into the final designs.

```
import cv2
import numpy as np

# Load the image
image = cv2.imread("path/to/image.jpg")

# Get image dimensions
height, width, _ = image.shape

# Define the number of rows and columns for partitioning
rows = 4
cols = 4

# Calculate the size of each tile
tile_height = height // rows
tile_width = width // cols

# Initialize a blank canvas to store results
output_image = image.copy()

for i in range(rows):
    for j in range(cols):
        # Calculate tile coordinates
        y_start = i * tile_height
        y_end = (i + 1) * tile_height
        x_start = j * tile_width
        x_end = (j + 1) * tile_width

        # Crop the tile
        tile = image[y_start:y_end, x_start:x_end]

        # Convert tile to HSV and create a binary mask
        hsv = cv2.cvtColor(tile, cv2.COLOR_BGR2HSV)
        lower_bound = np.array([0, 50, 50])      # Adjust as needed
        upper_bound = np.array([10, 255, 255])     # Adjust as needed
        mask = cv2.inRange(hsv, lower_bound, upper_bound)

        # Find contours in the tile
        contours, _ = cv2.findContours(mask,
                        cv2.RETR_EXTERNAL, cv2.CHAIN_APPROX_SIMPLE)
       # Draw contours on the original image
       for contour in contours:
           # Translate contour coordinates
           # to match the full image
                contour += [x_start, y_start]
                cv2.drawContours(output_image,
                    [contour], -1, (0, 255, 0), 2)

       # Display the final segmented image
       cv2.imshow("Segmented Image", output_image)
       cv2.waitKey(0)
       cv2.destroyAllWindows()
```

5.6 Quilt Design

In the quilt design module, our focus shifted towards unleashing creativity and innovation in transforming the partitioned textile pieces into visually stunning quilt patterns. Drawing inspiration from traditional quilting techniques as well as contemporary design principles, we explored a myriad of layout arrangements, color palettes, and pattern generation strategies. This involved experimenting with different combinations of quilt pieces, arranging them in geometric or organic patterns, and incorporating elements of symmetry, balance, and contrast to create dynamic and harmonious compositions. Our goal was to push the boundaries of conventional quilt design, leveraging the unique characteristics of the children's clothes to craft captivating and expressive works of art.

```
# Combine tiles into a quilt design
quilt = np.zeros_like(image)\selectlanguage{ngerman}
# Create a blank canvas

index = 0
for i in range(rows):
    for j in range(cols):
        # Calculate the position of the current tile
        y_start = i * tile_height
        y_end = (i + 1) * tile_height
        x_start = j * tile_width
        x_end = (j + 1) * tile_width

        # Place the transformed tile into the quilt
        quilt[y_start:y_end, x_start:x_end] = tiles[index]
        index += 1
```

5.7 Display and Evaluation

The final stage of our workflow involved showcasing the quilt designs digitally and subjecting them to rigorous evaluation based on various criteria including aesthetic appeal, coherence, and sustainability impact. Through digital visualization platforms and interactive presentations, we presented the quilt designs to stakeholders, collaborators, and the broader community for feedback and critique. This feedback loop was integral to refining and iterating upon our designs, ensuring that they resonated with audiences and met our overarching objectives of sustainability and creativity. Additionally, we conducted systematic evaluations of the sustainability impact of our designs, considering factors such as material usage, resource efficiency, and environmental footprint. By aligning our designs with principles of sustainability and circularity, we aimed to demonstrate the transformative potential of repurposing children's clothes into art (Fig. 5).

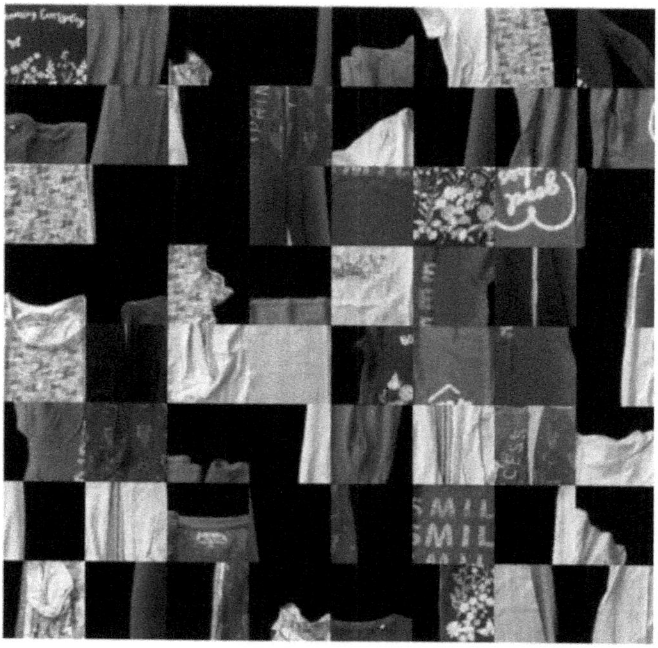

Fig. 5. Quilt Design Image

6 Conclusion

QuiltCraft stands as a pioneering fusion of technology and artistry, offering a sustainable solution to the pressing issue of children's clothing. Through the innovative integration of advanced image processing techniques facilitated by OpenCV and Python libraries, QuiltCraft transforms discarded textiles into visually captivating quilt patterns. By automating processes such as image segmentation and piece partitioning, QuiltCraft streamlines the creation of quilt designs, minimizing manual intervention while maximizing efficiency. Moreover, QuiltCraft serves as a catalyst for environmental consciousness, inspiring individuals to rethink their relationship with textiles and embrace sustainable practices. Through its interdisciplinary approach and community engagement efforts, QuiltCraft seeks to empower individuals to become agents of change in the global effort to reduce waste and promote sustainability.

6.1 Future Enhancement

QuiltCraft aims to harness the power of artificial intelligence (AI) to further augment its capabilities and drive continuous improvement in removing the background and detecting the size of the clothes. By integrating AI algorithms into its image processing pipeline, QuiltCraft seeks to enhance precision and efficiency in tasks such as image segmentation and pattern recognition. Additionally, QuiltCraft plans to exploit AI technologies to optimize quilt design suggestions based on user preferences and input. Furthermore, machine learning models could be employed to analyse extensive datasets of quilt patterns, generating innovative designs automatically. Through these advancements, QuiltCraft envisions a future where AI plays a pivotal role in revolutionizing sustainable textile design, fostering creativity, and advancing environmental stewardship.

References

1. Aghaei, M., Hosseini, R., Saeedi, P.: Fabricnet: a hybrid deep learning model for fine-grained fabric classification. Expert Syst. Appl. **192**, 116384 (2022)
2. Arbelaez, P., Pont-Tuset, J., Barron, J.T., Marques, F., Malik, J.: Multiscale combinatorial grouping. In: Proceedings of the IEEE Conference on Computer Vision and Pattern Recognition (CVPR) (2014)
3. Efros, A.A, Freeman, W/.T.: Image quilting for texture synthesis and transfer. In: Proceedings of the 28th Annual Conference on Computer Graphics and Interactive Techniques (SIGGRAPH), pp. 341–346. ACM (2001)
4. Hiramatsu, K., Shibata, Y., Tanaka, N.: Deep learning-based clothing classification from cluttered backgrounds for automated textile recycling. J. Fash. Technol. Textile Eng. **9**(3), 123–131 (2021)
5. Kumar, A., Patel, R., Mehta, A.: A review of intelligent systems for automated sorting in textile recycling. J. Clean. Prod. **294**, 126303 (2021)
6. Li, Y., Breen, D.E., McCann, J., Hodgins, J.: Algorithmic quilting pattern generation for pieced quilts. In:Proceedings of Graphics Interface 2019, pp. 21–28 (2019)
7. Yang, L., Sun, X.: Toward real-time textile defect detection using convolutional neural networks. Comput. Ind. Eng. **149**, 106795 (2020)
8. Moradi, A., Mahdavi-Amiri, N.: Vision-based approaches for upcycling post-consumer textiles: A sustainable pathway. Sustainability **15**(2), 525 (2023)
9. Saif, YA., Yusof, Y., Ahmed, M.L., Pathan, Z.K., Latif, K., Kadir, A.Z.A.: A framework to develop intelligent system for capturing product features using open cv technique. J. Technol. Oper. Manage. **15**(Number 2), 42–51 (2020)
10. Yuan, J.: Automatic generation technology of scrap denim clothes in quilting art patterns by integrating computer vision. J. Electric. Syst. **20**(6s), 207–216 (2024)
11. Zhang, W., Liu, F., Chen, J.: Smart garment recycling: Combining computer vision and sensor fusion for automated sorting. Sustain. Comput.: Inform. Syst. **35**, 100760 (2022)

Automatic Fraud Insurance Claim Detection System

K. Neela[1], M. Rekha[2], R. Varsha[1(✉)], and B. Saveetha[1]

[1] Sri Sairam Engineering College, Sai Leo Nagar, West Tambaram, Chennai 44, India
neela.it@sairam.edu.in, varsharajendiran99@gmail.com
[2] R.M.K. Engineering College, Kavaraipettai, Gummidipoondi, TamilNadu 601206, India
mra.it@rmkec.ac.in

Abstract. Fraud in insurance companies has grown to be a serious problem for the auto insurance sector in recent years, costing honest policyholders a lot of money in lost revenue and raising premiums. The paper offers a novel solution to this problem that uses algorithms in machine learning (ML) and natural language processing (NLP) methods to detect insurance fraud in auto sector. This project also uses AES Encryption technique for storing all the customer details. To find fraudulent patterns in the insurance documents, a machine learning model is constructed applying the proper techniques for XGBoost, logistic regression, Naive Bayes, Random forest, and SVM. To increase accuracy and efficacy, these models are trained using historical data from both valid and fraudulent insurance claims. The models are trained using historical data encompassing both legitimate and fraudulent claims to enhance accuracy and effectiveness. The objective is to compare and determine the algorithm that most accurately differentiates between fraudulent and legitimate claims. Model performance is evaluated using standard metrics such as accuracy, precision, recall, and F1-score. Among all tested algorithms, XGBoost achieved the highest accuracy of 89.73%, demonstrating superior performance in fraud detection.

Keywords: Machine learning · NLP · XGBoost · Classification · Encryption

1 Introduction

Insurance fraud encompasses a wide array of dishonest activities aimed at deriving unfair financial benefits from insurance coverage. These deceitful schemes can range from minor exaggerations of claims to elaborate scams involving fabricated events or false information. Since the inception of insurance companies, fraudulent claims have posed a challenge, as they can yield considerable financial gains and are often difficult to detect. This growth indicates the sector's increasing importance in the Indian economy and its evolving ability to manage risks and provide financial stability [1]. In global rankings, India ranks 15th in non-life insurance products and 10th in total premiums for life insurance. These standings highlight India's prominent role in the global insurance market, attributed to its large and diverse population and the growing

R. Geetha et al. (Eds.): AAIMB 2024, CCIS 2654, pp. 150–161, 2026.
https://doi.org/10.1007/978-3-032-13029-7_11

demand for insurance. Fraudulent activities significantly burden the insurance industry, leading to annual waste of billions of dollars. As a result, the need for technology to combat these fraudulent practices is becoming ever more pressing. This not only ensures a stable and trustworthy environment for consumers but also greatly reduces the chances of fraud. It is crucial to understand that higher premiums for policyholders often compensate for losses incurred from fraudulent claims, negatively impacting society as a whole [2] A two-step approach for fraud detection is proposed initially, a "rule-based detection algorithm" identifies claims likely to be fraudulent. Following this, fraud specialists review the flagged claims for any indications of fraud, determining if further investigation is necessary. Therefore, insurance providers must develop strategies for accurately identifying risk factors and mitigating the damage caused by fraudulent claims as swiftly as possible [3]. Traditionally, fraud detection has been a manual process. However, advancements in data analysis and the adoption of machine learning and deep learning algorithms have automated the detection process, improving accuracy, efficiency, and systematic approaches. The processing speed of information is progressively enhanced through various machine learning algorithms. Thus, by utilizing data mining and machine learning or deep learning techniques, fraud detection becomes increasingly accurate, efficient, and systematic [4]. Effective fraud detection depends heavily on the quality of features extracted from both structured data (e.g., claim amounts, policyholder details) and unstructured data (e.g., textual claim descriptions, scanned documents). Advanced techniques like natural language processing (NLP) and optical character recognition (OCR) enable the transformation of raw data into actionable features, which machine learning models utilize to improve predictive accuracy. In this research, XGBoost is utilized as the primary classifier to build an effective machine learning system for auto insurance fraud detection. The increasing volume and complexity of insurance claims pose significant challenges to traditional manual fraud detection methods, which are often time-consuming and prone to errors. There is a critical need for automated, accurate, and scalable solutions that can analyze large datasets in real-time or near-real-time to promptly identify suspicious claims and minimize financial losses.

To address this, our study validates the application of machine learning techniques, specifically XGBoost, for automating fraud detection in insurance claims. Additionally, the system integrates natural language processing (NLP) to extract meaningful information from unstructured claim documents, while simultaneously ensuring customer data privacy through encryption. This combined approach not only enhances detection accuracy but also improves client satisfaction and strengthens the company's reputation, ultimately benefiting the business as a whole.

2 Literature Survey

The authors of the paper highlight that the substantial sums claimed by fraudsters can lead to critical issues, prompting various organizations to collaborate with the government to tackle these practices. Such fraudulent schemes are severe and pervasive throughout all types of insurance claims. For instance, vehicle insurance fraud—often characterized by filing false accident claims—illustrates the project's aim to analyze insurance claim datasets for the detection of inflated and fraudulent claims. Machine learning techniques are used to build a model for labeling, categorizing claims, with a confusion matrix serving as a tool to compare the accuracy, precision, recall, and other relevant metrics of each applied algorithm for classification purposes [5].

Insurance fraud involves submitting fictitious claims to obtain benefits or funds to which individuals are not rightfully entitled. Among the many types of insurance fraud, auto fraud is notably common, often carried out through fictitious accident claims. This study aims to efficiently identify instances of vehicle fraud using various machine learning techniques, while a confusion matrix is created to evaluate and compare different approaches' efficiencies. A comprehensive assessment of the machine learning methods utilized in this research will also include significant metrics like recall, precision, and accuracy. The intention of this approach is to enhance the capacity of the insurance industry to detect fraud effectively and, ultimately, to minimize further fraudulent claims [6].

Criticism of traditional claim investigations arises from their tendency to produce inaccurate results due to their labor-intensive and protracted nature. To address these issues, the researchers propose an automated forgery identification device system powered by machine learning techniques, particularly the XGBoost algorithm. Objective of the system was to accurately predict claim that are prone to fraud more quickly. Data analysis is crucial in this approach, as it facilitates the extraction, cleaning, and validation of relevant information at every stage. Implementing this system will not only enhance the operational efficiency of insurance companies but also bolster their reputation and nurture a trustworthy relationship with clients. This shift towards automation and sophisticated analytical techniques aims to strengthen the integrity of the claims process overall [7].

The insurance industry faces considerable obstacles due to fraud which escalates premiums for consumers. This article discusses managing vehicle insurance forgery through several models like KNN, Support Vector Machine. The analysis leverages diverse factors such as age, policy information, incident specifics, and claims data to improve the detection procedure. These ml methodologies are specifically crafted to efficiently process legitimate claims while accurately identifying fraudulent ones. By utilizing advanced techniques, the study aims to enhance both the accuracy and speed of differentiating fraudulent from legitimate claims, ultimately benefiting insurers and policyholders by mitigating the effects of fraudulent activities [8]. The auto insurance market has transitioned into a multimillion-dollar industry fueled by high premium costs and the millions of claims submitted annually. Auto insurance constitutes a significant portion of the overall vehicle ownership expenses. The insurance sector relies on data science to streamline operations amidst its unpredictable nature attributed to risk assessment. This research conducts a comparative examination of prevalent ML algorithms for fraud detection. Each algorithm is assessed for reliability and effectiveness in detecting fraud. Additionally, various performance metrics, including the F1 score, AUC score, efficiency, recall, and precision, are evaluated to provide a comprehensive understanding of these ML techniques' capabilities concerning auto insurance fraud detection [9].

Obtaining auto insurance is mandatory in many countries, serving to transfer risk and provide financial stability. Given the growing scope of the automobile insurance sector, effectively addressing insurance fraud has become increasingly critical. Predicting fraud in auto insurance claims are especially challenging because of the imbalanced characteristics, which often results in insufficient classification models in most fraud detection systems. To rectify this imbalance and enhance the effectiveness of fraud detection, this study employs Adaptive Synthetic Sampling and Synthetic Minority Oversampling Techniques to boost model performance and the representation of fraudulent cases within the dataset. The research scrutinizes the predictive performance of various machine learning models, which includes, Balanced Random Forest, and Extreme Gradient Boosting, using tenfold cross-validation to ensure a robust evaluation and accurate detection of insurance fraud [10].

3 Challenge Overview

Fraud on auto insurance is a serious issue that impacts 20–36% of claims, significantly more than the estimated 10–20% of false claims overall. This high fraud rate puts a substantial financial strain on insurance companies, drives up premiums for legitimate customers, and results in considerable losses. Therefore, new and more effective approaches are needed to address these challenges. One promising solution is the use of automated fraud detection systems, that employ intelligent algorithms and data analytics to flag unusual activity potentially indicating fraud. These technologies are essential in combating fraud because they improve response times, reduce manual labor, and enhance detection accuracy. By adopting these cutting-edge tools, insurance firms can alleviate financial burdens and better protect lawful policyholders.

4 Existing System

The current strategy for identifying vehicle insurance fraud predominantly relies on manual approaches and basic software tools. Historically, insurers have utilized rudimentary data analysis and manual document assessments to flag potential fraud cases. However, these methods are labor-intensive, prone to errors, and frequently miss more sophisticated fraudulent schemes. Furthermore, the limitations of existing systems in adopting cutting-edge technologies like machine learning and natural language processing impede their effectiveness in detecting and preventing fraud. This insufficiency complicates insurers' ability to accurately assess risk and defend against fraudulent claims, resulting in financial losses for honest policyholders and heightening their rates. Clearly, there is an urgent need for a more advanced and automated approach to assess insurance documentation, uncover fraudulent tendencies, and enhance the overall security and integrity of the insurance sector.

5 Proposed System

The present study suggests a novel method for detecting car insurance fraud that apply ML methods, AES encryption, NLP techniques. The user of the system first uploads insurance claim documents in PDF format, which provide important details about auto insurance coverage. The system then extract and analyse important information from these documents, including coverage terms, vehicle specifications, and policy details, using sophisticated natural language processing algorithms. This procedure guarantees precise and effective data extraction, even from papers that lack structure. The framework is divided into 5 modules as given below (Fig. 1):

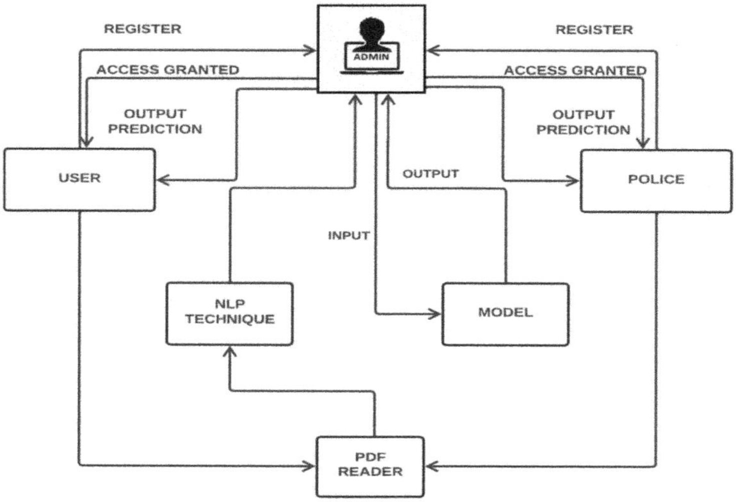

Fig. 1. System Flow

5.1 Data Collection

The initial step in optimizing a machine learning algorithm involves acquiring the required dataset. Data can be gathered through various approaches, such as manual input or automated web scraping, managed by a model coordinator. In this study, the dataset utilized is from Kaggle. It provides detailed customer-related information, such as the date of the incident, the insurance policy's binding date, the policy identification number, the type of incident, the involvement of authorities, the severity of the event, whether a police report was submitted, and the availability of witnesses, among other attributes.

5.2 Data Analysis and Preprocessing

Before data can be effectively utilized to detect auto insurance fraud, it must be meticulously cleansed and prepared. This step involves collecting data from various sources, including policy documents, insurance claims, and historical records. To streamline the dataset, redundant or irrelevant data points will be removed. If it's impossible to fill in missing values accurately, those entries will either be eliminated or replaced using statistical methods. Standardizing the data format will ensure consistency among variables, making comparison and analysis more manageable.

5.3 Feature Selection and Fraud Detection Model Building

Feature selection and the creation of a fraud detection model entail dividing the dataset into two segments: a target variable, y (indicating fraud presence), and independent variables, X (features). The fraud detection module reviews and verifies claims. This analysis utilizes five algorithms XGBoost, Random Forest-Ensemble Decision Tree Model, Logistic Regression- Binary Classification Algorithm, SVM, and Naive Bayes Probabilistic Classifier Based on Bayes' Theorem. The fraud detection model will be refined by selecting the algorithm that yields the highest accuracy.

5.4 Learning Phase and Testing:

The training phase involves using preprocessed historical data to train models. These models learn patterns from features such as insurance policy's binding date, the policy identification number, the type of incident, the involvement of authorities, the severity of the event. Hyperparameter tuning is applied to optimize model performance. In the testing phase, the trained models are evaluated on new data to assess their effectiveness using metrics like accuracy, precision, recall, and F1-score. This ensures unbiased performance evaluation. The results are analyzed to compare the models and identify the most accurate and reliable algorithm for fraud detection.

5.5 Natural Language Processing:

The proposed method for identifying auto insurance fraud utilizes NLP to enhance the analysis of claim documents. NLP extracts and analyses essential data from unstructured documents, such as PDF files. This section outlines the steps and techniques used for preprocessing and data extraction through NLP [11].

Parsing PDFs

The first step in this process is converting PDF documents into a machine-readable format. This step is crucial since most insurance claims are submitted in PDF format, which lacks structure. Tools like PyMuPDF and pdfminer are instrumental in this conversion, as they are designed to handle various PDF layouts, ensuring effective text extraction. PyMuPDF is used initially to handle complex and irregular PDF layouts that may contain tables, forms, or multiple columns. It provides high precision in locating and extracting text blocks with positional information, which helps maintain spatial context in claims with complex formatting. Pdfminer is then used as a secondary tool to extract text while preserving the flow and structure of the original document. This helps in maintaining sentence continuity and context, especially in simpler, text-heavy documents. PyMuPDF excels in managing complex layouts, while pdfminer is adept at maintaining the original text's structure, making both libraries advantageous for accurate data extraction. After extracting text from PDF files, further processing may be necessary to ensure clarity and suitability for analysis. This may involve removing extraneous information, correcting formatting errors, and organizing the content for subsequent NLP tasks. In addition to preparing the text for entity recognition, this preprocessed content is also passed into other NLP methods such as keyword extraction and text classification, which support downstream analysis.

Recognition of Named Entities (NER)
After parsing the PDFs, the next crucial step is recognizing and categorizing vital information within the extracted text, known as Named Entity Recognition (NER). This technique identifies critical elements such as policy numbers, dates, personal information, and other relevant identifiers that could signify potential fraud. Pre-trained models from the spaCy library are used for this process. SpaCy's robust framework and extensive training on diverse datasets make it an excellent choice for accurately identifying entities within text. Additionally, these models can be fine-tuned to focus specifically on insurance-related terminology and patterns, enhancing the entity recognition accuracy. The accuracy and effectiveness of the NER model are evaluated using standard metrics such as precision, recall, and F1-score, based on a labeled dataset of insurance claim documents where model outputs are compared with manual annotations. Beyond entity recognition, NLP is further leveraged to extract keywords commonly associated with fraudulent behavior and to classify the nature of claims based on their textual descriptions (e.g., accident, theft). NER is vital for converting raw text data into structured information that can be further analyzed. By categorizing the extracted data, insurers can better assess claims for possible fraud and streamline the claims processing workflow.

The outputs from NLP—including identified entities, extracted keywords, and claim-type classifications—are converted into structured features (e.g., flags, counts, or categorical values) and combined with numerical and categorical data in the dataset. These features are then directly used as inputs in the machine learning model for fraud prediction, ensuring that the insights gained from unstructured documents contribute meaningfully to the overall detection process.

5.6 Encryption Technique - AES

Safeguarding client privacy and security is essential for any system managing sensitive information, such as insurance claims. In this system, the Advanced Encryption Standard (AES) is implemented to securely store all client data online. This section explains the AES encryption process, its utilization, and the benefits it offers for data security [12].

Implementation of AES
Generate a Key and Initialization Vector (IV): Both the key and IV are essential for the AES encryption process. The IV ensures that blocks of plaintext that are similar produce separate blocks of ciphertext, but the key must be kept private.

Encrypt Client Data: This step involves using the AES technique to convert client data into ciphertext.

Store the Encrypted Data: The encrypted information is then securely stored in a database.

Decrypt Data: When needed, the data can be restored to its original state using the same key and IV.

5.7 Website Development

The Auto Insurance Fraud Detection website is developed with Flask, emphasizing a smooth and user-friendly experience. With a clean, responsive design, the site is accessible on various devices, ensuring usability for all users. The front-end employs HTML, CSS, and JavaScript to create an engaging and interactive interface, where essential features like claim submissions and fraud analysis results are easily navigable.The website caters to three main user groups: Insurance Company Admins, who can manage and review claims along with fraud analysis; Police Authorities, who can access relevant fraud reports for investigation; and Customers or Policy Holders, who can submit and monitor their claims. The layout is well-organized, with clear sections, enabling rapid access to essential tools and reports for each user group. A strong emphasis on usability ensures smooth interaction, while contemporary design principles provide a professional appearance. The site is optimized for accessibility, offering intuitive navigation and clear calls to action that make it straightforward for all users to handle and review insurance claims.

5.8 XGBoost

An enhanced ensemble learning technique called Extreme Gradient Boosting (XGBoost) generates decision trees in a sequential fashion. Each subsequent tree aims to correct the mistakes of its predecessors, focusing on wrongly classified cases. This methodology is based on gradient boosting, which progressively adds trees to minimize a specified loss function, thereby enhancing model accuracy over time. XGBoost improves computational efficiency through parallel processing, which accelerates training. It also employs regularization techniques to reduce the likelihood of overfitting, resulting in a more robust and generalizable model. Significantly, it can manage missing data internally, ensuring that model performance is maintained even when certain values are not present. Additionally, XGBoost automates feature selection, simplifying the preprocessing phase and minimizing the need for extensive feature engineering. Leveraging the strengths of individual decision trees, XGBoost emerges as a potent predictive model that excels in both accuracy and performance, particularly for complex datasets. Its capabilities make it especially valuable for tasks like fraud detection, where intricate patterns and anomalies need to be swiftly and reliably identified. The dataset will be sampled into smaller subsets . Decision trees were sequentially constructed to predict residuals (errors) from previous trees. Each tree provides a prediction. These predictions were summed to reduce errors, and the final output is obtained by aggregating the results, improving overall accuracy.

6 Results and Discussion

XGBoost outperformed all other models in accuracy and overall performance. It achieved an accuracy of 89.73%, making it highly effective in identifying fraudulent claims while maintaining a balance between precision and recall. In addition to accuracy, XGBoost achieved a high F1-score of 0.88, indicating strong balance between precision and recall. Its false positive rate was significantly lower compared to other models, reducing the likelihood of legitimate claims being flagged incorrectly. The Area Under the Precision-Recall Curve (AUPRC) further highlighted its robustness, particularly in handling class imbalance typical in fraud detection. Random Forest showed strong performance but not as consistently high as XGBoost. Logistic Regression and Naive Bayes performed at lower levels, with accuracy ranging below 70%. These models struggled with the dataset's complexity, leading to higher false positives and missed fraud cases. Their lower F1-scores and elevated false positive rates reflected their difficulty in accurately distinguishing fraudulent claims from legitimate ones.

SVM demonstrated decent performance but did not match XGBoost, especially in handling large datasets with diverse features. XGBoost's superior performance can be attributed to its ability to handle complex, large datasets and its resistance to overfitting. Its tree-boosting technique focuses on hard-to-classify cases, improving accuracy in fraud detection. It also achieved a favorable trade-off between precision and recall as seen in its AUPRC value, which underscores its reliability under real-world fraud detection conditions. Additionally, XGBoost's effectiveness without requiring extensive data preprocessing gives it an edge over other models.

While the models tested performed well, the study's dataset was not exhaustive. Future research should explore larger, more diverse datasets that reflect real-world fraud scenarios. Incorporating time-series analysis could also help capture evolving fraud patterns, leading to more adaptable fraud detection systems.

In conclusion, XGBoost demonstrated the highest accuracy and reliability for fraud detection in auto insurance. Its high F1-score, low false positive rate, and strong AUPRC reinforce its suitability for real-world deployment (Figs. 2, 3 and 4).

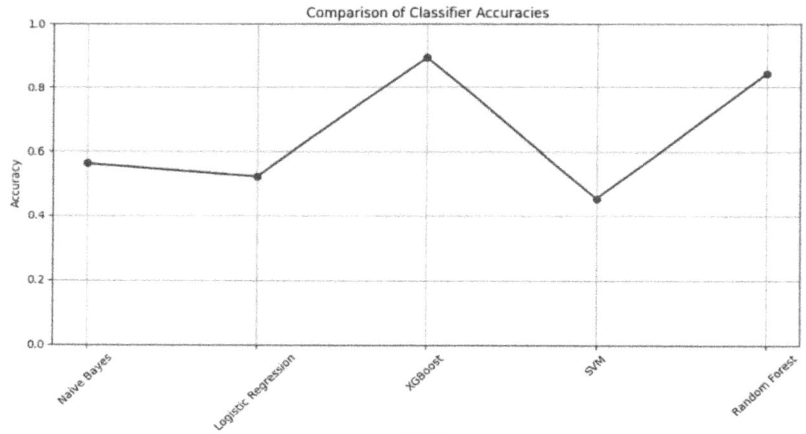

Fig. 2. Graph

AUPRC for XGBoost is: 0.93

CLASSIFIER	ACCURACY	PRECISION	RECALL	F1 SCORE
XGBOOST	89.73%	0.9	0.92	0.91
RANDOM FOREST	82.78%	0.9	0.71	0.81
SVM	52.31%	0.63	0.34	0.44
LOGISTIC REGRESSION	56.53%	0.61	0.61	0.61
NAÏVE BAYES	56.29%	0.58	0.77	0.66

Fig. 3. Comparison Table

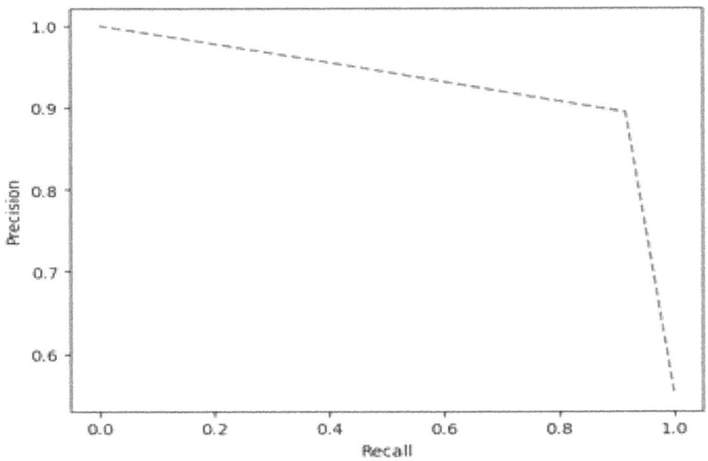

Fig. 4. Precision-Recall Graph

7 Conclusion

To identify fraud in auto insurance claims, this study examined a number of machine learning methods, such as Naive bayes, logistic regression, XGBoost, Support Vector Machines (SVM), and Random Forest. Of these techniques, XGBoost performed the best on all evaluated measures, proving its efficacy in handling the challenges of fraud detection.To graphically evaluate each model's performance and highlight its advantages and disadvantages, a line graph was employed. This study also demonstrated the importance of careful feature engineering and preprocessing to improve model effectiveness. The results indicate that machine learning models, particularly ensemble methods like XGBoost, hold significant potential to enhance the accuracy and efficiency of fraud detection in the insurance domain.

8 Future Enhancement

Future research would focus on enhancing the XGBoost by incorporating additional relevant features extracted from both structured and unstructured data sources. Fine-tuning hyperparameters through advanced optimization techniques, such as Bayesian optimization or grid search, could further improve model performance. Employing sophisticated approaches like ensemble methods, model stacking, or blending multiple algorithms may increase predictive accuracy and robustness.

Moreover, integrating temporal or sequential data analysis could help capture evolving fraud patterns over time. Enhancing feature engineering using domain-specific knowledge and automated feature selection methods may also boost the model's ability to discriminate between fraudulent and legitimate claims. These improvements would collectively strengthen the model's capacity to accurately detect false claims, leading to more efficient and reliable fraud detection systems in the insurance sector.

References

1. Vyas, S., Serasiya, S.: Fraud detection in insurance claim system: a review. In: 2022 Second International Conference on Artificial Intelligence and Smart Energy (ICIAIS), pp. 922–927. IEEE (2022)
2. Jain, A., Dutta, T., Senthil Kumar, T.: Fraud detection and analysis system for car insurance claim using random forest classifier. In: 2023 International Conference on Recent Advances in Science and Engineering Technology (ICRASET), pp. 1–5. IEEE (2023)
3. Okagbue, H.I., Oyewole, O.: Prediction of automobile insurance fraud claims using machine learning. The Scientific Temper 14, no. 03, vol. 756–762 (2023)
4. Kini, A., Chelluru, R., Naik, K., Naik, D., Aswale, S., Shetgaonkar, P.: Automobile insurance fraud detection: an overview. In: 2022 3rd International conference on Intelligent Engineering and Management (ICIEM), pp. 7–12.IEEE (2022)
5. Urunkar, A., Khot, A., Bhat, R., Mudegol, N.: Fraud detection and analysis for insurance claim using machine learning. In: 2022 IEEE International Conference on Signal Processing, Informatics, Communication and Energy Systems (SPICES), vol. 1, pp.406–411. IEE, (2022)
6. Roy, R., Thomas George, K.: Detecting insurance claims fraud using machine learning techniques. In: 2017 international conference on circuit, power and computing technologies (ICCPCT), pp. 1–6. IEEE (2017)
7. Tongesai, M., Mbizo, G., Zvarevashe, K.: Insurance fraud detection using machine learning. In: 2022 1st Zimbabwe Conference of Information and Communication Technologies (ZCICT), pp. 1–6. IEEE (2022)
8. Chauhan, R., Negi, R., Verma, D.K.: Analysis of machine learning algorithms for insurance fraud detection. In: 2023 IEEE 11th Region 10 Humanitarian Technology Conference (R10-HTC), pp. 649–655. IEEE (2023)
9. Singhal, A., Singhal, N., Sharma, K.: Machine learning methods for detecting car insurance fraud: compartive analysis. In: 2023 3rd International Conference on Intelligent Technologies (CONIT), pp. 1–5. IEEE (2023)
10. Rijanto, KP., Putri, D.A., Calvin, P., Edbert, I.S., Suhartono, D.: Machine learning algorithms with oversampling methods for automobile insurance fraud detection. In: 2023 IEEE 21st Student Conference on Research and Development (SCORed), pp. 138–143. IEEE (2023)

11. Sun, P., Yang, X., Zhao, X., Wang, Z.: An overview of named entity recognition. In: 2018 International Conference on Asian Language Processing (ILAP), pp. 273–278. Bandung, Indonesia (2018). https://doi.org/10.1109/IALP.2018.8629225
12. Kaur, J., Lamba, S., Saini, P.: Advanced encryption standard: attacks and current research trends. In: 2021 International Conference on Advance Computing and Innovative Technologies in Engineering (ICACITE), pp. 112–116. IEEE (2021)

Deep Learning

Classification of Alzheimer's Disease Through Deep Learning Approaches with Contextual and Inter-channel Features from Brain MRI

T. Illakiya(✉), B. Vidhya, and K. Jashvanth

Department of Computational Intelligence, School of Computing, Faculty of Engineering and Technology, SRM Institute of Science and Technology, Kattankulathur 603203, India
{illakiyt,vidhyab,jk4451}@srmist.edu.in

Abstract. Alzheimer's Disease (AD) is a neurological disorder defined by a gradual deterioration in cognitive abilities and memory impairment. It significantly impacts individuals' quality of life and imposes a considerable burden on healthcare systems. Prompt detection, especially during the Mild Cognitive Impairment (MCI) stage, is necessary for earlier interventions and enhancing disease classification. This phase includes Early MCI (EMCI), representing the initial stages of cognitive decline, and Late MCI (LMCI), which indicates progressive deterioration. Deep learning (DL) has the capacity to extract features automatically from image data, presenting significant potential to enhance the diagnosis of AD and MCI. However, challenges such as imbalanced datasets, highdimensional features, and the difficulty of capturing subtle changes in early disease stages have limited the effectiveness of existing models. To address these challenges, this work presents a deep learning-based model using DenseNet-169 architecture integrated with two attention mechanisms—Global Context Network and Triplet Attention. The Global Context Network captures long-range dependencies and contextual information, the Triplet Attention Module extracts cross-dimensional and position-sensitive features. Therefore, the proposed model concentrates on extracting relevant, multi-scale features, which enhances the capacity of the model to classify AD and its early phases more accurately. The proposed model gives an accuracy of 98.8% when evaluated on the ADNI dataset. The conclusions show the efficacy of custom-designed DL models in enhancing the early detection of AD.

Keywords: Classification · MRI · Alzheimer's disease · Attention network · Convolution neural networks · MCI

1 Introduction

Alzheimer's disease (AD) is a degenerative neurological condition marked by a gradual deterioration in memory and difficulties in performing routine activities, arising from impairments in behaviour, reasoning, and cognitive functions [1]. Cognitive decline, memory impairments, language difficulties, disorientation, and behavioural changes are

R. Geetha et al. (Eds.): AAIMB 2024, CCIS 2654, pp. 165–175, 2026.
https://doi.org/10.1007/978-3-032-13029-7_12

major symptoms of AD. The phase of changeover from typical aging and greater acute cognitive deterioration, as that observed in AD, is known as mild cognitive impairment (MCI). It is characterized by milder cognitive impairments, particularly in attention, memory, and executive functions. Despite these challenges, individuals with MCI are typically able to maintain their independence in daily activities. MCI can be categorized into two types: Late MCI (LMCI) and Early MCI (EMCI). EMCI represents an earlier stage of cognitive decline, where the impairments are subtle and often only detectable through detailed neuropsychological assessments. At this stage, individuals may face minor challenges with memory or problem-solving. However, these difficulties are not substantial enough to interfere significantly with daily activities. In contrast, individuals with LMCI experience more cognitive difficulties, particularly in memory and reasoning, making everyday tasks more challenging. The possibility of progressing to AD becomes considerably higher in LMCI patients. Early diagnosis and intervention at the EMCI or LMCI stage can be essential for alleviating symptoms and slowing the progress of the disease. The main factors include age, genetics—specifically the existence of the APOE ε4 gene—brain changes such as the buildup of amyloid plaques and tau tangles, along with lifestyle factors like diet and exercise. AD imposes substantial economic burdens on families and healthcare systems, increases dependency on caregivers, and leads to social isolation and emotional strain. Hence, it is important to diagnose the disease in early stage which helps in therapeutic intervention. Recent research suggests that by 2050, AD may affect 1 in every 85 individuals. Several techniques, such as MRI, PET, and CT scans, are available for diagnosing and predicting this form of dementia. Recent statistics reveal that the disease affects more than 6.5 million individuals worldwide. Traditional approaches to diagnose AD involve clinical evaluations, neuroimaging methods like MRI and PET scans, biomarker analysis, and neuropsychological assessments [2]. While these approaches have been foundational, the need for automation is increasingly evident. Manual assessments are labour-intensive, susceptible to human error, and often lack to scale effectively when applied to large populations. Additionally, early detection is important for managing AD, and automated systems offer more accurate, and timely analysis. Ultimately, the incorporation of automated systems with traditional classification methods holds great promise for optimizing resource utilization in healthcare and improving accuracy of AD diagnosis.

Machine learning (ML) and Deep Learning (DL) methods are emerging as powerful tools for detecting AD. ML algorithms can process large amounts of complex data from neuroimaging, biomarkers, and cognitive tests. It also detects patterns and anomalies that could signify the early stages of the disease [3]. These models can be trained to differentiate between patients, those with MCI, and cognitively normal individuals, enabling more precise classification. Computer-aided ML techniques provide an effective approach to developing complex, automated classification models that can handle large datasets and extract intricate patterns [4]. But traditional ML approaches typically involve multiple pre-processing steps, often requiring specialized information and are usually based on manual or partial automated feature learning. The proposed DL model incorporates the pre-trained model and customized CNN for effective classification of AD stages. By incorporating advanced DL architectures, the model automates the extraction of relevant features from neuroimages, and enhances classification accuracy.

2 Literature Survey

In AD studies, various techniques are developed to diagnose the disease automatically at an early stage, leveraging advancements in ML, DL, and neuroimaging technologies. This section explores the existing works that deal with AD classification, focusing on the various methods and approaches employed to diagnose and predict the progression of the disease. Dar et al. used the MobileNet model which was developed specifically to operate efficiently on mobile and embedded devices [5]. It utilized depthwise separable convolutions to minimize computational complexity and memory usage while maintaining high accuracy for applications such as object detection and image classification. Al-Zharani et al. used DenseNet121, which belongs to a category of CNN recognized as densely connected networks [6]. In this model, every layer is connected to all other layers in a feedforward configuration. This architecture reduces the number of parameters, mitigates vanishing gradient issues, and improves computational efficiency in deep networks. Aparna et al. used a DL model, Xception-Fractalnet which enhances inception by using depthwise separable convolutions to facilitate more efficient feature extraction [7]. FractalNet is a fractal like network structure that is designed to improve learning without the need for deep architectures. Pallikonda et al. incorporated HHOSqueezeNet, which integrates the Hawk-Hybrid Optimization method with the lightweight SqueezeNet architecture [8]. It optimizes the model parameters while preserving a minimal footprint, making it effective for tasks such as image categorization in resource-limited settings.

Pallikonda et al. utilized Mask Region-based CNN (Mask R-CNN), an object detection framework that integrates region proposal methods with CNN-based feature extraction for accurate classification and localization [9]. Sanjay et al. incorporated a DL model, GM ABC-NN, which combines Generative Models (GM), Approximate Bayesian Computation (ABC), and Neural Networks (NN) to improve inference in complex models [10]. The framework enhances image preprocessing, and feature selection, followed by AD stage classification, achieving better accuracy and performance. Naveen et al. used ResNet-50, a deep CNN that utilizes residual learning via skip connections to reduce the vanishing gradient issue [11]. The utilization of 50 layers promotes feature extraction and elevates performance in picture recognition tasks. Shanmugam et al. incorporated a DL model, ResNet-18, a deep CNN using residual learning with skip connections to mitigate the vanishing gradient issue. Consisting of 18 layers, it balances model complexity and performance, enabling feature extraction while maintaining high accuracy in image classification tasks [12]. Illakiya et al. proposed a DL model using DenseNet-169 and custom CNN [13]. DenseNet uses densely connected layers to improve feature flow and reuse, enhancing efficiency and performance. CustomCNN integrates specialized layers tailored to the specific task, further optimizing feature extraction. This architecture is effective in classifying Mild Cognitive Impairment MCI and AD analysis. Illakiya et al. used a DL framework that uses the Hybrid Multi-Focus Attention Block to prioritize key features, while EfficientNet-B0 optimizes performance with fewer parameters. Additionally, the Group Shuffle Depth-wise Convolution enhances feature extraction for capturing contextual information. Together, these models improve feature extraction and efficiency in DL [14]. Despite significant advancements, the existing DL models tends to face challenge in extracting relevant features due to the complexity and variability in brain MRI data. Many models struggle with effectively capturing subtle differences in early-stage AD features, leading to reduced accuracy in early detection.

3 Proposed Model

The proposed system utilizes the DenseNet-169 in conjunction with two attention modules: The input images are initially processed using the DenseNet-169 architecture to acquire important features. The resulting features are propagated to the Global Context Network (GCN) and subsequently processed by the Triplet Attention Module. The GCN captures long-range dependencies through contextual aggregation, improving the capacity of the model to comprehend relationships over the entire image. Simultaneously, the Triplet attention block acquires cross-channel, direction-sensitive (local attributes), and position-aware information allowing the network to concentrate on key spatial regions within the image. Figure 1 illustrates a detailed representation of the proposed architecture.

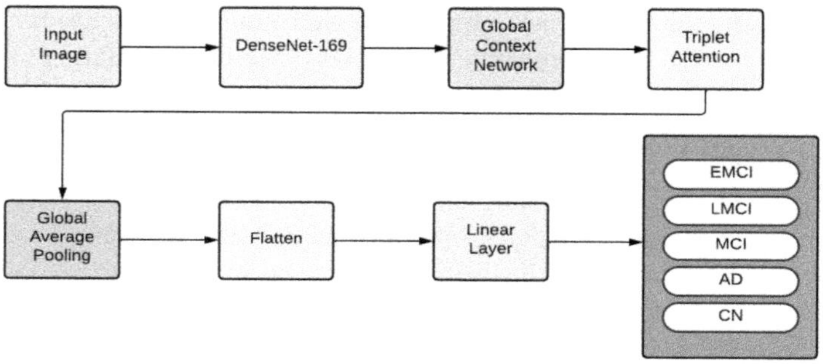

Fig. 1. Outline of the proposed customized DL network for AD classification.

3.1 Global Context Network

The Global Context Network (GCN) capture contextual information from the feature map [15]. Its main objective is to gather data in a broader perspective and offer a global perspective of the given input. This allows the model to assess associations between different regions and elements in the feature map, thereby enhancing its capability to generate predictions informed by contextual understanding.

Figure 2 presents the block diagram to understand the functioning of the GCN. The figure depicts the information flow within the GCN, emphasizing its links to the previous and subsequent layers of the block. This module comprises three stages: Contextual Modelling, Transformation, and Fusion. Long-range dependencies define connections between components and features within the input. Traditional CNNs obtain features by combining data from local pixel neighborhoods, successfully capturing local details but possibly overlooking long-range features or relations among distant elements in the input data. The non-local network addresses these dependencies by calculating the output at

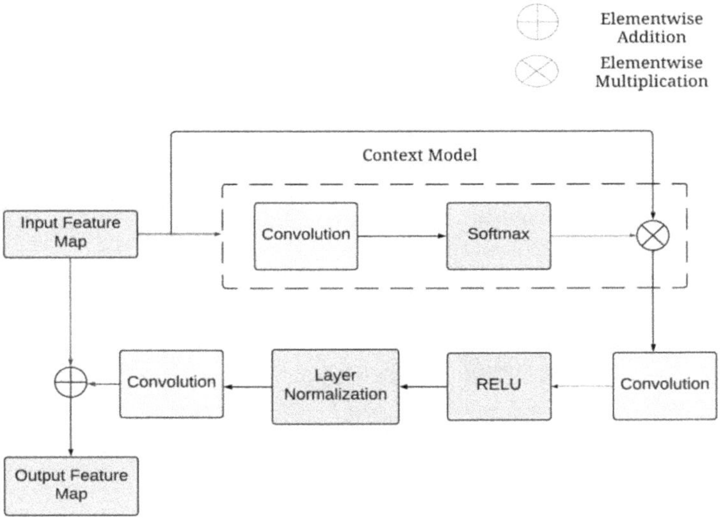

Fig. 2. Architecture of the Global Context Network.

each point through a weighted combination of feature maps. The proposed model's non-local block employs a softmax function and 1x1 convolution to derive global context characteristics and attention weights via global pooling. The features are next processed through a 1x1 convolution. In the GC block, the transform module performs squeeze-and-excitation on the feature map produced by the non-local network, enabling channel attention. A 1x1 convolution is applied in this module to spatially compress the feature map using a bottleneck ratio, creating a channel descriptor. This is followed by ReLU activation, layer normalization, and another 1x1 convolution to generate an attention map. The generated map is utilized to refine the feature maps, emphasizing the most significant characteristics.

3.2 Triplet Attention

A triplet attention mechanism extracts cross-dimensional relationships between the spatial dimension W or H and the channel dimension C [16]. Averaging is employed to aggregate the results from all three branches. Interdependence across the dimensions of the input is recorded using the three branches: (C, W), (C, H), and (H, W). The first two branches are responsible for maintaining cross-dimensional interactions between channels and either breadth or height, respectively. Spatial attention is fostered by the third branch. The schematic representation of the triplet attention block is illustrated in Fig. 3.

This attention block processes the input through 3 distinct branches, each corresponding to the Width W, Channel C, and Height H dimensions. In order to facilitate interactions between the C and H dimensions, the feature map is rotated with respect to the H axis. The output is generated by applying batch normalization layer and a convolutional layer, then the output map is reduced to $2 \times H \times C$ using Z-pooling. To preserve

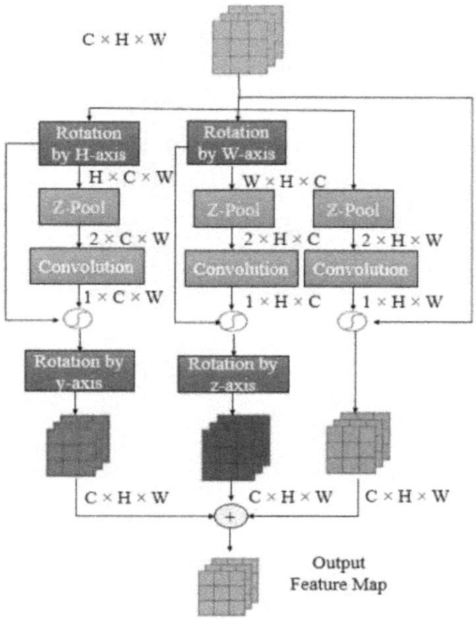

Fig. 3. A visual representation of the triplet attention module.

the original configuration, the final attention weights are rotated with respect to the height dimension. The second branch follows the same procedures as the first, but rotates along the W axis. The final component of the model employs Z-pooling to diminish the input tensor's channels to two. The subsequent steps adhere to a same pattern as observed in branch 1 and branch 2. Subsequently, the revised tensors generated by each branch are averaged. Traditional methods for calculating channel attention generally include allocating a singular weight to each channel in an input tensor and equally adjusting the feature mappings according to these weights. This method results in a considerable loss of interdependence among channels and their respective pixel positions. To address this limitation, the cross-dimension interaction was employed in the triplet attention block. By integrating spatial and channel attention mechanisms into the triplet module, the proposed model captures interdependencies across the spatial and channel dimensions. This results in enhanced feature representation, improved representation of meaningful features, and eventually, increased performance in classification of AD.

3.3 Classification

Following the extraction and refinement of prominent features through the DenseNet169 model, GCN, and Triplet Attention module, the model progresses to the classification phase. The revised feature maps are passed to Global Average Pooling (GAP), which lowers the dimensions by calculating the average of each feature map, thereby preserving essential data and mitigating overfitting. This creates a lower-dimensional tensor that incorporates the most salient features from the input image. Subsequently, the output from GAP is transformed into a one-dimensional vector to facilitate the final phase. The flattened vector is subsequently transmitted through a fully connected linear layer, which correlates it to the output classes. The classification performance is assessed with the categorical cross-entropy loss function. It measures the alignment among the true class and predicted probability scores, thereby ensuring accurate predictions.

4 Experimental Results

The contents of this section include the description of the dataset and the employed augmentation techniques. It also explains the comparison and analysis of the proposed architecture with state-of-the-art networks and existing methodologies.

4.1 Dataset Description

The dataset utilized in this research is derived from MRI scans obtained through the ADNI. The original 3D images underwent conversion to 2D format, followed by preprocessing and reformatting into JPG images. The dataset is grouped into five distinct categories: AD, LMCI, MCI, EMCI, and Cognitive Normal (CN). Every image measure 256×256 pixels. The complete dataset comprises 1296 images, including 171 from AD, 580 from CN, 240 from EMCI, 72 from LMCI, and 233 from MCI.

4.2 Data Augmentation

The 2D images experience online augmentation to improve the model's generalizability and incorporate fresh alterations throughout the training process. The dataset was diversified by employing geometric adjustments, including image flipping and rotational transformations. The range of parameters for the random rotation function was defined between 0° and 90°, and the probabilities for vertical and horizontal flips were fixed to 50%.

4.3 Analysis of the Proposed Model

The effectiveness of the proposed DL network was thoroughly analyzed, integrating the DenseNet-169, GCN and Triplet Attention modules. By combining DenseNet's strong capabilities for feature extraction with the attention mechanisms of the custom CNN, the DL model exhibited better classification accuracy, better feature representation, and enhanced generalization. The accuracy and loss curves for both training and validation were plotted to analyze the performance across the training process. These graphs help finding the identification of trends such as convergence and overfitting and are presented in Fig. 4.

Table 1 presents the results of ablation studies conducted to evaluate the effectiveness of different architectural configurations on classification accuracy. The baseline architecture, DenseNet-169, achieved an accuracy of 87.20%. By incorporating a Graph Convolution Network (GCN) with DenseNet-169, the accuracy improved to 94.6%, showing the added benefit of GCN for feature extraction and structural learning. The proposed

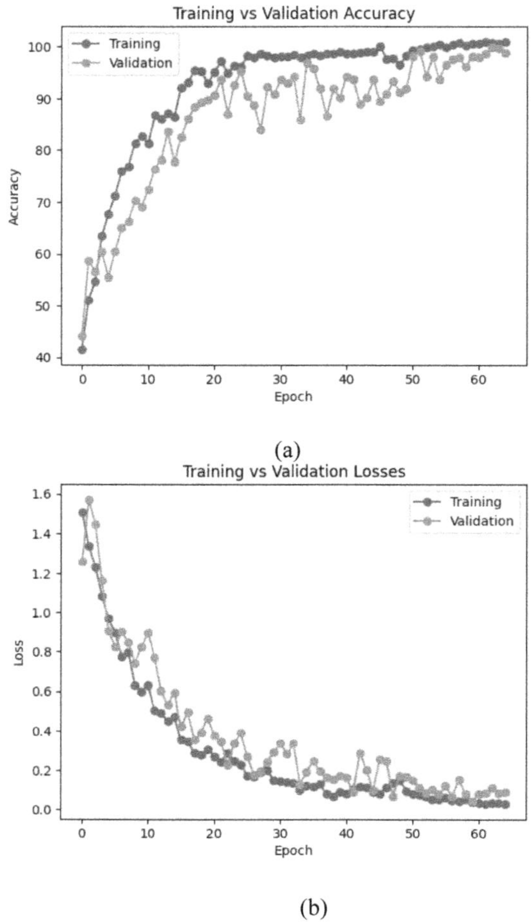

Fig. 4. Observations made on training and validation a) Accuracy b) Loss

Table 1. Evaluation of the ablation studies.

Architecture	Accuracy in (%)
DenseNet-169	87.20
DenseNet-169 with GCN	94.6
Proposed model - DenseNet-169 with GCN and triplet attention model	**98.8**

model, which integrates both GCN and a triplet attention module with DenseNet-169, significantly outperforms the previous configurations, achieving an accuracy of 98.8%.

This substantial improvement highlights the effectiveness of combining attention mechanisms for enhancing feature representation and capturing complex relationships in the data.

4.4 Comparison with the State-of-the-Art Networks

The effectiveness of the proposed model against various state-of-the-art architectures for AD and MCI classification was evaluated using the ADNI dataset. As shown in Table 2, traditional convolutional models like VGGNet-16 and ResNet-50 achieved classification accuracies of 76.4% and 78.2%, respectively. AlexNet and ResNet-152 performed slightly better with accuracies of 79.1% and 79.5%, respectively. EfficientNet-B0 and DenseNet-201 offer more efficient parameterization and deeper learning, with an improved accuracy of 84.7% and 85.9%.

VGGNet-16, ResNet-50, AlexNet, and ResNet-152 struggle with subtle feature extraction while EfficientNet-B0 and DenseNet-201, though efficient, do not capture contextual or cross-dimensional dependencies, leading to suboptimal performance. DenseNet-169, utilized as the backbone in our proposed architecture, achieved 87.2% accuracy as a standalone model. However, by incorporating it with the GCN and Triplet Attention mechanisms, the proposed model achieved a significant performance improvement, reaching an accuracy of 98.8%. This improvement highlights the impact of combining dense feature extraction with cross-dimensional and global attention module, allowing for better localization of disease-relevant brain regions. Therefore, the proposed model outperforms all compared architectures in terms of accuracy.

Table 2. Analysis of proposed models with respect to the pre-trained models.

S. No	Model Trained	Accuracy in (%)
1	VGGNet-16	76.4
2	ResNet-50	78.2
3	AlexNet	79.1
4	ResNet-152	79.5
5	EfficientNet-B0	84.7
6	DenseNet-201	85.9
7	DenseNet-169	87.2
8	**Proposed Model**	**98.8**

4.5 Comparative Analysis of the Existing Works

The effectiveness of the proposed approach is evaluated against prior research on AD detection with brain MRIimaging. Table 3 presents a comparison between the proposed model and existing approaches. Using DL techniques on the ADNI dataset. The existing works employed various architectures, including CNN, MobileNet, DenseNet-121, and GM-ABC-NN, with performance varying based on the chosen architecture and training methods.

Table 3. Evaluating the efficacy of the suggested method in comparison to existing studies.

Source	Method	Accuracy in (%)
Mahmud et al. [1]	CNN	79.0
Dar et al. [5]	MobileNet	96.6
Zharani et al. [6]	DenseNet121	97.33
Sanjay et al. [10]	GM-ABC-NN	97.56
Proposed Model	**Custom CNN**	**98.8**

These studies explored different techniques to improve classification accuracy, with the MobileNet and DenseNet121 models. However, the proposed model achieves 98.8% accuracy, outperforming all previous methods. This shows the strength of customized DL models in handling complex multi-class classification tasks for AD detection.

5 Conclusion

In conclusion, the proposed DL framework, which integrates the DenseNet-169 architecture with advanced attention mechanisms such as the Global Context Network and Triplet Attention module. The model shows significant improvements in AD detection across five classes: AD, MCI, LMCI, EMCI, and CN. These attention modules enable the model to extract both global contextual information and local spatial features, leading to enhanced classification performance. The proposed network gives an accuracy of 98.8% using the ADNI dataset, surpassing other models in AD classification. This highlights the potential of the customized DL architecture in supporting accurate and early diagnosis of AD and its stages. Future work could focus on further optimizing the model for scalability and testing its performance on other neuroimaging datasets to validate its generalization capabilities.

References

1. Mahmud, S.M., et al.: Detection of different stages of Alzheimer's disease using CNN classifier. CMCComput. Mater. Continua **76**(3), 3933–3948 (2023). https://doi.org/10.32604/cmc.2023.039020
2. Arya, A.D., Verma, S.S., Chakarabarti, P., et al.: A systematic review on machine learning and deep learning techniques in the effective diagnosis of Alzheimer's disease. Brain Inf. **10**, 17 (2023). https://doi.org/10.1186/s40708-023-00195-7
3. Shukla, A., Tiwari, R., Tiwari, S.: Review on Alzheimer disease detection methods: Automatic pipelines and machine learning techniques. Sci. **5**(1), 13 (2023). https://doi.org/10.3390/sci5010013

4. Illakiya, T., Karthik, R.: A dimension centric proximate attention network and Swin Transformer for age-based classification of mild cognitive impairment from brain MRI. IEEE Access **11**, 128018–128031 (2023). https://doi.org/10.1109/ACCESS.2023.3332122
5. Mohi ud din Dar, G., et al.: A novel framework for classification of different Alzheimer's disease stages using CNN model. Electronics **12**(2), 469 (2023). https://doi.org/10.3390/electronics12020469
6. Al-Zahrani, M., Ansarullah, S.I., Al-Eissa, M.S., Dar, G.M., Alqahtani, R.A., Alkahtani, S.: Exploring the efficacy of deep learning techniques in detecting and diagnosing Alzheimer's disease: a comparative study. J. Disabil. Res. **3**(6), 20240064 (2024). https://doi.org/10.57197/jdr-2024-0064
7. Aparna, M., Rao, B.S.: Xception-Fractalnet: hybrid deep learning based multi-class classification of Alzheimer's disease. Comput. Mater. Continua **74**(3) (2023). https://doi.org/10.32604/cmc.2023.034796
8. Pallikonda, A.K., Varma, P.S.: Multi-class classification of Alzheimer's disease stages using SqueezeNet based approach for automated diagnosis. Int. J. Comput. Inform. Syst. Indust. Manage. Appl. **15**, 11–11 (2023). https://cspub-ijcisim.org/index.php/ijcisim/article/view/549
9. Pallikonda, A.K., Varma, P.S., Vivekanandam, B.: Mask region-based convolution neural network (Mask R-CNN) classification of Alzheimer's disease based on magnetic resonance imaging (MRI). Int. J. Image, Graph. Signal Process. **15**(6), 54–66 (2023). https://doi.org/10.5815/ijigsp.2023.06.05
10. Sanjay, V., Swarnalatha, P.: Effective classification of Alzheimer disease based on image tractography framework utilizing GM-ABC-NN. Alex. Eng. J. **93**, 336–347 (2024). https://doi.org/10.1016/j.aej.2024.03.008
11. Naveen, N., Nagaraj, G.: Cholli: enhancing early Alzheimer's disease detection: leveraging pre-trained networks and transfer learning. Int. J. Intell. Syst. Appl. **16**(1), 52–69 (2024). https://doi.org/10.5815/ijisa.2024.01.05
12. Shanmugam, J.V., Duraisamy, B., Simon, B.C., Bhaskaran, P.: Alzheimer's disease classification using pre-trained deep networks. Biomed. Signal Process. Control **71**, 103217 (2022). https://doi.org/10.1016/j.bspc.2021.103217
13. Illakiya, T., Ramamurthy, K., Siddharth, M.V., Mishra, R., Udainiya, A.: AHANet: Adaptive hybrid attention network for Alzheimer's disease classification using brain magnetic resonance imaging. Bioengineering **10**(6), 714 (2023). https://doi.org/10.3390/bioengineer-ing10060714
14. Illakiya, T., Karthik, R.: Alzheimer's Disease Neuroimaging Initiative: A deep feature fusion network with global context and cross-dimensional dependencies for classification of mild cognitive impairment from brain MRI. Image Vis. Comput. **144**, 104967 (2024). https://doi.org/10.1016/j.imavis.2024.104967
15. Cao, Y., Xu, J., Lin, S., Wei, F., Hu, H.: Global context networks. IEEE Trans. Pattern Anal. Mach. Intell. **45**(6), 6881–6895 (2020). https://doi.org/10.1109/TPAMI.2020.3047209
16. Misra, D., Nalamada, T., Arasanipalai, A.U., Hou, Q.: Rotate to attend: convolutional triplet attention module. In: Proceedings of the IEEE/CVF Winter Conference on Applications of Computer Vision, pp. 3139–3148 (2021). https://doi.ieeecomputersoci-ety.org/10.1109/WACV48630.2021.00318

Early Prediction of Alzheimer's Disease Using Deep Learning Models

S. Deeparani(✉), G. Archana, R. Deepika, and D. Dhanrithii

Department of Computer Science and Engineering, Sri Venkateswara College of Engineering, Sriperumbudur, Tamil Nadu, India
{deeparanis,2021cs0007,2021cs0533,2021cs0300}@svce.ac.in

Abstract. Alzheimer's Disease (AD) is a type of dementia that progresses with time. This disease impacts cognitive functions of elderly patients affecting their daily life. This paper provides a cutting-edge approach to early AD detection using advanced deep learning system, specifically targeting the challenge of timely and accurate diagnosis. This is done by feeding neurodiagnostic imaging data, especially Magnetic Resonance Imaging (MRI) and Positron Emission Tomography (PET) scans, through the proposed hybrid 3D CNN-LSTM model. The methodology encompasses comprehensive data preprocessing, including image normalization, augmentation, and skull stripping, to enhance model performance. The research achieves a hybrid 3D CNN-LSTM model with 94% accuracy in classifying Alzheimer's disease stages, successfully processing and analyzing neuroimaging data from the ADNI dataset. The proposed model proves to be effective in handling spatial data such as shape, size, position etc. as well as temporal data such as trends, patterns and sequences. The study enables precise early detection across four cognitive stages: Alzheimer's Disease (AD), Cognitive Normal (CN), Early Mild Cognitive Impairment (EMCI), and Mild Cognitive Impairment (MCI). A user-friendly web interface further supports scan upload and diagnostic classification, contributing a promising computational methodology for identifying early onset of Alzheimer's Disease and potentially revolutionizing diagnostic approaches in clinical practice.

Keywords: Alzheimer's Disease · Early Detection · Machine Learning · Deep Learning · 3D CNN-LSTM · Neuroimaging Analysis

1 Introduction

Alzheimer's Disease is a progressive disorder impacting vital parts of the brain that poses a significant challenge to global healthcare, particularly affecting elderly populations by progressively impairing memory and cognitive functions. Traditional diagnostic methods such as clinical evaluation are limited by complex manual interpretation and potential human error, often detecting the disease at advanced stages when interventions are less effective.

Hence, there exists a crucial demand for more precise and early diagnostic tools which ensures driven significant research interest in advanced computational approaches.

R. Geetha et al. (Eds.): AAIMB 2024, CCIS 2654, pp. 176–191, 2026.
https://doi.org/10.1007/978-3-032-13029-7_13

Machine learning and deep learning technologies offer unprecedented opportunities to transform AD detection by providing more objective, accurate, and timely diagnostic capabilities. This research establishes a modern framework to enhance Alzheimer's Disease detection by developing a sophisticated deep learning model that combines 3D Convolutional Neural Networks (CNN) with Long Short- Term Memory (LSTM) networks.

The proposed methodology leverages the unique strengths of both 3D CNNs and LSTM networks. Convolutional layers aids in extracting spatial features from complex 3D neuroimaging data, capturing intricate structural details of brain anatomy, while LSTM networks provide robust temporal sequence modeling, enabling the detection of progressive changes and subtle patterns that might indicate early-stage cognitive impairment.

By integrating neuroimaging data from MRI and PET scans with advanced preprocessing methods such as image normalization, augmentation, and skull stripping, the research aims to overcome existing limitations in AD diagnosis. The main goal is to deploy a computational methodology for identifying the onset of Alzheimer's disease, providing timely treatment to prevent severe after effects of the disease.

This study makes substantial contributions to the field by developing a sophisticated hybrid deep learning model, implementing comprehensive data preprocessing techniques, and creating a user-friendly diagnostic interface. By addressing the critical challenges in early AD detection, the research seeks to support healthcare professionals in providing more proactive and precise care for patients likely to develop Alzheimer's disease.

2 Related Works

In the recent past, various researchers have implemented different machine learning and deep learning-based approaches for providing timely detection in order to prevent the progression of the disease.

M. El-Geneedy et al. [1] formulated a CNN-based method in order to detect AD. The dataset used in this research is 2D MRI images from OASIS- 3, which is used to attain a testing accuracy of 99.68%. The architecture included five 2D convolutional layers, to reach a validation accuracy of 99.27% and a training accuracy of 100%.

Diogo et al. [2] introduced a machine learning framework that utilized structural MRI data from ADNI and OASIS to differentiate between MCI and AD. The balanced accuracy of 90.6% along with a Matthew's Correlation Coefficient (MCC) of 0.811 were achieved for differentiating healthy controls (HC) from AD.

X. Jiang et al. [3] constructed a CNN model consisting of 8 layers for AD detection that incorporated batch normalization, dropout, and data augmentation methods. Trained on a dataset of 7399 AD patients and 7399 HC, the method achieved a 97.77% sensitivity, 97.76% specificity, 97.79% precision, and 97.76% accuracy.

A. B. Tufail et al. [4] explored various deep 2D convolutional networks (2D-CNNs) to extract and combine local brain features for AD diagnosis. Utilizing transfer learning models like Inception v3 and Xception, the study showed improved performance on the OASIS-1 dataset, which included 416 subjects.

M. Liu et al. [5] introduced a deep learning approach that utilized a CNN for simultaneous hippocampal segmentation and AD classification. The model achieved a Dice coefficient of 87%, 88.9% accuracy, and 92.5% AUC for distinguishing between AD and normal controls (NC).

D. A. Arafa et al. [6] presented a deep learning method that utilized both a custom CNN and a fine-tuned VGG16 model for classifying AD. The 2D CNN achieved impressive accuracies of 99.95% and 99.99% across various stages of AD, while the VGG16 model reached 97.44%, incorporating dropout techniques to reduce overfitting. Arya et al. [7] examined the combined dataset of PET and MRI scans for diagnosing AD, employing CNNs for feature extraction and recurrent neural networks (RNNs) to forecast disease progression. The study also highlighted the importance of Explainable AI (XAI) for enhancing model transparency.

S. Bringas et al. [8] developed a 1D CNN aimed at classifying stages of AD using time-series data. This model effectively extracted relevant features and tackled issues related to imbalanced data, and utilized various layers along with softmax activation for multiclass classification, further improved by batch normalization and dropout techniques.

S. E. Sorour et al. [9] introduced AD-DL, a deep learning technique developed to detect AD through brain MRI images. This model assessed five different architectures, including a 2D CNN-LSTM hybrid that achieved an 99.92% accuracy and an 99.70% F1-score. It effectively captures spatial as well as temporal data, thereby improvising the classification to detect and prevent AD.

H. A. Helaly et al. [10] investigated the use of CNNs for detecting and classifying various stages of AD through brain MRI scans. The research combined straightforward CNNs for both 2D and 3D convolution with transfer learning using VGG19, optimized using the Adam optimizer with 0.0001 learning rate.

R. Franciotti et al. [11] utilized random search for hyperparameter optimization across Random Forest (RF), Gradient Boosting (GB), and Extreme Gradient Boosting (XGB) methods. The observations indicated that RF outperformed both GB and XGB, while GB and XGB showed superior performance after applying voting methods. Although the voting system enhanced stability, it did not consistently surpass the performance of individual models.

Patil et al. [12] assessed AD and assessed two ML methods for early detection, highlighting an 18-layer CNN achieving 98% accuracy, outperforming 3D CNNs. Transfer learning with a customized VGG19 architecture further enhanced accuracy. The study encourages exploring different feature sets for earlier detection, affirming that deep learning, especially CNNs, is effective for early AD detection.

P. Matlani [13] proposed a hybrid deep learning framework for AD diagnosis using IAWF for image enhancement, PCA-NGIST for feature extraction without segmentation, IWHO for feature selection, and BiLSTM-ANN for classification. This was done using ADNI and OASIS dataset, procuring accuracies of 99.22% and 98.96% respectively.

F. Hajamohideen et al. [14] proposed a Siamese Convolutional Neural Network (SCNN) using a triplet-loss function to represent MRI images in the form of k- dimensional embeddings. The CNNs, pre-trained as well as non-pretrained, were utilized for

4-way AD classification. This classifier used both ADNI as well as OASIS dataset with accuracies of 91.83% and 93.85% respectively.

M. Leela et al. [15] proposed the HEMRDTL model, which uses EEG and CT-MRI data using techniques such as transfer learning, deep VGG-19 and robust principal component analysis (RPCA) for early AD detection, achieving an accuracy of 98.96%. The model extracts structural features from fused CT-MRI via a fine-tuned VGG-19 and functional features from EEG using modified RPCA.

3 Dataset

This research utilized ADNI and OASIS database. The two main modalities used for model training are MRI and PET scans. The MRI scans were of NifTi format while PET scans were of DICOM format which were transformed to NifTi format for uniformity. The scans are preprocessed and further classified into 4 classes: AD, CN, EMCI and MCI.

3.1 MRI Scans

(See Fig. 1)

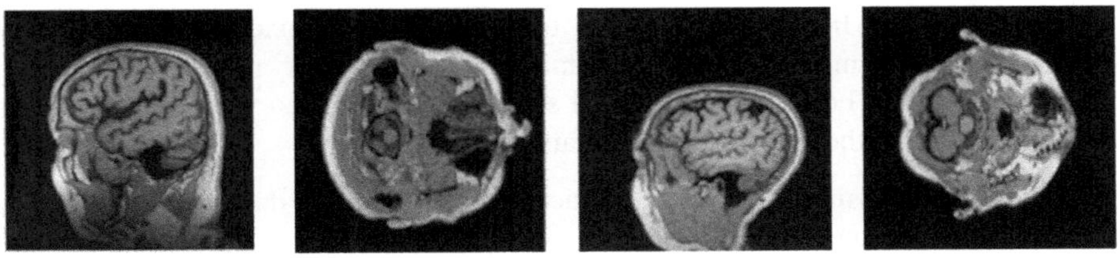

Fig. 1. The dataset of MRI scans of different classes AD, CN, EMCI and MCI.

3.2 PET Scans

(See Fig. 2)

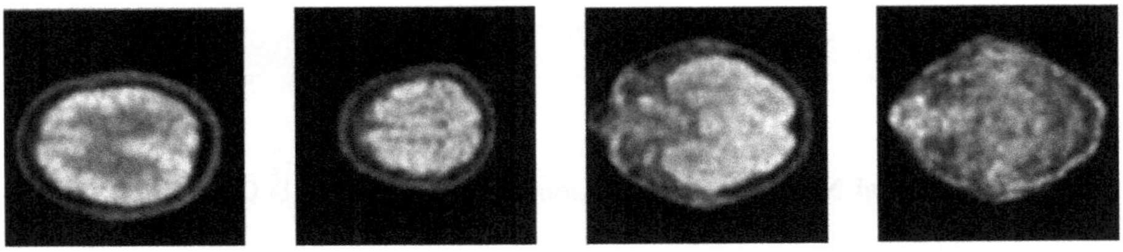

Fig. 2. The dataset of PET scans of different classes AD, CN, EMCI and MCI.

4 Methodology

4.1 Data acquisition

For this study, a balanced dataset was curated with 20 MRI scans for each class, representing AD, CN, MCI and EMCI. The total of 60 scans are stored in NIfTI format (nii or nii.gz), preserving high-resolution 3D brain images. In the case of Positron Emission Tomography (PET) scans, approximately 21 scans per class were collected. These DICOM slices of 64 scans were stacked and converted into NIfTI format to maintain consistency with the MRI data.

4.2 Data Preprocessing

Several preprocessing steps were used on the MRI and PET scans to assure uniformity and improve model performance. The key preprocessing operations included:

- **Image Normalization**: Intensity normalization was applied to standardize the intensity values across all scans.
- **Image Augmentation**: To improve model generalization and prevent overfitting, various augmentation techniques were applied: Rotation by 90°, 180°, and 270° and flipping along both axes. Hence the dataset size increased to 300 MRI scans and 320 PET scans.
- **Image Resizing**: Images were resized to either 128x128 pixels or 64x64 pixels, depending on the model's input requirement.
- **Skull Stripping**: For MRI scans, skull stripping was done to filter out non- brain tissues, directing the analysis toward brain regions only.

These preprocessing steps were critical to maintaining data consistency and enhancing the robustness of the model.

4.3 Comparative Models

The various Machine Learning (ML) models used are:

1. 3D CNN
2. 3D CNN LSTM
3. 3D CNN SVM
4. 3D CNN RNN

3D CNN

A **3D Convolutional Neural Network** extends the traditional 2D CNN by processing volumetric data such as MRI and PET scans. This model extracts spatial features from 3D brain images, preserving relationships between slices in the volumetric data.

- **Input Layer**: The model receives 3D volumes of data, with each voxel representing a point in the 3D space.

- **3D Convolutional Layer**: 3D convolutional filters slide across all three spatial dimensions (height, width, and depth) to extract feature maps that capture spatial patterns and structural abnormalities in the brain, such as atrophy or lesions.
- **Pooling Layer**: It helps in reducing the dimensions of the input feature maps using pooling layers, ensuring the model retains the most informative features while reducing computational complexity. Max-pooling or average pooling techniques are commonly employed.
- **Fully Connected Layer**: The resultant flattened representations are passed through this layer, where the model learns high-level representations necessary for classification.
- **Output Layer**: The final output layer provides probabilities of each class i.e. AD, CN, MCI and EMCI using various activation functions, loss functions and optimizers.

3D CNN LSTM

The **3D CNN + LSTM** network combines the spatial and temporal features extracted by 3D CNN and LSTM networks respectively This hybrid model is particularly useful as it captures both spatial patterns and temporal dependencies across multiple frames or time points.

- **Input Layer**: The input consists of volumetric data representing 3D brain scans.
- **3D Convolutional Layer**: The spatial features were extracted by the model from the volumetric data by applying 3D convolutional layers. These layers learn high-dimensional spatial features, such as abnormal brain structures or regions of interest.
- **Pooling Layer**: Pooling layers downsample the feature maps, emphasizing the most relevant spatial information.
- **LSTM Layer**: The 3D CNN displays the output, reshaped into a sequential format and passed through the LSTM layers. The LSTM units learn temporal relationships in the data, tracking changes in the features over time (e.g., progression of brain atrophy). This step is crucial for detecting subtle, time- dependent patterns that could indicate early stages of Alzheimer's.
- **Fully Connected Layer**: The output is then passed through fully connected layers for further integration of spatial and temporal features, leading to refined predictions.
- **Output Layer**: The final output layer generates class probabilities, classifying the input scans into AD, CN, MCI, or EMCI.

3d CNN SVM

The **3D CNN + Support Vector Machine (SVM)** architecture integrates the spatial attribute extraction functionality of 3D CNNs with the powerful classification technique of SVMs. The CNN component learns spatial features, while the SVM classifier is trained to separate these features into distinct classes.

- **Input Layer**: The model receives 3D volumetric scans as input.
- **3D Convolutional Layer**: Spatial characteristics from the MRI and PET scans are captured from the convolutional layers.
- **Pooling Layer**: These layers help in minimizing the feature map's dimensions while preserving key features.

- **Feature Extraction**: After convolution and pooling, the final output from the 3D CNN is reshaped into feature vectors, which are then passed to the SVM for classification.
- **SVM Classifier**: The SVM helps in finding the optimal hyperplane to separate data points into different classes, and hence the margin between the classes is maximized. The support vectors, which are the data points nearest to the hyperplane, define the boundary, ensuring precise classification.
- **Output Layer**: The SVM classifier predicts the class labels based on the learned decision boundary.

3D CNN RNN

The **3D CNN + Recurrent Neural Network (RNN)** model combines the spatial and temporal features extracted by 3D CNN and RNN networks respectively. This hybrid model is suited for sequential data analysis, where temporal dependencies within the volumetric data must be considered.

- **Input Layer**: The model receives 3D volumetric data as input.
- **3D Convolutional Layer**: The spatial features are captured from input data through convolutional layers.
- **Pooling Layer**: They help in reducing the feature map's dimensions while preserving key features.
- **RNN Layer**: The 3D CNN reshapes its output into a sequence and passes to RNN layers, which learn temporal dependencies across the sequential frames of the neuroimaging data. This is important for modelling structural brain changes over time, which could indicate progress of the disease.
- **Fully Connected Layer**: Flattened output from the RNN layers passes through this layer to integrate spatial and temporal information for classification.
- **Output Layer**: The output layer provides class probabilities for the target classes (AD, CN, MCI, and EMCI), making the final decision based on the learned features.

Proposed Model: 3D CNN + LSTM

Among the previously described models, we propose the **3D CNN + LSTM** hybrid model as the most efficient approach for Alzheimer's classification. This model excels in capturing both spatial features (through 3D CNN layers) and temporal dependencies (via LSTM layers), making it well-suited for analyzing neuroimaging data.

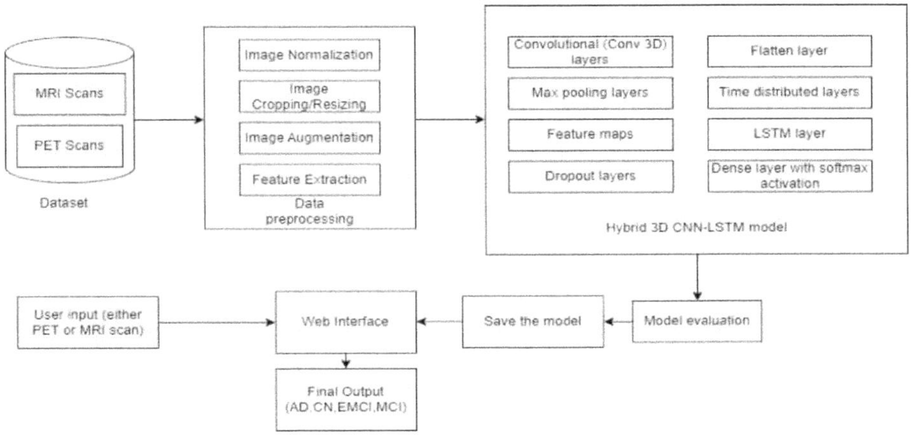

Fig. 3. The proposed architecture diagram of hybrid 3D CNN LSTM model

Based on the performance comparison of multiple deep learning approaches, the 3D CNN + LSTM hybrid architecture was selected as the proposed model due to its superior ability to capture both spatial and temporal aspects of neuroimaging data. While 3D CNNs are effective in learning structural features from brain scans, LSTM networks excel in modelling time-dependent patterns and progression making this combination highly suitable for identifying different stages of AD. As shown in Fig. 3, the architecture integrates these components into a single, end-to-end trainable pipeline that classifies brain scans into four diagnostic categories: AD, MCI, EMCI, and CN.

The 3D CNN component comprises several convolutional layers with 3 × 3 × 3 kernels that extract local features indicative of neurodegeneration, such as cortical thinning or hippocampal atrophy. After each convolution, batch normalization is applied to stabilize training, followed by ReLU activation to introduce non-linearity. Max pooling layers with 2 × 2 × 2 windows reduce spatial dimensions and retain essential structural patterns. These layers work together to encode the anatomical features of the brain into compressed, high-level representations.

To transition from spatial to temporal analysis, the output from the CNN is reshaped into a sequence using a Time Distributed layer. This format allows each volumetric slice or time step to be analysed chronologically, preparing the data for temporal modelling. The Time Distributed wrapper ensures that the CNN processes each frame independently while preserving its position in the sequence.

The reshaped sequence is then fed into LSTM layers, which are designed to capture dependencies across the time steps. The LSTM unit maintains a memory of previously seen frames and updates this memory using gates (input, forget, output) that determine which information to keep or discard. This allows the model to detect subtle progression trends such as increasing asymmetry or structural decline over time which are key to early-stage AD diagnosis.

The LSTM's output is flattened and passed through fully connected dense layers. These layers synthesize spatial and temporal features into a final decision- making representation. The output layer, activated by a softmax function, consists of four neurons representing the target classes: AD, MCI, EMCI and CN. The final predicted class is chosen by finding the maximum probability.

To prevent overfitting, dropout layers (with a rate of 0.5) are placed after the LSTM and dense layers. The model is trained using the Adam optimizer, 0.001 learning rate and categorical cross-entropy loss function. As highlighted in the results section, this hybrid model achieved 94% classification accuracy on both MRI and PET datasets, outperforming alternative architectures and proving to be highly effective for early AD detection.

Results and Discussion

The below tables and diagrams demonstrate the efficiency of the proposed model using evaluation metrics such as accuracy and confusion matrix for both MRI and PET scans (Fig. 4, 5, 6, 7 and Tables 1, 2).

Table 1. Accuracy of models Trained on MRI images

S.No	Model	Accuracy
1	3D CNN	50%
2	3D CNN-RNN	56%
3	3D CNN-SVM	19%
4	3D CNN-LSTM	94%

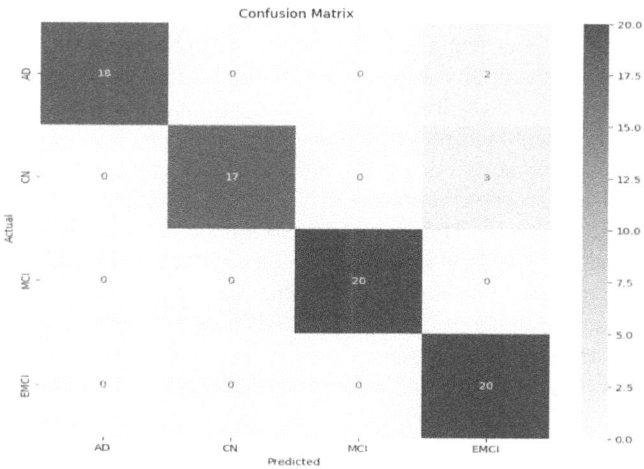

Fig. 4. The confusion matrix of 3D CNN LSTM model on MRI scan

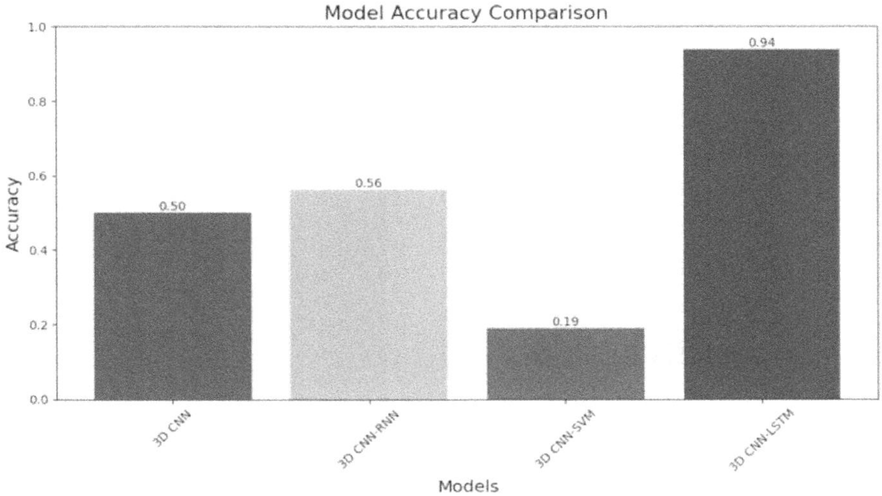

Fig. 5. The comparison of accuracies across all models trained on MRI scans

Table 2. Accuracy of Models Trained on PET scans

S.No	Model	Accuracy
1	3D CNN	78.43%
2	3D CNN-RNN	84.31%
3	3D CNN-LSTM	94.12%
4	3D CNN-SVM	64.75%

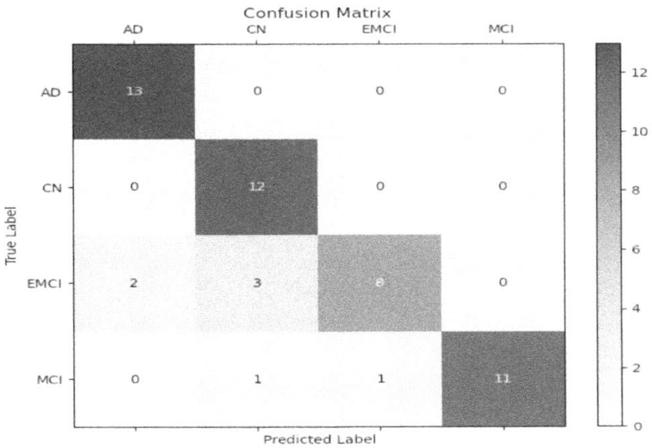

Fig. 6. The confusion matrix of 3D CNN LSTM model on PET scan

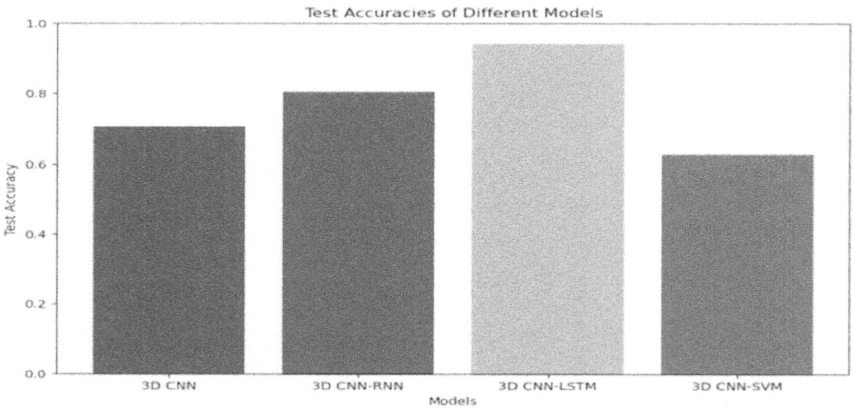

Fig. 7. The comparison of accuracies across all models trained on MRI scans

As observed in above figures and tables, the 3D CNN LSTM model achieves around 94% overall accuracy for both MRI and PET scans which is higher compared to other models. This is due to the exceptional features of 3D CNN LSTM which includes spatial and temporal feature extraction, improved generalization, sequential dependence handling and non-linearity of the model. This model works as the backend of web interface.

Table 3. The comparison with state of the art models

Study	Dataset	Methodology	Accuracy	Remarks
M. El-Geneedy et al. [1]	OASIS-3 (MRI)	2D CNN with 5 layers	99.68%	High accuracy but limited to 2D images; lacks 3D spatial relationship modeling
D. A. Arafa et al. [6]	Multiple (MRI) (self – created)	Custom CNN and VGG16 (transfer learning)	99.99%	Uses fine-tuned VGG16; overfitting risk with small datasets
S. E. Sorour et al. [9]	Brain MRI (self – created)	2D CNN-LSTM hybrid	99.92%	Captures spatial and temporal features; MRI- based, not PET-specific
Arya et al. [7]	PET + MRI (self – created)	CNN + RNN	91.2%	Multimodal data predicts MCI to AD transition

(*continued*)

Table 3. (*continued*)

Study	Dataset	Methodology	Accuracy	Remarks
M. Liu et al. [5]	ADNI (MRI)	Multi-model CNN for hippocampal analysis	88.9%	Focuses on hippocampal segmentation; lacks end- to-end volumetric analysis
P. Matlani [13]	ADNI, OASIS (MRI)	Hybrid BiLSTM-ANN framework	99.22%	Extensive preprocessing and feature engineering required
F Hajamohideen et al. [14]	ADNI, OASIS (MRI)	Siamese CNN with triplet-loss function	91.83%-93.85%	Lower accuracy than our model, despite advanced loss function
Our Model	ADNI, OASIS (MRI and PET)	3D CNN-LSTM	**94%**	Captures 3D spatial and temporal features, achieving competitive accuracy

Table 3 shows the comparison of our proposed model with other state of the art models. Prior works like El-Geneedy et al. and Arafa et al. obtained high accuracy but relied on 2D CNNs or transfer learning, lacking 3D spatial understanding while Sorour et al. and Arya et al., captured temporal features but used limited or unimodal data. Some models required high preprocessing and produced lower accuracy. In contrast, our 3D CNN-LSTM leverages both MRI and PET data to model spatial and temporal features, achieving robust performance (94%) with a more comprehensive and scalable approach.

The web interface offers a user-friendly experience, including pages for registration, login, and scan upload, where users can securely enter patient details, scan dates, and optional comments. The "Choose a Scan" page guides users through the upload process, ensuring MRI and PET data are classified into AD, CN, MCI and EMCI.

Each uploaded scan is preprocessed through noise reduction and normalization and analyzed by a hybrid CNN-LSTM model for feature extraction and classification. CNN extracts spatial features from MRI and PET scans, while LSTM captures sequential data, enabling precise diagnostic predictions. Testing on the ADNI and OASIS datasets demonstrated accurate classification across AD stages, showcasing the model's capability to detect early cognitive decline effectively.

The screenshots of the developed web application are given below (Figs. 8, 9, 10, 11 and 12):

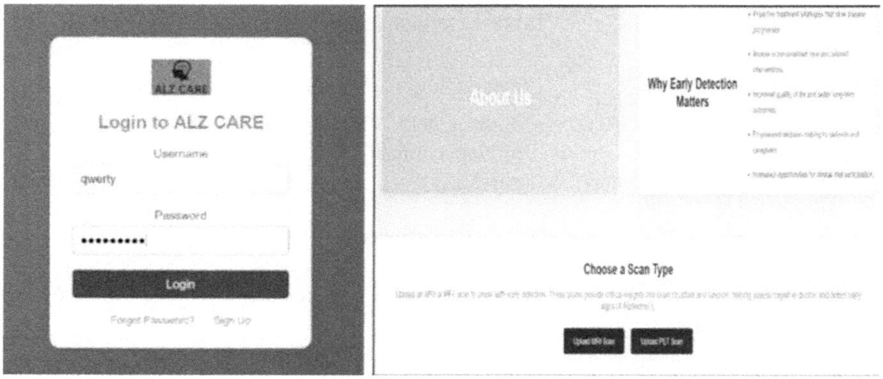

Fig. 8. The login page and the home page of the web interface.

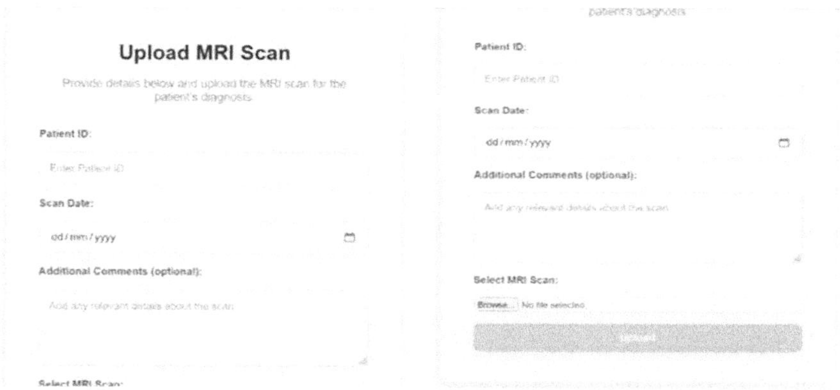

Fig. 9. The interface to upload MRI scans

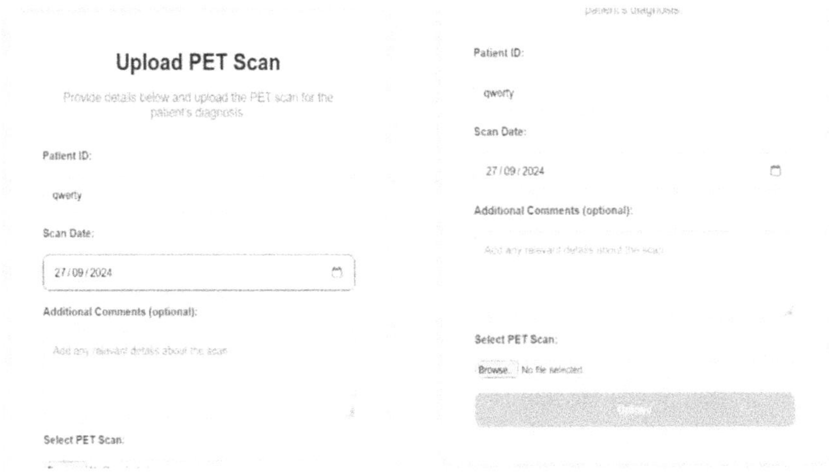

Fig. 10. The interface to upload PET scans.

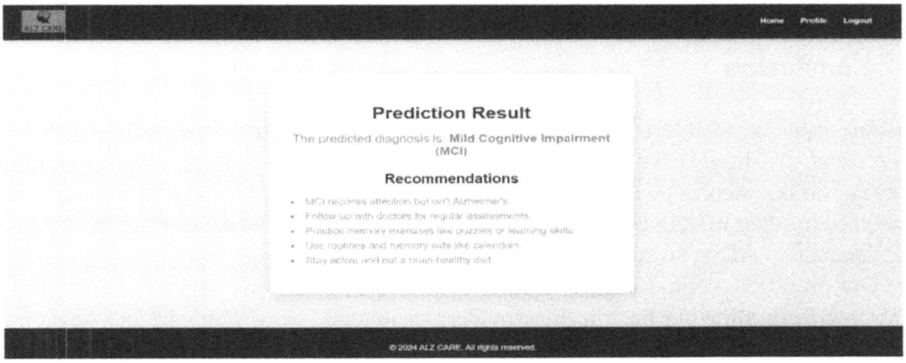

Fig. 11. The prediction for MRI scan.

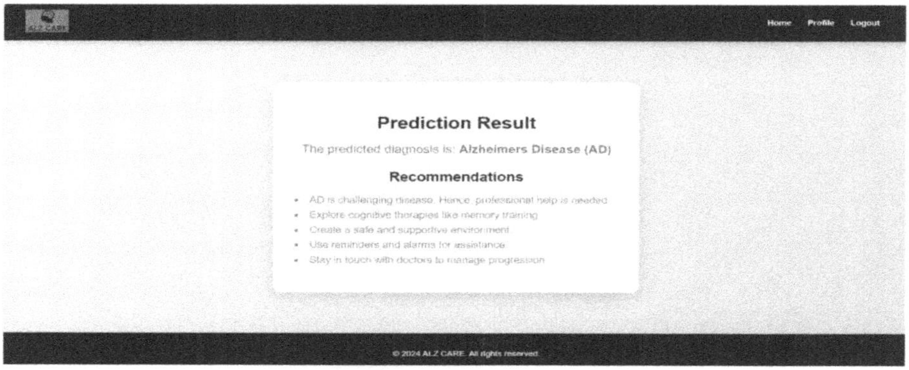

Fig. 12. The prediction for PET scan.

Built with TensorFlow and Flask, the backend supports model deployment, allowing users to access classification results and receive preventative care recommendations based on individual risk profiles. Clinical validation through real-world testing in medical settings will be pursued in future work to ensure the model's reliability and usefulness in actual diagnostic workflows.

5 Conclusion

In this paper, a reliable early detection model for Alzheimer's Disease is developed to prevent the progressive nature of the disease. Through a comprehensive literature survey, various neural learning models integrating 3D CNN, RNN, LSTM, etc. were analyzed for their effectiveness in processing MRI and PET scan data, which are critical for diagnosing AD at an early stage.

Our proposed methodology integrates neuroimaging analysis within a user-friendly web interface, allowing healthcare providers to upload scans, input patient data, and receive diagnostic classifications such as AD, MCI, EMCI and CN. This web interface, with its backend support for seamless integration and data processing, aims to empower clinicians and caregivers with timely, accurate diagnostic insights.

Future enhancements include implementing Phase2Phase noise reduction techniques for cleaner MRI and PET data, further improving diagnostic accuracy. Additionally, monitoring disease progression through sequential data modeling will enable predictive capabilities, allowing clinicians to better track AD development and tailor interventions accordingly.

References

1. El-Geneedy, M., Moustafa, H.E.-D., Khalifa, F., et al.: An MRI-based deep learning approach for accurate detection of Alzheimer's disease. Alex. Eng. J. **63**, 211–221 (2022)
2. Diogo, V.S., Ferreira, H.A., Prata, D.: Early diagnosis of Alzheimer's disease using machine learning: a multi-diagnostic, generalizable approach. Alzheimer S Res. Therapy **14** (2022)

3. Jiang, X., Chang, L., Zhang, Y.-D.: Classification of Alzheimer's Disease via eight-layer convolutional neural network with batch normalization and dropout techniques. J. Med. Imag. Health Inform. **10**, 1040–1048 (2020)
4. Tufail, A.B., Ma, Y.-K., Zhang, Q.-N.: Binary classification of Alzheimer's disease using SMRI imaging modality and deep learning. J. Digit. Imaging **33**, 1073–1090 (2020)
5. Liu, M., Li, F., Yan, H., et al.: A multi-model deep convolutional neural network for automatic hippocampus segmentation and classification in Alzheimer's disease. Neuroimage **208**, 116459 (2019)
6. Arafa, D.A., Moustafa, H.E.-D., Ali, H.A., et al.: A deep learning framework for early diagnosis of Alzheimer's disease on MRI images. Multimedia Tools Appl. **83**, 3767–3799 (2023)
7. Arya, A.D., Verma, S.S., Chakarabarti, P., et al.: A systematic review on machine learning and deep learning techniques in the effective diagnosis of Alzheimer's disease. Brain Inform. **10** (2023)
8. Bringas, S., Salomón, S., Duque, R., et al.: Alzheimer's Disease stage identification using deep learning models. J. Biomed. Inform. **109**, 103514 (2020)
9. Sorour, S.E., El-Mageed, A.A., Albarrak, K.M., et al.: Classification of Alzheimer's disease using MRI data based on Deep Learning Techniques. J. King Saud Univ. – Comput. Inform. Sci. **36**, 101940 (2024)
10. Helaly, H.A., Badawy, M., Haikal, A.Y.: Deep learning approach for early detection of Alzheimer's disease. Cogn. Comput. **14**, 1711–1727 (2021)
11. Franciotti, R., Nardini, D., Russo, M., et al.: Comparison of machine learning-based approaches to predict the conversion to Alzheimer's Disease from mild cognitive impairment. Neuroscience **514**, 143–152 (2023)
12. Patil, V., Madgi, M., Kiran, A.: Early prediction of Alzheimer's disease using convolutional neural network: a review. Egyptian J. Neurol. Psych. Neurosurg. **58** (2022)
13. Matlani, P.: BiLSTM-ANN: early diagnosis of Alzheimer's disease using hybrid deep learning algorithms. Multimedia Tools Appl. **83**, 60761–60788 (2024)
14. Hajamohideen, F., Shaffi, N., Mahmud, M., et al.: Four-way classification of Alzheimer's disease using deep Siamese convolutional neural network with triplet-loss function. Brain Inform. **10** (2023)
15. Leela, M., Helenprabha, K., Sharmila, L.: Prediction and classification of Alzheimer Disease categories using Integrated Deep Transfer Learning Approach. Measure. Sens. **27**, 100749 (2023)

Emotion Recognition in Text and Images Using Deep Learning

R. Vinston Raja[1], Sakthitharan Subramanian[1], Manoj Kushwaha[1], V. Kaliraj[2(✉)], Dharaniya[3], and N. Shree Makesh[3]

[1] SRM Institute of Science and Technology, Kattankulathur, Chennai, Tamil Nadu, India
[2] S. A. Engineering College, Chennai, Tamil Nadu, India
kallis@saec.ac.in
[3] Easwari Engineering College, Chennai, Tamil Nadu, India

Abstract. Emotion recognition involves predicting or extracting feelings based on human reactions. Predicting emotions is a crucial task even without facial expressions. Human depression can lead to severe health consequences. By using a deep learning algorithm, companies can predict emotions through emotional analysis to better understand their clients' responses to their products. Emotional analysis helps to clarify human feelings. Supervised algorithms like Support Vector Machines, Random Forests, and Naive Bayes are employed to classify emotions in the early stages. However, these algorithms often struggle with accuracy and cannot handle large volumes of data efficiently. Various companies utilize emotional analysis to enhance their understanding of clients' responses to their products. The existing system operates with CNN and RNN models for emotion classification. While CNN does not take previous inputs into account, RNN considers the most recent input, despite its short-term memory limitations. This issue can be addressed by utilizing LSTM. The proposed system employs LSTM for text emotion recognition and CNN to analyze human facial expressions. This text classification method assesses incoming text and determines whether it falls into categories such as happy, angry, surprised, sad, or fearful. The CNN algorithm identifies facial expressions related to feelings such as sadness, happiness, or joy. Performance analysis is based on comparing the accuracy rates of both CNN and LSTM.

Keywords: Emotion recognition · RNN · LSTM · CN

1 Introduction

Emotional recognition is a text mining technique that identifies emotions without facial cues. Various industries utilize emotional analysis to better understand customer reactions. This analysis clarifies emotions in the text. Supervised algorithms like Naive Bayes, Support Vector Machines, and Random Forests classify emotions initially, but they often lack accuracy and struggle with large datasets. Neural Networks and other deep learning methods (CNN, RNN, ANN) have gained prominence due to their superior performance. Computer technology detects human emotions using images and messages to enhance

R. Geetha et al. (Eds.): AAIMB 2024, CCIS 2654, pp. 192–209, 2026.
https://doi.org/10.1007/978-3-032-13029-7_14

human-machine interaction. Recent advancements in pattern recognition and machine learning improve emotional understanding. Facial expression recognition serves as a vital element in real-time applications for intelligent human-machine interaction, with uses in medical assistance, online education, gaming, and public security.

Facial emotion and expression recognition analyzes features from facial images to classify human emotions such as happiness, sadness, surprise, anger, aversion, crying, and neutrality. It is essential for researching emotional reorganization. Facial expression mining utilises machine learning and algorithms to detect the emotions behind facial expressions and is also applied in healthcare. Understanding depression goes beyond mere sadness; it can significantly impact a person's mental health, causing fatigue, distress, and despair. Accurate drug assessments can be based on consistent records of facial expressions. Autism spectrum disorder (ASD) is a developmental disability affecting social communication, interaction, and behavior. Facial expression recognition can aid in interpreting the emotions of autistic children, enabling doctors to understand their psychosomatic changes better and develop more effective treatments. Facial expression recognition in education can allow the teaching and learning system to capture and record students' subtle changes in their studies, thereby improving the situation for instructors to teach students about emotions.

Online chat refers to internet communications that allow real-time text messaging between friends. Unlike traditional text messages, chat messages are typically short and informal, enabling quick replies. However, informal messages complicate sentiment analysis due to poor formatting, limited size (usually a few sentences), and language that includes grammatical errors, slang, and abbreviations. Friends often use emojis, like the Smiling Face with Smiling Eyes, to express emotions, typically indicating happiness or positivity [3]. The meaning of words in chat can differ significantly from their usual meanings, and emotions can relate to previous messages. While many emotional features have been studied, few studies have effective strategies for emotion recognition in text. Emotion classification algorithms generally perform well at the page or paragraph level but may struggle at the sentence level, making it difficult to apply sentiment analysis methods directly to online chat. To test our approach, we conducted experiments with participants engaging in online chats, who provided very positive feedback afterward.

2 Literature Review

Deep learning algorithms yield efficient results in applications such as speech recognition, object detection, and data classification. Human emotions are communicated through nonverbal cues

This challenge can be overcome using deep learning models for emotional recognition. RNN with LSTM is used to predict emotions with different feelings. The main objective is to classify the unstructured emotions in text. Machine learning techniques struggle to handle text interpretation due to a lack of labelled data. They use the Twitter US Airlines sentiment dataset for the classification deep learning model. [3, 4] Twitter is one of the most instant messaging applications.

Users can express their opinions on Twitter in text and share their thoughts and opinions. Here, the research areas and gaps in sentiment analysis are addressed, along with their future scope. [5] Nowadays, people mostly use E-commerce for purchases and consumption. In an E-commerce platform, emotion recognition can increase user satisfaction by using emotion recognition in many user reviews. In this paper, a new model, SLCABG, is proposed for emotion recognition, combining a Bidirectional Gated Recurrent Unit with a CNN. [6, 7] It aims to predict the emotion in a piece of text, helping to understand the opinion and sentiment of the text in natural language processing. Using emotion recognition, we can extract useful information from the vast amount of data flooding the internet. [8, 9] Different studies were conducted on completely automatic methods that execute feature extraction from datasets without human mediation

[10] Analysis of emotion is a crucial task. The classification was performed at the spoken level, utilising new features such as DAL scores and n-grams. Nevertheless, this

The Method had a disadvantage; it needed a correct expression bound to capture the exact emotion prediction. Furthermore, it does not adequately address the uncertainty of meaning. [11] The main disadvantage is the accuracy compared to other training algorithms. The classification process utilised Tree Tagger for POS (part of speech) tagging, as well as Naive Bayes and N-Grams, which have various dispositions for sentiment classification. Still, the multilingual group of data is not contained in the classification.

Emotion classification with Twitter data [12] by utilising opposition predictions to classification reviews captured from various sites and described as boisterous labels. Labels are used as an evolving practice model. Still, 1000 tweets are classified manually, and another 1000 are captured for testing. Nevertheless, they did not analyse the info collection for testing. However, Twitter features are extracted using patterns such as retweets and links to replies. Extra alternatives like the prior polarity of conversation and POS tags were used.

[13] In the emotion recognition dataset collected from the survey. They set an evaluation to recognise the significance of features in NLP. They acted on selecting feature solids and analysis, which helped illustrate that studying text features provides accuracy for the classifier.

[14] The authors proposed a core by using texts from online blogs for emotion study and utilising the emotions expressed in posts as indicators of end users' moods. The classification is emotional despair at the sentence level, which uses the reasoning phase through SVM and CRF. Still, some investigations were constructed using many methodologies for solving the emotion of text at the text level. [15] The usual ML classification algorithms, LSTM, had tested for emotional recognition to achieve reasonable accuracy. If the classification is finished utilising RNN, there appear to be few exploding and decaying of result veracity, all the while the points of backpropagation. This question is rectified utilising an advanced RNN, LSTM, with its more complicated makeup that contains memory units that recollect the instructions stored in memory cells long ago.

3 Working Methodology

The dataset contains valuable comments reflecting various emotions [16]. This work examines a wide range of emotions, not limited to just positive and negative. Users often provide reviews, comments, emojis, or product suggestions on Twitter, which serve as a basis for review analysis. User-generated content effectively captures sentiment about products and services. Compared to other platforms, Twitter [17] is favored for product reviews and political statements. Tweets and retweets reflect public emotions about a company's products, aiding companies in addressing product issues and enhancing sales and growth. However, the large volume of Twitter data is inaccessible to humans, making machine learning and deep learning essential. These technologies allow for efficient processing of vast amounts of data and quicker conclusions [18]. Nonetheless, overly brief content lacks the information needed for valid analysis. This motivated us to investigate deep learning methods for developing a robust classification model [19]. User-generated content, such as reviews, comments, blogs, chats, images, and videos, frequently appears, providing insights into customer perspectives and helping address business and societal challenges [20]. While there are many methods for classifying emotional sentences, we utilize the advanced LSTM networks technique for emotion prediction, as illustrated in Fig. 1 [21, 22]. This model effectively performs sentiment analysis on text. Previous models, which employed CNN and RNN for emotional recognition, had limitations: CNN overlooks prior outputs [23], and RNN offers only short-term memory. To address this, we implement LSTM for emotion recognition, improving efficiency and reducing errors [24].

4 Proposed Approach

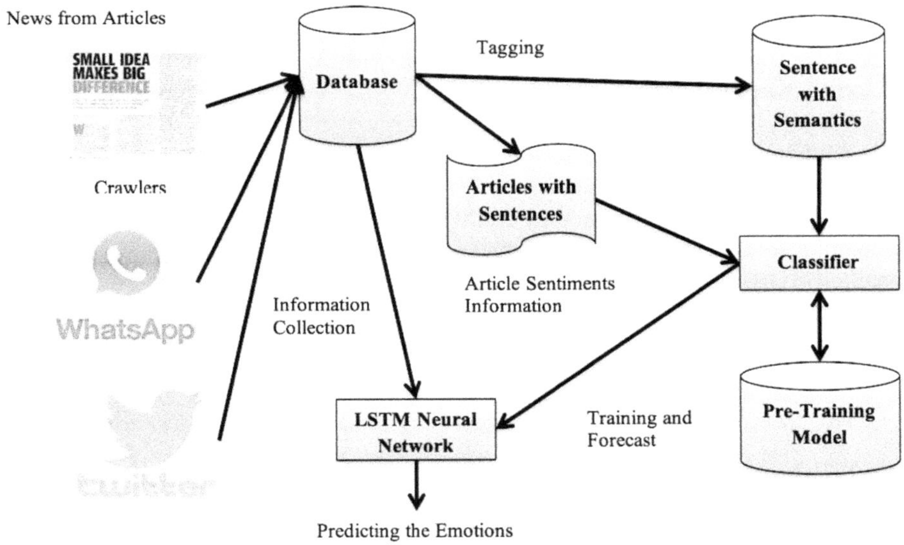

Fig. 1. Architecture Diagram of Emotion Prediction

Collecting the sentences, text messages, emojis and images of users to store in a shared database, as mentioned in Fig. 1. The Architecture Diagram of Emotion Prediction consists of an input dataset. The tagging process is initiated for Sentences with semantics

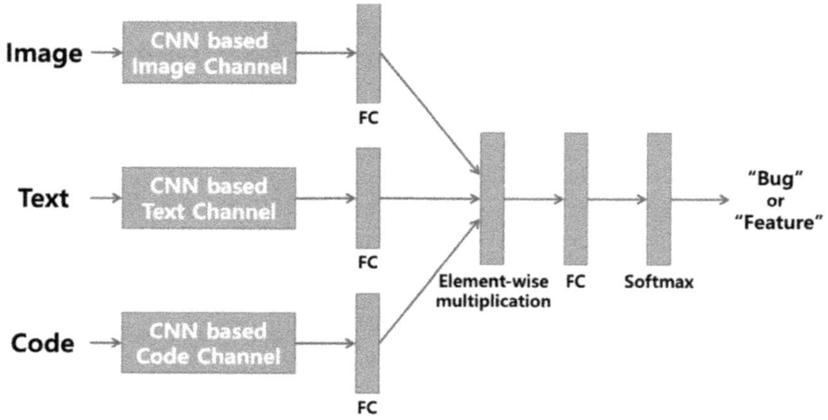

Fig. 2. Proposed Model of CNN-based text identification

that are used to split the words using the tokenisation method. This module is used to check the semantics of text messages. Once the semantic processes are complete, the data will move to the classifier to generate the training dataset. The training dataset will be shaped using the Pre-Training Model. Article sentences will be fetched from the Article with Sentences module and applied to the filtering to generate a training dataset. The trained dataset will serve as the input to the LSTM Neural Network, which will predict both sentences and emotions.

5 Long Short-Term Memory

RNN predicts the next word based on previous words in the chat window but faces gradient exploding and vanishing issues. Long Short-Term Memory (LSTM), depicted in Fig. 2, enhances text prediction by modifying RNN's structure. LSTM retains information in memory cells for future use and addresses RNN's gradient challenges. Additionally, LSTM is effective for sentiment analysis as its variations capture long-term dependencies.

$$[\![oneh]\!]_{(}t-1) = Hiddenstateatprevioustimestamp\ t-1(Shottermmemory)$$

$$[\![onec]\!]_{(}t-1) = Cellstateatprevioustimestamp\ t-1(Long-termmemory)$$

$x_i = Input\ vectorat\ currentt\ imestamp\ t$

$h_i = Hidden\ state\ at\ current\ timestamp\ t$

$c_i = Cell\ state\ at\ current\ time\ stamp\ t$

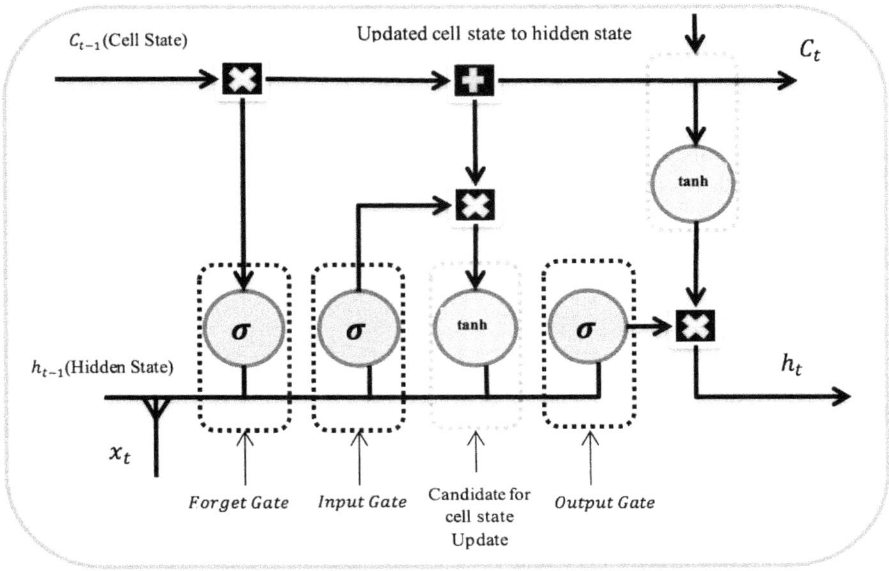

Fig. 3. LSTM Working Principle

6 Workflow of LSTM

Step 1: Initialization of the weights

Weights for different gates are :

Input gate: w_{xi}, w_{xg}, b_i, w_{hy}, w_g, b_g

Forget gate : w_{xf}, b_f, w_{ho}

Output gate: w_{xo}, b_o, w_{ho}

Step 2: Passing through different gates.

Inputs tx, h_{t-i}, and c_{t-1} are given to the LSTM cell.

Passing through the input gate:

$Z_g = w_{xg} * x + w_{hg} * h_{t-1} + b_g$

$g = \tanh(Z_g)$

$Z_j = w_{xi} * x + w_{hi} * h_{t-1} + b_i$

$i = \text{sigmoid}(Z_i)$

Input_gate_out = g*i

Passing through forget gate:

$Z_f = w_{xf} * x + w_{hf} * h_{t-1} + b_f$

$f = \text{sigmoid}(Z_f)$

Forget_gate_out = f

Passing through the output gate:

$Z_o = w_{xo} * x + w_{ho} * h_{t-1} + b_o$ o

= sigmoid(zo) Out_gate_out

Step 3: Calculate the output h_t and current cell state c_t.

Calculating the current cell state c_t :

$c_t = (c_{t-1} * \text{forget\_gate\_out}) + \text{input\_gate\_out}$

Calculating the output gate ht:

$h_t = \text{out\_gate\_out} * \tanh(ct)$

Finally, the latest cell state (c_t) and the hidden state (h_t) return to the recurrent unit, and the **process repeats at time step t+1**. The loop continues until we reach the end of the sequence.

First, import the training, testing, and validation datasets. Next, clean up the data by removing stop words like a, an, the, etc. After removing them, feed the output to the LSTM model for classification.

7 Flow of LSTM processing for classifying and predicting Emoji and Text

The LSTM process flow for Emoji and Text classification and Prediction consists of input, hidden, and output layers. The input layer fetches data like sentences and text messages from various sources and emojis. The hidden or Training layer extracts emotions from text and emojis. The next hidden layer concatenates the emotion features into emojis, as mentioned in Fig. 3.

7.1 Data Preprocessing

Before we predict emotions, the data must be preprocessed. Data preprocessing is a way of cleaning the data. It is classified into two steps: Stop word removal and tokenisation.

Initially, there are numerous unwanted words, articles, blank spaces, taglines, and hashtags, and some content may be repeated. Removing those doesn't harm; they don't carry any meaning. By removing them, we can reduce the runtime and increase efficiency.

7.2 Stop Word Removal

In the collected datasets, discarding unwanted or unnecessary words is referred to as stop word removal. Generally, articles and pronouns are classified as stop words and have no meaning. Stop words are words that are discarded before processing user inputs. By removing those phrases, we remove the low-level data from our textual content, allowing us to focus more attention on the crucial information. Examples of stop words are a, an, the, the, the, that, which, etc.

The stop word removal process discards these kinds of words. We have to import a separate stop word library to remove them. Only the words containing

Emotions remain after the stop words are removed, significantly decreasing the training time.

7.3 Tokenisation

It is the process of splitting the sequence of strings into pieces. The Words are separated using white space, punctuation marks or line breaks. Normally, hashtags are an important part of tweets; removing those hashtags and cleaning the punctuation marks inside tweets are important processes we must follow in tokenisation. Discard all the special characters, like ' ' '! ()-[]{};:'", <>. ? @%^* " ' / from the tweets. Replacing sensitive data with a token, an algorithmically generated number, is referred to as tokenisation in Fig. 3. Emoji and Text Classification and Prediction.

According to the LSTM architecture, the input dataset will consist of a sequence of text and emoji data. The text data will be retrieved from various user information stored in the database, including chat and email conversations. The LSTM layer will then use the information from the database to sum up all text t1, t2, t3,...tn and emojis e1,e2,e3,....en (Fig. 4).

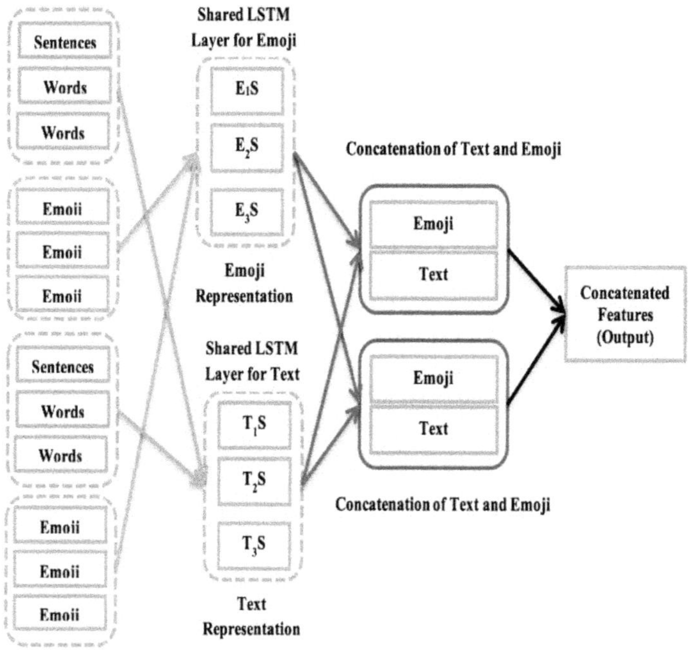

Fig. 4. Emoji and Text Classification and Prediction

The text information and emojis will be processed separately, and the features from text to emoji and emoji to text will be predicted with the help of a concatenation layer. The top layer is to predict the final results of emotions.

$$CText = \sum_{(} i = 1) \llbracket (t_i) = t_1, t_2, t_3, \ldots . t_n \rrbracket \; Emojis = \sum_{(} i = 1) \llbracket (e_i) \rrbracket$$
$$= e_1, e, e_3, \ldots . e_n$$

Each text will be converted into a vector.

$$T = t_1, t_2, t_3, \ldots t_n, \in V^{(}t * d)$$

$$E = e_1, e_2, e_3, \ldots e_n, \in V^{(}e * d)$$

8 Text Classification

The last layer's attention head in the text-only model predicts the label for this sample incorrectly. The lines show attention scores for the text tokens, with intensity indicating weight. Higher attention scores are circled. In Fig. 5

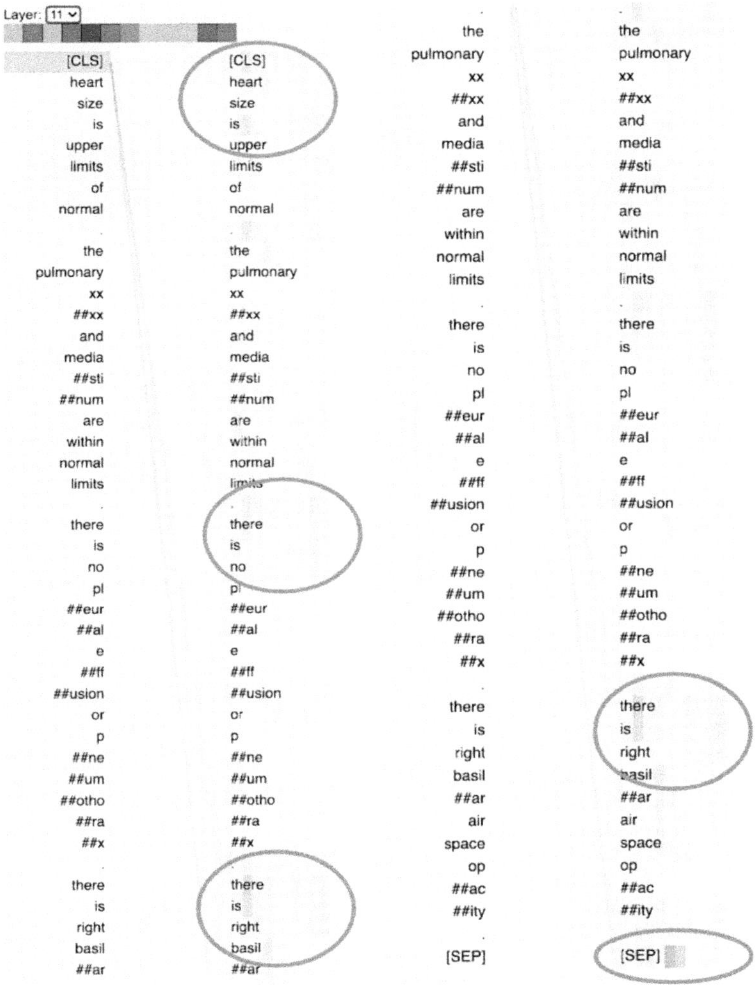

Fig. 5. Text Classification and Prediction

Figure 6 shows the study's flowchart. The experiment aims to normalise raw infrared spectra of pasta, extract features through principal component analysis (PCA), process the data, and input it into traditional and deep learning models for analysis.

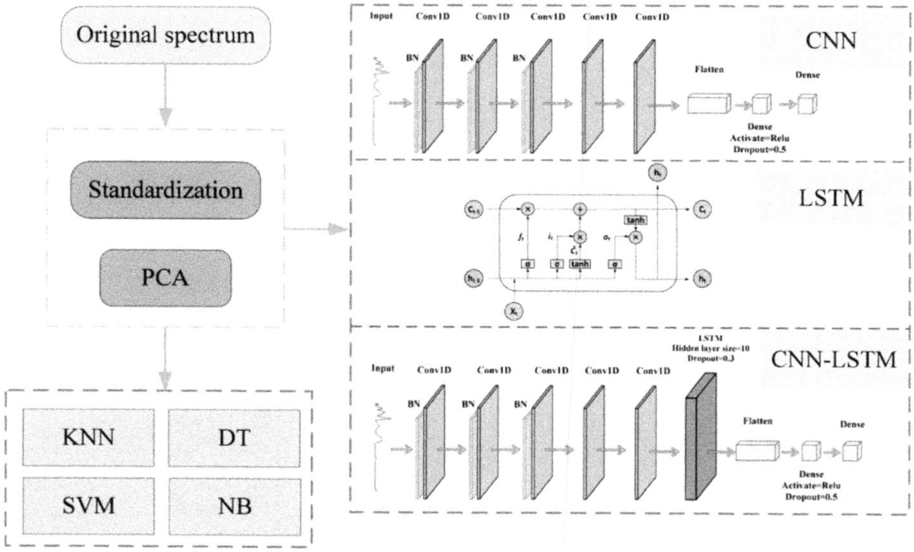

Fig. 6. Emoji and Text Classification and Prediction using LSTM and CNN

9 Implementation

The dataset has two parts for training and testing the LSTM model. To build this classifier, we follow these steps:

1. Conduct Exploratory Data Analysis (EDA).
2. Develop the classifier model.
3. Train and assess the model.

Step 1: Conduct Exploratory Data Analysis (EDA).

Load the dataset using pandas.

train.head()

"text/plain":

```
"                              0              1\n",
"0                 I want to eat\\t           4\n",
"1           he did not answer\\t           3\n",
"2               he got a raise\\t            \n",
"3         she got me a present\\t           0\n",
"4   ha ha ha it was so funny\\t           2."
]
```

Our dataset has five emojis: heart, baseball, smile, disappointed, fork and knife.

Let's store the above information in a dictionary for ease of use.

emoji_dict = { 0 : ":heart:", 1 : ":baseball:", 2 : ":smile:", 3 : ":disappointed:"4 : ":fork_and_knife:"}

Now, using the emoji module, see how these emojis turn out!

Import emoji

for ix in emoji_dict.keys():

```
print (ix,end=" ")
print (emoji.emojis(emoji_dict[ix], use_aliases=True))
"text": [
    "0 ♥\n",
    "1 ☹\n",
    "2 □n",
    "3 □\n",
    "4 □\n"
  ]
```

Unlike the formal method of using the train_test_split() function from sklearn, you can create the training and testing data from the datasets.

```
X_train = train[0]
Y_train = train[1]
X_test = test[0]
Y_test = test[1]
for ix in range(X_train.shape[0]):
X_train[ix] = X_train[ix].split()
for ix in range(X_test.shape[0]):
X_test[ix] = X_test[ix].split()
Y_train = to_categorical(Y_train)
```

Let's check if our above conversion to categorical labels is done.

```
print (X_train[0],Y_train[0])
np.unique(np.array([len(ix) for ix in X_train]) , return_counts=True)
In [14] "# Now checking the above conversion by printing train and test data at 0th index\n",
Our[14] "text": [    "['never', 'talk', 'to', 'me', 'again'] [0. 0. 0. 1. 0.]\n"    ]
In[15] "text/plain": ["np.unique(np.array([len(ix) for ix in X_train]) , return_counts=True)"]
Out[15]    "(array([ 1, 2, 3, 4, 5, 6, 7, 8, 9, 10]),\n",
           " array([ 4, 5, 26, 35, 20, 21, 11, 5, 1, 4], dtype=int64))"
```

Step 2: Develop the classifier model.

LSTM stands for Long Short-Term Memory Network is a type of recurrent neural network that processes sequential data like audio and text. LSTMs can learn long-term dependencies and work well with time-series data. The model is defined as follows:

```
from TensorFlow.keras.models import Sequential
from TensorFlow.keras.layers import Dense, Input,
Dropout, SimpleRNN, LSTM, Activation
model = Sequential()
model.add(LSTM(128, input_shape=(10,50),
return_sequences=True))
model.add(Dropout(0.5))
model.add(LSTM(128, return_sequences=False))
model.add(Dropout(0.5))
model.add(Dense(5))
model.add(Activation('softmax'))
```

Step 3: Train and assess the model.

Compile and fit our model to the embedding matrix from training data, using categorical cross-entropy as the loss function and Adam as the optimizer. We measure accuracy as a metric.

```
pred = model.predict_classes(embedding_matrix_test)
accuracy = float(sum(pred==Y_test)) / embedding_matrix_test.shape[0]
```

The following code prints the test dataset text and its predicted emoji by our classifier:

```
for ix in range(embedding_matrix_test.shape[0]):
   if pred[ix] != Y_test[ix]:
      print(ix)
      print(test[0][ix], end=" ")
      print(emoji.emojis(emoji_dict[pred[ix]], use_aliases=True), end=" ")
      print(emoji.emojize(emoji_dict[Y_test[ix]], use_aliases=True))
   "output_type": "stream",
   "text": [
    "6\n",
    "['I', 'am', 'upset'] □
    ♥n", "7\n",
    "['We', 'had', 'such', 'a', 'lovely', 'dinner', 'tonight'] □ ♥n",
    "11\n",
    "['work', 'is', 'hard'] □ □\n",
    "12\n",
    "['This', 'girl', 'is', 'messing', 'with', 'me'] ♥□\n",
    "13\n",
    "['are', 'you', 'serious', 'ha', 'ha'] □ □\n",

    "['you', 'brighten', 'my', 'day'] ♥□\n"]
```

Display the confusion matrix.

```
import itertools
from sklearn.metrics import confusion_matrix
conf_matrix = confusion_matrix(Y_test, pred)
print("Confusion Matrix")
print(conf_matrix
 ) "text": [
    "Confusion Matrix\n",
    "[[ 8 0 3 1 0]\n",
    " [ 0 5 0 0 0]\n",
    " [ 2 2 12 2 0]\n",
    " [ 3 2 2 7 1]\n",
    " [ 0 0 0 0 6]]\n"
   ]
```

Figure 7 confusion matrix for emoji prediction, when visualised, looks like this.

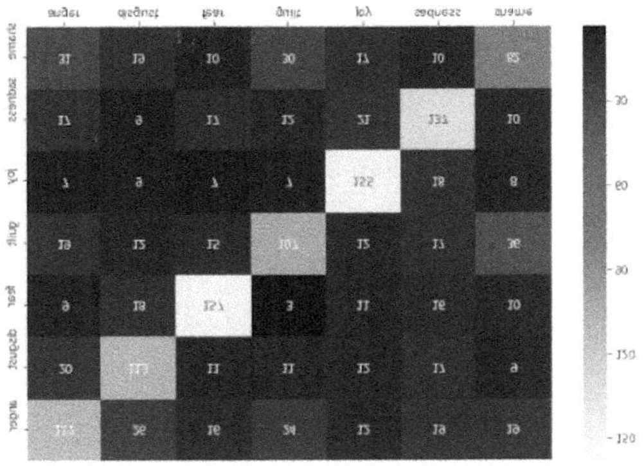

Fig. 7. Confusion matrix for emoji prediction

Figure 5 illustrates the loss and accuracy of our model on the dataset. This plot shows the loss and accuracy metrics clearly (Fig. 8).

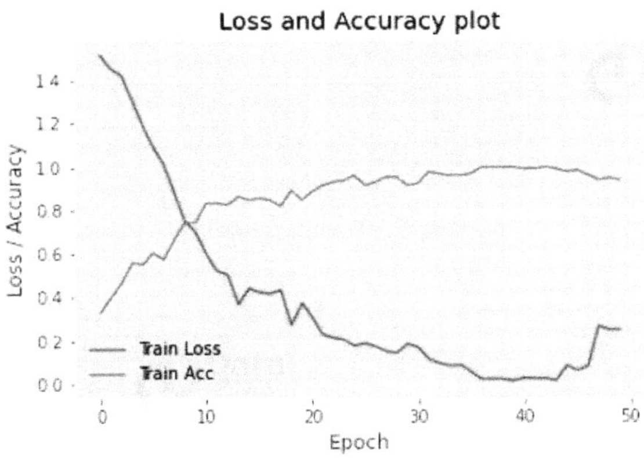

Fig. 8. Loss and Accuracy plot

In our methodology, we begin by importing datasets specifically for training, testing, and validation. The training dataset consists of 16000 rows, while both the test and validation datasets contain 4000 rows each, all previously reprocessed and labeled with various emotions. While most studies focus solely on positive and negative feelings within datasets, this paper incorporates a wider spectrum of emotions.

Table 1 represents the positive and negative emotions with numbers. Here, the feeling of sadness is higher than that of other feelings (Fig. 9).

Table 1. Total Number of Emotions

Dataset	Positive			Negative		
	Joy	Love	Surprise	Fear	Sadness	Anger
Train	5605	2156	650	2061	4698	2286
Test	739	248	76	243	582	291
Validation	730	293	96	225	553	288

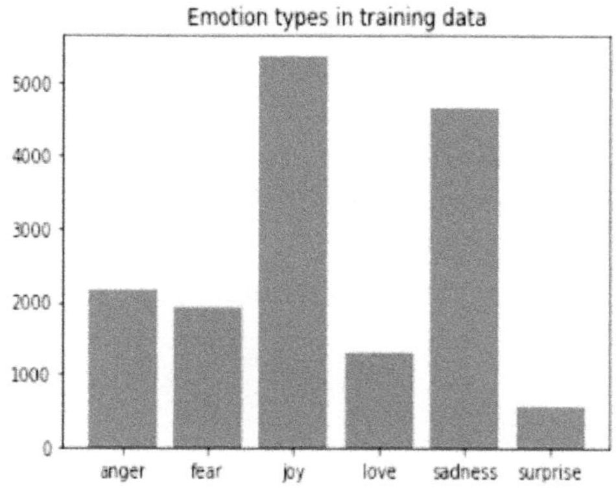

Fig. 9. Training data

10 Prediction

This sentence classification method examines the input sentence and decides whether the underscored emotion is positive or negative, as well as the liability associated with certain +ve or −ve statements. The intensity of a +ve or −ve argument is represented by probability. The model's delicacy is 86.68 per cent while using the LSTM networks algorithm working with Fig. 7. Value vs. Accuracy graphical representation of accuracy (Fig. 10).

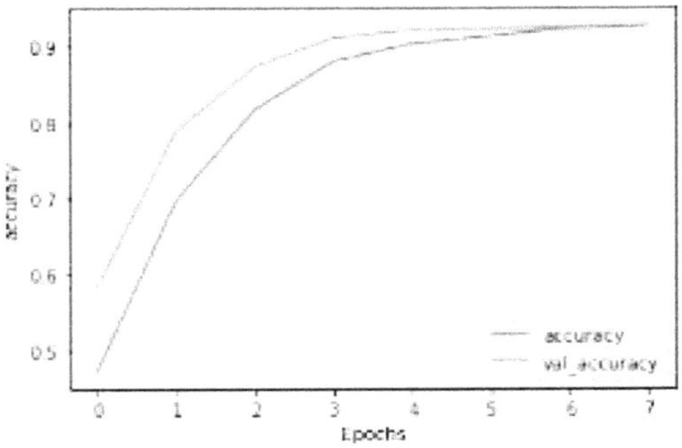

Fig. 10. Value vs. Accuracy graphical representation of accuracy.

Testing meets validation accuracy at six epochs, so model accuracy peaks at these epochs (Fig. 11).

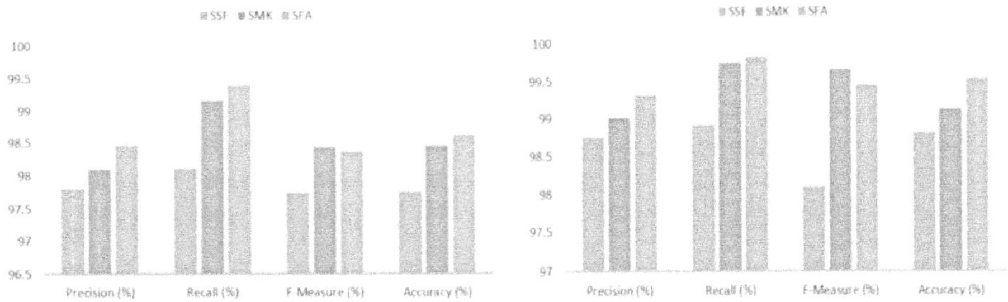

(a) CNN (b) CNN-LSTM

Fig. 11. Value vs. Accuracy graphical representation of accuracy.

11 Result and Future Work

This model begins with training data followed by emotion prediction. It classifies the dataset based on this training. To prevent overfitting, a large dataset and clear training process are essential. The model learns classification during training, and its performance is assessed with a test set.

I am feeling bad Message:[' I am feeling '] Emotion: Sadness □

12 Conclusion

In the proposed system, we recognised emotions without facial expressions and voice modulation, and we achieved an analysis rate of 92.8% using LSTM.

Future Work

We plan to extend our project in the future by recognising emotions in emojis, voice messages, and videos. We also plan to increase accuracy further and decrease the error rate.

References

1. Zhang, H., Jolfaei, A., Alazab, M.: A face emotion recognition method using convolutional neural network and image edge computing. IEEE Access **7**, 159081–159089 (2019). https://doi.org/10.1109/access.2019.2949741
2. Karpov, N., Lyashuk, A., Vizgunov, A.: Sentiment analysis using deep learning. In: Kalyagin, V.A., Pardalos, P.M., Prokopyev, O., Utkina, I. (eds.) NET 2016. SPMS, vol. 247, pp. 281–288. Springer, Cham (2018). https://doi.org/10.1007/978-3-319-96247-4_20
3. Chen, C.-H., Lee, W.-P., Huang, J.-Y.: Tracking and recognizing emotions in short text messages from online chatting services. Inf. Process. Manage. **54** (2018). https://doi.org/10.1016/j.ipm.2018.05.008
4. Kumawat, S., Yadav, I., Pahal, N., Goel, D.: Sentiment analysis using language models: a study. In: 2021 11th International Conference on Cloud Computing, Data Science & Engineering (Confluence). IEEE (2021)
5. Raisa, J.F., Ulfat, M., Al Mueed, A., Reza, S.M.S.: A review on twitter sentiment analysis approaches. In: 2021 International Conference on Information and Communication Technology for Sustainable Development (ICICT4SD). IEEE (2021)
6. Khan, L., Amjad, A., Ashraf, N., Chang, H.-T., Gelbukh, A.: Urdu sentiment analysis with deep learning methods. IEEE Access **9**, 97803–97812 (2021). https://doi.org/10.1109/access.2021.3093078
7. Liu, Y., Lu, J., Yang, J., Mao, F.: Sentiment analysis for e-commerce product reviews by deep learning model of Bert-BiGRU-Softmax. Math. Biosci. Eng. **17**, 7819–7837 (2020). https://doi.org/10.3934/mbe.2020398
8. Vinston Raja, R., Ashok Kumar, K.: Financial derivative features based integrated potential fishing zone (IPFZ) Future forecast. J. Intell. Fuzzy Syst. **45** (2023). https://doi.org/10.3233/jifs-231447
9. Srivastava, S., Nagpal, A., Bagwari, A.: Various approaches in sentiment analysis. In: 2020 12th International Conference on Computational Intelligence and Communication Networks (CICN). IEEE (2020)
10. VinstoncSend mail to Raja R.V. Innovative time series-based ECG feature extraction for heart disease risk assessment
11. Devi, S., Saran, M., Yadav, R.K., Maurya, P., Tripathi, U.N.: An optimized approach for emotion detection in real- time for twitter sentiment analysis with natural language processing. Int. J. Innovat. Res. Eng. Manage. **11** (2024). https://doi.org/10.55524/ijirem.2024.11.3.19
12. Subramanian, R.R., Akshith, N., Murthy, G.N., Vikas, M., Amara, S., Balaji, K.: A survey on sentiment analysis. In: 2021 11th International Conference on Cloud Computing, Data Science & Engineering (Confluence). IEEE (2021)
13. Hu, M., Liu, B.: Mining and summarizing customer reviews. In: Proceedings of the teNth ACM SIGKDD International Conference on Knowledge Discovery and Data Mining. ACM (2004)
14. Chikersal, P., Poria, S., Cambria, E., Gelbukh, A., Siong, C.E.: Modelling public sentiment in twitter: using linguistic patterns to enhance supervised learning. In: Computational Linguistics and Intelligent Text Processing, pp 49–65. Springer International Publishing (2015)

15. Miles, C., et al.: Correlator convolutional neural networks as an interpretable architecture for image-like quantum matter data. Nat. Commun. **12** (2021). https://doi.org/10.1038/s41467-021-23952-w
16. Vinston Raja, R., Ashok Kumar, K.: Fisher scoring with condition-based ensemble supervised learning classification technique for prediction in PFZ. J. Uncertain Syst. **15** (2022). https://doi.org/10.1142/s1752890922410094
17. Günalan, E., Turhan, S., Çavak, B.Y., Cebioğlu, İK., Çonak, Ö.: The evaluation of videos about branched-chain amino acids supplements on YouTube™: a multi-approach study. Int. J. Environ. Res. Public Health **19** (2022). https://doi.org/10.3390/ijerph192416659
18. Automatic Identification of Hurricane Damage Using a Transfer Learning Approach with Satellite Images | International Journal of Intelligent Systems and Applications in Engineering
19. Study of ECG Analysis based Cardiac Disease Prediction using Deep Learning Techniques | International Journal of Intelligent Systems and Applications in Engineering
20. Khamphakdee, N., Seresangtakul, P.: An efficient deep learning for thai sentiment analysis. Data **8** (2023). https://doi.org/10.3390/data8050090
21. Hochreiter, S,.: The vanishing gradient problem during learning recurrent neural nets and problem solutions. Int. J. Uncert. Fuzz. Knowl.-Based Syst. **06** (1998). https://doi.org/10.1142/s0218488598000094
22. Bibi, M., et al.: Class association and attribute relevancy based imputation algorithm to reduce twitter data for optimal sentiment analysis. IEEE Access **7**, 136535–136544 (2019). https://doi.org/10.1109/access.2019.2942112
23. Almagambetov, A., Velipasalar, S., Casares, M.: Robust and computationally lightweight autonomous tracking of vehicle taillights and signal detection by embedded smart cameras. IEEE Trans. Industr. Electron. **62** (2015). https://doi.org/10.1109/tie.2015.2400420
24. Diwan, T., Tembhurne, J.V.: Sentiment analysis: a convolutional neural networks perspective. Multimedia Tools Appl. **81** (2022). https://doi.org/10.1007/s11042-021-11759-2

Applying Inception-V3 for Skin Disease Classification: A Deep Learning Approach Using the ISIC Dataset

K. P. Aishwarya, P. Bharath, Edwin Mathews, R. Monika(✉), and B. Nidharshana

Department of AI and DS, Sri Sairam Engineering College, Chennai, Tamil Nadu, India
aishwarya.ai@sairam.edu.in, {sec21ad057,sec21ad010,sec21ad084,
sec21ad003}@sairamtap.edu.in

Abstract. Early identification of skin diseases is crucial for effectual patient management along with treatment, yet traditional diagnostic methods often face limitations in consistency and speed. This manuscript traverses on the application of Inception-V3, a Deep learning technique aimed at automated skin disease prediction using ISIC which is the dataset, containing diverse and annotated skin lesion images. Inception-V3, renowned for its multi-scale feature extraction capabilities and efficient use of computational resources, is employed to intensify the accomplishment of automated skin disease diagnosis. Our architecture of Inception-V3 is fine-tuned to accommodate the precise characteristics concerning skin images, involving modifications to the final layers of network and hyperparameters to optimize performance for this classification regarding various skin conditions. Our experiments reveal that Inception-V3 achieves notable advancements in this classification precision compared to baseline models. This model's performance is assessed through various metrics such as precisions, F1- scores, confusion matrices, and recall, demonstrating its potential for reliable skin disease classification. We also address challenges encountered during the implementation, including class imbalance and overfitting, and discuss strategies employed to mitigate these issues.

Keywords: Convolutional neural network · Machine learning · Inception · Skin disease prediction · Softmax · Tensorflow

1 Introduction

Deep learning is a subset of artificial intelligence, employing various statistical models which use data as a primary source for classifying, recognizing patterns and autonomously making anticipations or decisions even to the extent that they are not programmed to do so. It has various applications in practice, including video and audio recognition, language processing, content recommendation, as well as in identifying fraud, impacting various sectors like finance, healthcare, e-commerce, and marketing. Advances in computing power, data storage, and processing techniques have accelerated the development of ML, which continues to make progress with new models and applications coming out frequently.

R. Geetha et al. (Eds.): AAIMB 2024, CCIS 2654, pp. 210–224, 2026.
https://doi.org/10.1007/978-3-032-13029-7_15

In India, the healthcare sector has seen significant advancements but still faces challenges such as inadequate infrastructure, limited access to services, and a shortage of healthcare professionals. Machine learning and AI offer solutions to these issues by enhancing diagnostic accuracy, detecting early disease indicators, predicting treatment outcomes, and improving overall efficiency in healthcare delivery.

Skin diseases, including various different types of cutaneous conditions including nevus, melanoma, keratosis and rosacea hold considerable health challenges globally. Early and accurate discovery is critical for essential treatment and better patient results. Traditional characteristic methods often depend on medical examination and histopathological analysis, which, while effective, can be rigorous and susceptible to human error. This increasing availability of image synthesis and evolution in machine learning algorithms provide new opportunities for automating and enhancing the diagnostic process.

Recent developments in deep neural network technology, particularly CNN confirmed noteworthy achievement through image classification type tasks, including skin image analysis. Among these, inception- V3 stands out due to its innovative architecture, which combines deep convolutional layers with multiple resemblant adulterants to internee features at different scales. This design let inception- V3 to achieve high delicacy while keeping computational effectiveness, by accomplishing an ideal candidate to complex type problems analogous to skin complaint discovery. Our ISIC dataset serves as a precious resource on training and assessing skin complaint bracket models. It comprises a large collection of affected areas of skin, representing a wide diapason of dermatological conditions. Our dataset provides a foundation for developing and validating ML algorithms, as it encompasses a diverse selection of skin types, complaint stages, and image rates.

In this manuscript, we traverse the operation of Inception - V3 for skin disease classification using the ISIC dataset. Our ideal is to examine the model's efficacy in directly distinguishing different cutaneous conditions and to tackle challenges associated with cutaneous image analysis. We begin by detailing the preprocessing way needed to prepare the dataset, including data addition ways to intensify the model's strength. We also describe the variations made to the commencement- V3 armature to optimize its performance for skin complaint brackets.

Through comprehensive trial and evaluation, we aim to demonstrate the eventuality of Inception - V3 in perfecting individual delicacy and effectiveness. Our results will impart to the developing body of exploration on automated dermatological results, offering perceptivity into the strengths and limitations of using deep learning technology for skin disease detection. Also, we will bandy the counter accusations of our findings for unborn exploration and the expansion of practical tools for dermatologists.

2 Related Works

The classification of cutaneous conditions is a popular exploration area with numerous researches which focuses on developing machine learning models on precise classification of skin conditions. Also there are numerous recent literature checks on detection of skin diseases.

Rimi T et al., in their study titled "Derm NN: Skin Disease Detection Using Machine Learning Creed, Convolutional Neural Network," demonstrated the efficacy of Convolutional Neural Networks (CNNs) in classifying skin lesions. The research highlighted that CNN-based models could achieve diagnostic accuracy comparable to that of dermatologists. By utilizing a robust dataset comprising over 100,000 skin lesion images, the model exhibited outstanding performance, particularly in melanoma detection. This study underscores the potential of CNNs in dermatological applications, showcasing their ability to analyze complex image features and provide reliable predictions. The findings emphasize the role of AI in enhancing early detection and diagnosis of critical skin conditions, paving the way for scalable and accessible healthcare solutions. The limitation faced was that the accuracy was better only to limited number of datasets [1].

According to the study which was carried out by Waweru et al. (2020), they were able to create a system that could detect skin cancer and carcinoma with the use of DenseNet201 architecture multitask. It applies DenseNet201 to a certain physiognomic of deep convolutional neural networks which is favored by many for its ability to perform categorical feature extraction due to feature reuse across various layers of the network. This helps to both enhance the models capacity to detect minute changes in the skin lesion image and increase the classification accuracy. When multitask learning is added to this problem it was also beneficial as the model was able to perform more than one task such as lesion segmentation and classification at the same time. Because the tasks were closely related seamless feature transfer was employed so that the models accuracy and robustness were improved in the diagnosis of the several varieties of skin cancers which include melanoma and basal cell carcinoma. This multitask learning framework also helped reduce the problem of overfitting and improved the generalization of the model to

unseen data. The system has considerable usefulness in supplementing strategies for the pores and skin cancer detection since it provides an integrated approach for detecting not only the type of lesion but also the part where it is detected. Thus the findings of this study will suffice the developing trends in automated skin cancer diagnostics. The DenseNet architecture contains numerous layers which affects the overall computation time unlike Inception V3 model [2].

The research work by Prakriti et al. (2020) examined the application of Fast R-CNN in the prediction of skin diseases with a particular emphasis on the model's ability to detect and classify some skin diseases. An effective skin lesion demarcation and classification process were achieved through the use of Fast R-CNN, which is an object detection model. This method incorporates the feature learning aspects of convolutional neural networks as well as region based convolutional networks to enhance the speed and precision of detecting skin related diseases. The model successfully differentiates between the various

types of skin malady indicating a propitious clinical application of this model where a fast and accurate diagnosis of skin disease is needed. As a result of the new feature of focusing on notable areas of the image, the processing time and the classification success rate were remarkably increased, which satisfactorily met the needs of real-time skin disease detection. This study offers insights into the ongoing work done aimed at automating the diagnosis of skin diseases. The results also affirm the use of complex deep learning methods in enhancing diagnostic accuracy [3].

The research by Roy et al. (2020) aimed at the use of segmentation techniques in skin disease imagery, focusing more specifically on the use of image analysis of skin diseases from the data set. They employed sophisticated algorithms to the previously specified methods in order to remove overlying images of skin lesions and thus facilitate the analysis of the characteristics of the lesions more accurately. By focusing on specific details, the model was able to discriminate various skin disease conditions with greater accuracy. The segmentation methods developed and addressed in this study were important in improving the performance of the model since they enhanced the capability of handling noise and background information in the images. This suggested that image processing techniques and feature extraction methods are critical in enhancing the performance of skin disease diagnosis systems thereby adding to the knowledge base in dermatology image analysis. The study showed that image processing complemented by machine learning approaches can be effective for the classification of several skin conditions, and thus automated diagnostic systems will be more advanced and accurate [4].

The paper by Chakraborty et al. (2020) aimed at the diagnosis of skin ailments employing a single metaheuristic approach combined with ANN for effective image classification. The study employed the use of ANN to distinguish between various skin ailments after optimizing the feature selection through the use of metaheuristic approaches such as genetic algorithms or particle swarm optimization. This technique was used in classifying images of skin lesions, particularly angioma, lentigo simplex and one more's images. The results also showed that the use of metaheuristic strategies improved the performance of the network in that it enhanced its capacity to remember the essential pointers in the skins of the images and their efficiency in terms of the time spent in diagnosis. [5].

Kolkur et al. (2020) concern themselves with a comprehensive pattern driven feature extraction approach for skin disease detection. The study sheds light on the importance of recognizing certain distinctive patterns associated with skin lesions for better diagnostic accuracy. Hence, it is recommended that these pattern-based properties may lead to better classification of different types of skin problems. To this end, certain skin photograph patterns are intended to facilitate the differentiation between different types of cutaneous diseases. This special methodology also creates opportunities for advanced feature extraction techniques as additives to current skin disease diagnosis methods because it helps the more detailed and systematic analysis of dermatological images. The results of the study also help move forward development of dermatology in the area of automated skin disorders diagnosis by pattern recognition for improvements in classification systems [6].

The research paper published by Navarro et al. (2020) aims at the detection and clustering of skin diseases by making use of the enhanced Bag of Features (BoF) paradigm. The investigation employs the Speeded-Up Robust Features (SURF) algorithm in feature engineering, a process known for its efficacy in identifying and extracting interesting and salient features from skin images. Afterward, the K-means clustering algorithm is applied so that visually-comparable images of skin lesions are grouped with other similar images into clusters.

Such integration in the schemes of Surge and K-means enhances the capability of the system to view differences and similarities among different forms of skin disorders, thereby improving classification and identification of new forms of patterns in the dataset being used. This method presents a new way of classifying skin diseases as well as increasing the degree of precision and relative ease of detecting skin disorders and even finding new or rare ones [7].

The research done by Anthal J. (2020) centers on the recognition of the non segmental type of vitiligo through deep learning based approaches. It is possible through this research to determine the affected areas of skin and characterize the areas affected by vitiligo (depigmented skin). The LVQ algorithm along with its focus of Classification is integrated within a multilayer perceptron in order to boost classification performance of the target skin regions, through skin vitiligo Weichen power neural machine. This study provides an architecture in which such a system (LVQ-MLP) can be used to train on images of skin lesions and then performs faster detection of vitiligo then localization of the affected body parts allowing for timely treatment of the disease. The method proves that complex pattern recognition can be limited to spatially restricted regions which helps to streamline vitiligo in clinical practice [8].

According to the study of Rathod, another interesting aspect of the use of Convolutional Neural Networks (CNNs) that merits further attention is their application to the early diagnosis or prognosis of skin diseases. In this study, CNNs are used for the automatic selection of features from images of skin and therefore eliminate the need for defining the relevant features manually. This approach will greatly improve the precision of the task while reducing the time taken to carry out the automated interventions. The study emphasizes the possibilities of the use of CNNs for practical tasks in the field of Dermatology, which will help to efficiently and quickly detect and classify a wide range of skin diseases [9].

According to the work of Suganiya Murugan et al. (2020. The objective of their work is to recommend and treat various types of skin diseases by using appropriate types of machine learning systems. This study examines the use of image data and clinical details to classify skin diseases using several machine learning models. The increasing changing nature of the data as well as the desire to increase the accuracy of the ML model skin disease prediction which is the second aim of the study which will help in the case that the model is able to tell the different types of skin diseases above. The practical part of the study includes the description of treatment and cure recommendation approaches based on the specific diagnosis made, which enables a more integrated view starting from diagnosis to treatment. This particular work demonstrates the use of ML in detection of skin diseases but also suggests how the ML can help in dealing with treatment suggestions which would help a practitioner MD in taking better decisions regarding the patients [10].

According to Suparyati et al. (2020), their study predicted lumpy skin disorders with the guidance of Random Forest Algorithm. The study also employs Random Forest, an ensemble machine learning technique to classify and predict the occurrence of lumpy skin disorders based on clinical data and image data features from various perspectives.

The study utilized the Random Forest algorithm which is appropriate for high dimensional problems since it can handle complexity by reducing the dimensionality of the problem space by extracting key descriptive features of the skin including the texture, shape and size to assess the magnitude of lumpy skin disorder in a person. The study explains how this technique can be used for effective screening in order to provide enhanced opportunities for correct diagnosis and useful approaches for the detection and categorization of such skin diseases [11].

Trstenjak et al. have investigated the possibility of predicting the risk of skin cancer with the help of a mobile application utilizing an Enembedded Convolutional Neural Network (CNN). This paper describes the design of a mobile application intended to predict the type of skin lesions with the help of photographs that have been uploaded into it. This application has become widely available which allows many people all over the world to determine their risk for skin cancer. The mobile version of the application guarantees adequate performance and accuracy of CNN in the classification of lesions. The paper also highlights the importance of embedded artificial intelligence technologies in improving the time of diagnosing and in developing portable methods of diagnosis of skin-related diseases [12].

Mauro Picardo and Co-authors in their work "Detection and feature extraction of skin disease and microbiome", emphasize the co occurrence of skin disorders and commensal microbiota. A number of points are raised, especially how the architecture of microbial parasites and their host and environmental factors might steer the complex formation of not directly infectious acquired dermatological diseases and conditions. Microbial factors however, specifically imbalance and dysbiosis are believed to be the general causative elements that disrupt homeostatic mechanisms of the skin and set forth disease processes. These things provide additional evidence on the importance of microbiota in understanding health and disease and in this case, skin disorders [13].

Maryamah et al attempt to make use of deep learning approaches in the area of fashion analysis. The first methodology consisted of data scraping from official websites with the aim to create a complete and high quality dataset for the analysis. The research used the model Inception V3 which is one of the best architectures for deep convolutional neural networks and image collocation and feature extraction techniques to locate and recognize structures and attributes in fashion. This approach clearly not only confirmed the effectiveness of Inception V3 in dealing with large scale image datasets but also showed its capacity to reconstruct numerous details such as the patterns, colours, and design elements. The study emphasizes the need of advanced AI to be deployed for more developed applications of the fashion spectrum, variation spotting, auto tagging, and suggestion mechanisms [14].

The research entitled "Deployment of Web Application for AI Skin Disease Prediction" was carried out by Manasvi Srivastava and others and aimed at the use of web technologies and artificial intelligence for enhancing the diagnosis of skin diseases. The research focused on devising methods for scraping websites to gather the requisite information for the datasets needed to train the predictive models. The study incorporated the real time use of the web-based application whereby users could upload images of their motorcycle for diagnosis and prediction of results using the relevant AI algorithms [15].

3 Methodologies

This particular study deals with applying the Inception-V3 deep learning architecture on skin disease classification using the ISIC dataset. The methodologies employed are designed to prepare the data, train the model, and evaluate its performance effectively. Below, we detail each component of our approach.

3.1 Data Preparation

Dataset Description: The ISIC dataset comprises thousands of annotated skin lesion images, representing various dermatological conditions including nevus, keratosis and acne. Every image is accompanied by a label indicating the presence and type of skin disease.

Preprocessing: Prior to providing the input data into the model, several processing steps are undertaken:

Resizing: Images are resized to the input dimensions required by Inception-V3 (299 × 299 pixels) to maintain consistency and ensure compatibility with the network.

Normalization: Pixel values are regularized to a range of [0, 1] by splitting by 255. This step improves our model's confluence throughout training by guaranteeing that all input values are on an analogous scale.

Data Augmentation: To improve model validity and generalization, data augmentation including rotation, flipping, scaling, also color adjustments are applied. This assists in pretending variations in real-time scripts and helps overfitting.

3.2 Model Architecture

Inception-V3 Overview: Inception-V3 is a CNN based algorithm noted for its correctness and performance in image classification. This network contains various layers such as convolutional layer, dense layer, pooling layer, and inception modules, where multiple filter sizes are usd and capture different levels of feature extraction at once. This architecture is designed to balance depth and computational complexity effectively.

Model Adaptation: To tailor Inception-V3 for skin disease classification, we utilize a trained model on the dataset to leverage the model's learned features from the dataset, which are fine-tuned to recognize specific patterns in skin images where the final layers are modified to equal with the subclasses count contained in the data source. Dense layer can be replaced by a softmax activation for multi-class classification.

3.3 Training Procedure

Training Setup: This model has been trained by supervised learning technique with these following settings:

Loss Function: For this case, the model has been trained to use categorical cross-entropy loss function which is appropriate for several class outputs.

Optimizer: The adaptive learning rate method is used and the Adam optimizer is used for a faster convergence.

Learning Rate: A default learning rate is set and it is changed during model training in order to minimize overfitting and achieve faster learning.

Epochs and Batch Size: Our model has been trained for a fixed number of epochs with certain batch size based on computational resources and size of the dataset.

Validation and Testing: This dataset has been splitted into various subsets such as training, validation, and also testing where the test set is mainly employed for supervising the execution of the model and fine tuning hyperparameters, while the test subtest assesses the final model's generalization ability (Fig. 1).

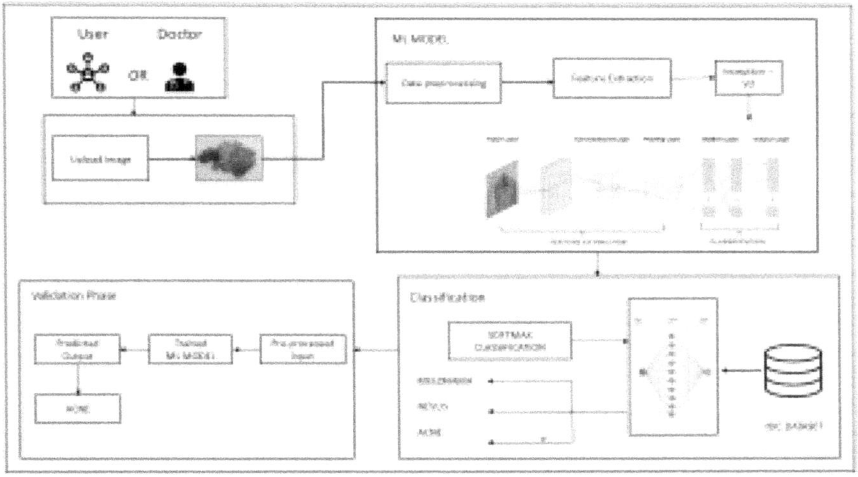

Fig. 1. Inception Model Architecture

3.4 Evaluation

Performance Metrics*:* Model implementation is guesstimated using several metrics: *Accuracy:* Measures the segment of correctly distinguished images among the total number of images.

Metrics: The metrics used to anticipate a detailed efficacy of the model's accomplishment are precision, recall and F1score.

Confusion Matrix: A confusion matrix is created to view the model's clustering efficiency across multiple categories.

Error Analysis: An analysis of misclassified images is performed to understand common patterns or issues that may contribute to classification errors. This analysis informs potential improvements to the model and preprocessing steps.

4 Algorithm Used

Inception-V3 is a special CNN technique which is intended for elevating image classification tasks through modular and efficient feature extraction. It integrates various mathematical techniques to optimize learning and performance.

4.1 Core Concepts

Convolutional Layers: Convolutional layers apply filters to input images to generate convolutional feature maps which are also called activation maps. The convolution operation is mathematically defined as:

$$y(i.j) = \sum_{m=0}^{k_h-1} \sum_{n=o}^{k_w-1} X(i+m, j+n) \cdot K(m, n)$$

where PPP is the input image, KKK is the filter applied which can also be the kernel, sss is the size of the filter, and (m,n)(m, n)(m,n) on the feature map represents the position.

Inception Modules: Inception modules use multiple types of convolutions in parallel to capture diverse features. Each branch in an inception module applies a different filter size. The results of these branches are concatenated along the depth dimension.

Factorization: Factorization is used to reduce computational complexity. For instance, a 4x4 convolution can be reduced into 2 × 2 convolutions (Figs. 2 and 3):

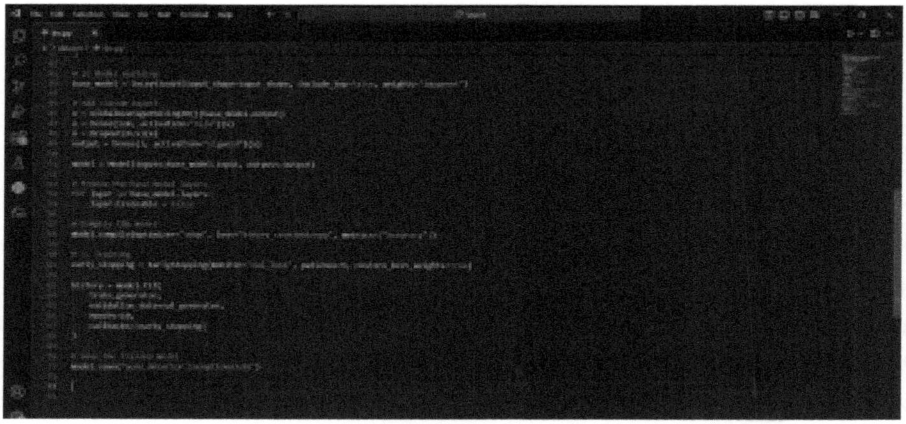

Fig. 2. Code for Model Training

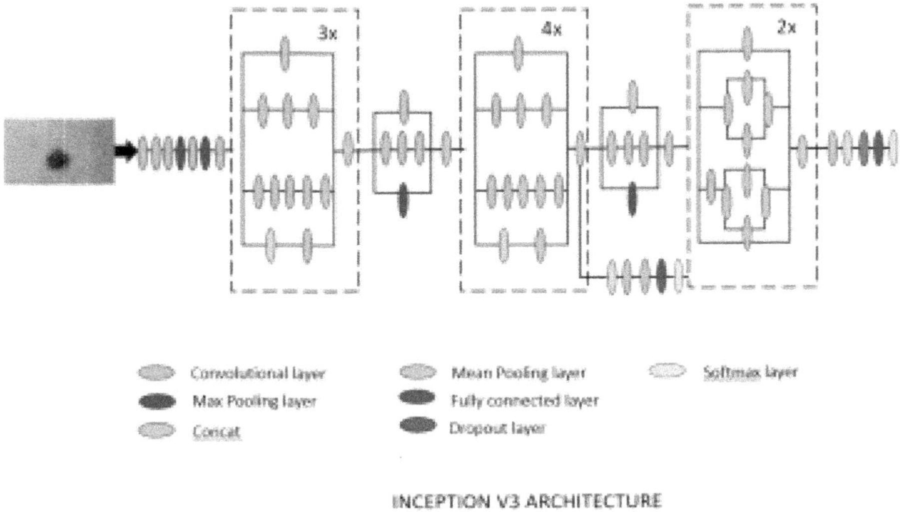

Fig. 3. Inception V3 architecture

4.2 Architecture Components

Stem: The stem consists of initial convolutional layers that process the image. The convolution operation in the stem is typically followed by max-pooling:

$$\text{MaxPooling}(x, y) = \max(I(x, y))$$

where dimensions of these feature maps are reduced by the process of Max-Pooling by consideration of the highest value in every particular region of the feature map.

Inception Modules: Inception modules are composed of several branches:
1 × 1 Convolution
3 × 3 Convolution
5 × 5 Convolution
4 × 4 Max-Pooling Convolution.

Auxiliary Classifier: The auxiliary classifier is a secondary network added to intermediate layers to improve training stability. It is defined as:

$$Y_c(i,j) = b_c + \sum_{d=1}^{D_{in}} \sum_{m=1}^{k_h} \sum_{n=1}^{k_w} X_d(s \cdot i + m - p, s \cdot j + n - p) \cdot K_{c,d}(m, n)$$

where the Dense layer outputs logits, and Softmax converts these logits into probabilities.

5 Result and Analysis

An image of a person diagnosed with Acne is given as input for analysing the efficacy of the model and it predicts the issue of the skin (Fig. 4 and Table 1).

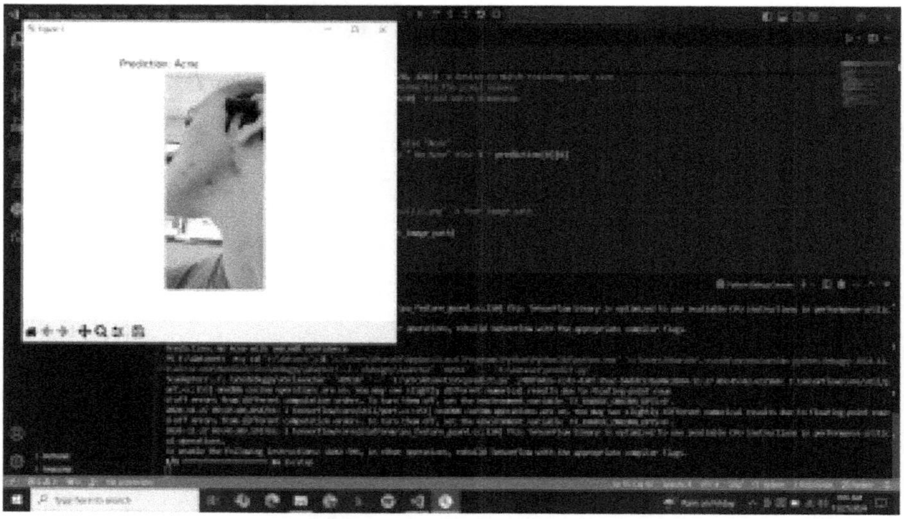

Fig. 4. Predicted Output

Table 1. Confusion Matrix

Binary Cross-entropy Loss	0.2271
Confusion Matrix	[[3,0],[0,4]]

The following features were observed in the input image during feature extraction stage (Fig. 5)

- Presence of pustules and pimples.
- Inflammation of skin and redness in certain areas.
- Visible blackheads and whiteheads.

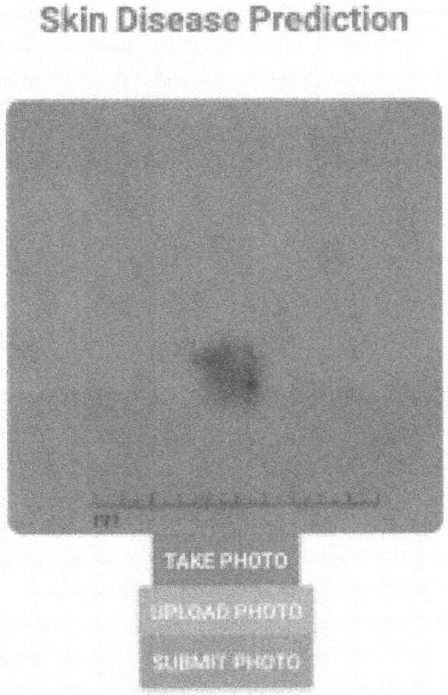

Fig. 5. The image is given to the model for prediction and it displays the result.

6 Application Interface

This novel approach was attempted to make the skin disease prediction application UX outstanding and easy to use. The home screen is simple and has a neat layout with buttons for image uploading, old reports viewing and learning. Users can also take good quality photographs of their lesions, which are later examined and strong predictions including the fact as to what skin diseases are most probably present are provided. The results page manages to display the findings alongside the recommendations and they're in most instances trouble free. The app provides students with materials concerning skin diseases, it traces old predictions and observes other security measures relevant to users data. This interface guarantees equal ease and accuracy of AI-based accurate predictions (Fig. 6).

Fig. 6. The affected skin image is uploaded in the application.

7 Future Scope

The application of Inception-V3 for skin disease classification opens several avenues for further research and improvement:

Enhanced Model Architectures: Exploring advanced models such as Inception-V4 or EfficientNet could provide improved accuracy and efficiency. Additionally, hybrid models that integrate other machine learning techniques or attention mechanisms might offer enhanced feature extraction capabilities.

Data Diversity and Augmentation: Incorporating multi- modal data, including clinical metadata and patient demographics, could enhance model performance and expand the dataset with more diverse images.

Model Interpretability and Explainability: Enhancing the interpretability about the model is important for clinical acceptance. Implementing explainable AI techniques and developing decision support systems can help healthcare professionals better understand and trust the model's predictions.

Clinical Validation and Deployment: Conducting prospective clinical trials and combining the model into existing medical systems are essential for real-world validation and adoption. Ensuring user-friendly interfaces and interoperability with electronic health records (EHRs) will facilitate practical deployment.

8 Conclusion

Inception-V3 has demonstrated strong performance in skin disease classification, effectively leveraging its sophisticated convolutional architecture to analyze and identify various dermatological conditions with high accuracy. Utilizing the ISIC dataset, this study highlights the model's capability to enhance diagnostic processes by providing reliable and automated support for dermatologists.

The results emphasize greatly on the eventuality of deep learning in advancing dermatological insights and increasing the overall percentage of patient care. However, to fully realize the benefits of such technologies, subsequent studies should focus on several key areas: exploring more advanced neural network architectures, incorporating diverse and multi-modal data sources, and enhancing model interpretability to ensure transparency and trust in clinical settings. Additionally, real-world validation through clinical trials and integration into existing healthcare systems are crucial for practical application.

Addressing ethical considerations and ensuring data privacy will also be vital in the responsible deployment of these technologies. Overall, continued innovation and refinement in these areas will help in creating more effective, well organized, and equitable diagnostic tools, ultimately contributing to better health outcomes and more accessible dermatological care.

References

1. Sultan, R.T., Ahmed, F., Derm, N.N.: Skin disease detection using convolutional neural network. In: 2020 4th International Conference on Intelligent Computing and Control Systems (ICICCS)
2. Waweru, A., Miao, K., Kawan, P.: Deep learning in skin lesion analysis towards cancer detection. In: 2020 24th International Conference Information Visualisation (IV)
3. Prakriti, D., Akbar, A., Amit, G., Subodh, D.: A deep learning based approach for automated skin disease detection using Fast R-CNN. In: 2021 Sixth International Conference on Image Information Processing (ICIIP)
4. Roy, K., Chaudhuri, S., Ghosh, Dutta, S., Chakraborty, K., Sarkar, R.: Skin Disease detection based on different Segmentation Techniques. In: 2019 International Conference on Opto-Electronics and Applied Optics (Optronix)
5. Chakraborty, S., Mali, K., Chatterjee, S., Banerjee, S., Mazumdar, K.G., Debnath, R.: Detection of skin disease using metaheuristic supported artificial neural networks. In: 2017 8th Annual Industrial Automation and Electromechanical Engineering Conference (IEMECON)
6. Kolkur & Kalbande: Survey of texture-based feature extraction for skin disease detection. In: 2016 International Conference on ICT in Business Industry & Government (ICTBIG)
7. Navarro, M.C., Bustillos & Barfeh: Skin disease analysis using digital image processing. In: 2019 International Conference on Computational Intelligence and Knowledge Economy (ICCIKE)

8. Anthal, J., Upadhyay, A., Gupta, A.: Detection of vitiligo skin disease using LVQ neural network. In: 2017 International Conference on Current Trends in Computer, Electrical, Electronics and Communication (CTCEEC)
9. Rathod, J., Sodha, A., Bhavathankar, P.: Diagnosis of skin diseases using CNN. In: 2018 Second International Conference on Electronics, Communication and Aerospace Technology (ICECA)
10. Murugan, S., Srividhya, S.R., Pradeep Kumar, S., Rubini, B.: A machine learning approach to predict skin diseases and treatment recommendation system. In: 2023 5th International Conference on Smart Systems and Inventive Technology (ICSSIT)
11. Suparyati, E.U., Muhammad, A.: Lumpy skin disease prediction based on meteorological and geospatial features using random forest algorithm with hyperparameter tuning. In: 2022 5th International Conference on Information and Communications Technology (ICOIACT)
12. Trstenjak, B.M.: Mobile application for skin cancer risk prediction based on embedded convolution neural network model
13. Mauro, P.M.D., Ottaviani, M.: Skin microbiome and skin disease. These complex microbial communities may contribute even in the development of noninfectious pathologies
14. Maryamah, M., Najma, A.A., Liviani, M.R.: Image classification on fashion dataset using inception V3
15. Srivastava, M., Yadav, V., Singh, S.: Implementation of web application for disease prediction using AI
16. Calistri, P., Kris, D., Simon, G., Eyal, K.: Lumpy skin disease. 2019 EFSA Journal
17. Juereanor, M., The, R., Arshad, A.: Data mining application in telehealth settings and disease prediction. 2022 Open University Malaysia
18. Lilia, R., Heni, B., Amel, L.: Skin genetic disease genes: patterns and prediction
19. Divya, N., Deepthi, P., Hariprasad: cureskin-Skin disease prediction using MobileNet Model. July 5th 2021 International Journal of Scientific research in computer science
20. Gomez, D., Montero, J., Yanez, J.: A Colouring algorithm for image classification
21. Anthony, H., Milad, B.: Impact of data augmentations when training the inception model for image classification
22. Murali, K., Hitendra, S.: A 3D-inception CNN for hyperspectral image classification

Real-Time Text Prediction and Speech Synthesis for Accessibility: A Bidirectional LSTM Approach

Sornavalli Ganesan(✉), Selcia Selvam, Sasmitha Baskaran, and Shashwat Shivam

Department of Information Technology, Sri Sivasubramaniya Nadar College of Engineering, Kalavakkam, Tamil Nadu, India
sornavallig@ssn.edu.in

Abstract. In this paper, we present a new approach to text generation using a Bidirectional Long Short-Term Memory (LSTM) network. The model is trained on articles from Medium, focusing on predicting the next word in a given text sequence based on an initial input. Before training, the data is carefully processed by tokenizing the text and creating pairs of inputs and outputs for the model. The trained LSTM model delivers highly accurate predictions, showcasing its effectiveness in natural language processing tasks. Additionally, we integrate the text prediction system with Google Text-to-Speech (gTTS) to convert the generated text into spoken language that closely resembles human speech. This combination of text generation and speech synthesis enhances accessibility for visually impaired individuals and creates a more engaging experience in voice-enabled applications. Our system achieves high accuracy, operates efficiently with minimal computational resources, and performs in real-time. These qualities make it a promising solution for assistive technologies and other applications requiring low-resource natural language processing.

Keywords: Text generation · Bi-LSTM · Speech synthesis · Natural Language Processing · Accessibility

1 Introduction

Modern natural language processing (NLP) systems, such as predictive text and speech synthesis, have transformed how humans interact with technology. These tools play an essential role in communication, from typing suggestions on smartphones to assistive devices for people with disabilities. However, despite significant advancements, there is still a gap in creating NLP systems that are efficient, easy to use, and work well in real-time, especially in settings with limited resources or for accessibility needs.

This study aims to address that gap by combining the predictive abilities of Bidirectional Long Short-Term Memory (Bi-LSTM) networks with the speech synthesis capabilities of Google Text-to-Speech (gTTS). Our system is designed to support individuals with visual impairments, offering a lightweight and responsive solution for text prediction and speech synthesis. By training the Bi-LSTM model on a collection of Medium

R. Geetha et al. (Eds.): AAIMB 2024, CCIS 2654, pp. 225–235, 2026.
https://doi.org/10.1007/978-3-032-13029-7_16

article titles, the system predicts the next word in a sequence and converts it into clear, natural-sounding speech using gTTS. This approach ensures smooth interaction, even on devices with limited processing power.

To further improve accessibility, the system integrates voice-based input through tools like SpeechRecognition and pyttsx3. This allows users to interact with the application by speaking, making it easier for those who rely on voice communication or have difficulty using traditional input methods. By combining these technologies, our solution provides a simple, effective, and inclusive way for users to generate text and hear it spoken aloud in real-time.

The goal of this study is to create an NLP tool that meets practical needs while enhancing accessibility. By focusing on resource efficiency and ease of use, we aim to make advanced technology available to underserved populations, creating a positive impact on their daily lives and improving how everyone interacts with technology.

2 Literature Review

Recent progress in text generation has been driven by transformer-based models. These models are great at working with long pieces of text and can be customized for specific purposes. For instance, researchers found that including topic-related details in the input makes it easier to create focused reviews, like ones about specific topics in Turkish. This shows how flexible and effective transformer models can be [10].

Transformers, introduced by Vaswani et al. in 2017, are now widely used for many language tasks, including generating text. Their key feature is the "attention mechanism," which helps them understand how different parts of a sentence relate to each other, even if they're far apart. This solves some issues that older models like RNNs and LSTMs struggled with when handling longer texts [12].

Bidirectional LSTM (BLSTM) models are also quite popular, especially in text-to-speech (TTS) systems. These models read text in both directions—forward and backward—helping them capture more context. This leads to more natural-sounding speech. Deep BLSTM models have even been used to create realistic talking head systems [2, 8]. They're also effective for real-time TTS, making speech tools more accessible for people with disabilities [1, 5].

Improving accessibility is a key goal for TTS systems. For example, they're essential for people with visual impairments, as they convert written text into speech. Researchers are working on adding emotion-aware features to make the speech more expressive and user-friendly [3, 7]. However, challenges remain, like improving real-time performance and managing complex text inputs [4].

Some researchers are combining classification models with language models to better control the way text is generated. This approach allows for more personalized content, like tailored reports or articles [13, 14].

Another focus area is improving how text is represented in these models. Traditional methods sometimes miss the deeper meaning of phrases, which can affect tasks like classifying or clustering text. To address this, researchers have introduced techniques like knowledge graphs and parsing trees to enhance understanding. This has significantly improved model performance in these areas [15].

There's also exciting progress in combining text generation with speech synthesis. These integrated systems produce more natural-sounding speech, which is helpful for virtual assistants and screen readers. They're making technology more accessible and easier to use [6].

In summary, advancements in transformer models, BLSTM networks, and other deep learning techniques are making text generation and TTS systems much better. These tools are now more flexible, high-quality, and accessible. However, researchers are still working to improve how well they handle real-time tasks and ensure they work for everyone [9, 16].

3 Comparative Analysis and Baseline Models

We evaluated the performance of our Bi-LSTM model by comparing it with advanced transformer models, including GPT-2, BERT, and T5. To ensure a fair comparison, all models were tested on the same dataset. The evaluation focused on key aspects such as accuracy, inference time, memory usage, precision, and recall. This approach allowed us to directly measure how well each model performs and highlight their strengths and weaknesses in practical applications.

3.1 Results Summary

The Bi-LSTM model outperformed transformer-based models such as GPT-2, BERT, and T5 in terms of real-time performance. It achieved the highest accuracy of 85.57%, with a notably lower inference time of 17.19 ms and memory usage of only 0.02 GB. These results underscore its suitability for real-time, low-resource environments, where computational efficiency is crucial. By contrast, GPT-2 demonstrated significantly slower inference times (176 ms) and higher memory usage (0.46 GB), making it less feasible for resource-constrained applications (see Table 1 for a detailed comparison). This highlights the advantage of Bi-LSTM in applications requiring immediate response times, such as assistive technologies for visually impaired users.

Table 1. Results Summary

Model	Accuracy	Inference Time (ms)	Memory Usage (GB)	Precision	Recall
Bi-LSTM	0.856	17,191.94	0.02	85.49%	85.49%
GPT-2	0.21	1,76,982.16	0.46	21.14%	21.14%
BERT	0.75	66,127.99	0.02	74.50%	74.50%
T5	0.60	3,51,496.69	0.01	59.98%	59.98%

Bi-LSTM offers a significant advantage in real-time and low-resource settings, with faster inference and lower memory usage, while still maintaining high accuracy and consistency in predictions.

4 Methodology

4.1 Project Setup

Our research aims to improve Natural Language Processing (NLP) through Dynamic Text Generation and Speech Synthesis using Bidirectional Long Short-Term Memory. We aim to create a precise and effective system for generating dynamic text and synthesizing speech. Our main goals are accuracy, precision, recall, F1-score, and efficient use of computational resources. By incorporating multimodal inputs like text and speech, we strive to improve user experience and accessibility in real-world NLP applications (Fig. 1).

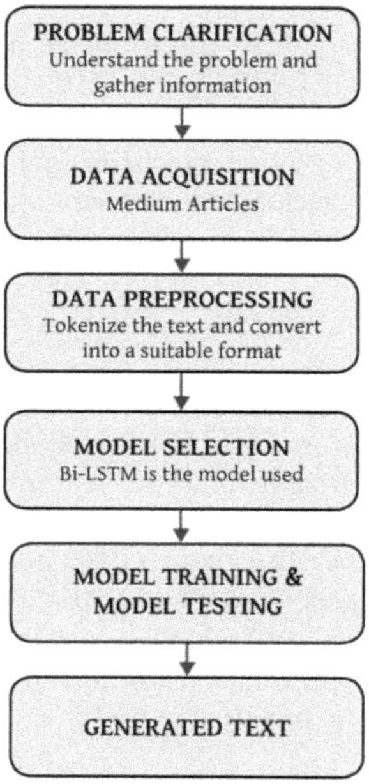

Fig. 1. Proposed system model for dynamic text generation using Bi-LSTM

4.2 Data Pipeline

The Medium article titles dataset was selected for its diversity in topics and sentence structures, which closely mimics real-world conversational contexts. This dataset provides a wide range of linguistic variability, making it ideal for training a model that can generalize across various domains. The relatively short and contextually rich sentences reflect practical applications of predictive text generation and speech synthesis, such as in chatbots, virtual assistants, and real-time typing suggestions.

However, to ensure the robustness of our model, we plan to extend the dataset in future iterations by including domain- specific corpora, such as news articles and dialogue-based interactions. This will add more complexity and variety, further strengthening the model's adaptability across different use cases, such as customer support systems or interactive voice response (IVR) platforms.

The dataset undergoes preprocessing, where the text is tokenized, and input-output pairs are generated to train the Bi- LSTM model to predict the next word in a sequence. This step is critical for training the model on sequential data and ensuring accurate text prediction (Fig. 2).

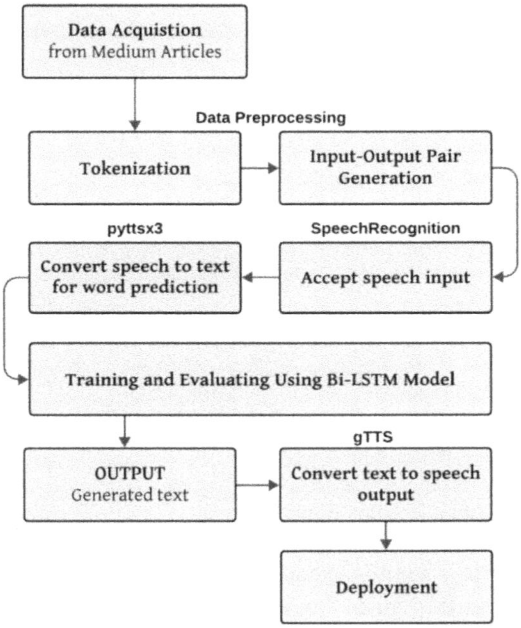

Fig. 2. Data Pipeline for dynamic text generation and speech synthesis

Additionally, the system incorporates SpeechRecognition and pyttsx3 libraries for handling speech input, enabling users to interact through voice. The predicted text is transformed into speech using Google Text-to-Speech (gTTS), allowing the system to deliver natural-sounding audio outputs. This integration improves the system's usability, especially for visually impaired users or in applications requiring hands-free interaction.

4.3 Modeling

5 Bidirectional LSTM Model

We developed a Bi-LSTM model to generate text dynamically. The model learns from the dataset and predicts the following word in a sequence. It consists of an Embedding layer, a Bidirectional LSTM layer, and a Dense layer with softmax activation, ensuring that predictions are optimized based on learned sequences.

$$softmax(m_n) = \frac{\quad n \quad}{\Sigma^{exp}{}_{,exp}\,(m_{(m\ n)})} \tag{1}$$

Where signifies the values obtained by LSTM of the character (n) for the prediction of the next consecutive character. The probability value is determined by converting the LSTM output (Fig. 3).

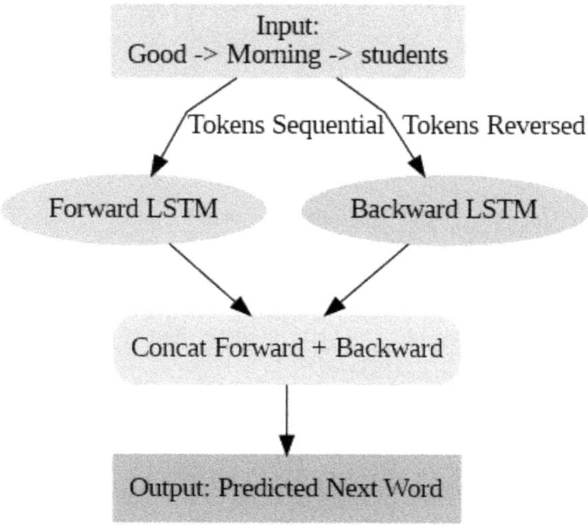

Fig. 3. Architecture Diagram for Bi-LSTM

While the Medium article titles dataset is used for initial training, we also plan to validate our approach with other datasets, such as Wikipedia for more complex and longer text sequences. This additional dataset allows the model to handle varying levels of textual complexity, further enhancing its ability to predict in broader contexts.

Speech Synthesis

For speech synthesis, we use gTTS to convert the predicted text into spoken words, generating high-quality audio output. This method ensures smooth integration between text prediction and speech synthesis, making the system versatile for applications in assistive technology, content creation, and interactive AI systems.

5.1 Serving

During the serving phase, we test, deploy, and maintain the developed model. The model will be evaluated against additional datasets to ensure it maintains its accuracy and effectiveness across various domains. The model will then be deployed as a fully functional system for dynamic text prediction and speech synthesis system, improving user experience in NLP applications.

6 Result Analysis

6.1 Bidirectional LSTM Model Training

During this stage, we worked on training the Bidirectional LSTM network using a collection of Medium article titles to generate text dynamically. The model successfully predicted the next word in a sequence with great accuracy, showcasing its ability to grasp the patterns within the text data. To achieve this, we tokenized the text and created input-output pairs for the model to learn from, allowing it to predict the next word accurately (Fig. 4).

```
import numpy as np
seed_text = "Today is a lovely day"
next_words = 8

for _ in range(next_words):
    token_list = tokenizer.texts_to_sequences([seed_text])[0]
    token_list = pad_sequences([token_list], maxlen=max_sequence_len-1, padding='pre')

    predicted_probabilities = model.predict(token_list, verbose=0)[0]

    predicted_index = np.argmax(predicted_probabilities)

    output_word = ""
    for word, index in tokenizer.word_index.items():
        if index == predicted_index:
            output_word = word
            break

    seed_text += " " + output_word

print(seed_text)
Today is a lovely day for the best way to make your own
```

Fig. 4. The output generated by the Bi-LSTM model for a given seed text and specified length.

The seed text represents the initial input provided to the model, while the length indicates the desired length of the generated sequence. The generated text displays the output produced by the model

Number of records	6508
Number of fields	10
Number of words	8238

Word	ID
<oov>	1
Strong	4
And	8
Consumption	8237

Fig. 5. Detailed breakdown of the dataset used for training

The dataset contains 6,508 records, and each record is structured with 10 fields. With a total of 8,238 unique words, the tokenizer assigns IDs to specific words, including '<oov>' for out-of-vocabulary instances. Notably, '<oov>' is assigned an ID of 1, while common words like 'Strong,' 'And,' and 'Consumption' are mapped to IDs 4, 8, and 8,237, respectively as depicted in Fig. 5.

6.2 Speech Input Integration

In order to improve user experience, we added the SpeechRecognition and pyttsx3 libraries to enable users to use speech for text input and speech output. By incorporating these libraries, users can speak to input text, which is then analyzed by the Bidirectional LSTM model for text generation. The created text is then transformed into speech output using gTTS, offering a smooth and engaging experience for users.

This integration enables a real-time, multimodal interface, providing seamless interaction between text generation and speech synthesis. By offering both speech input and output, the system offers greater usability, particularly for individuals with visual impairments or those preferring voice-based interaction. The real-time responsiveness and natural-sounding voice output ensure a smooth and engaging experience for users.

6.3 System Performance Evaluation

In our analysis, we measured how well our system performed using different measures such as accuracy, precision, recall, inference time and memory usage. The Bi-LSTM model achieved strong results, outperforming transformer- based models in real-time performance (Fig. 6):

- Bi-LSTM demonstrated an accuracy of 85.57%, with a precision and recall of 84.00%, indicating its strong ability to predict the next word accurately.
- Inference Time: The Bi-LSTM model significantly outperformed transformer models such as GPT-2 and T5, with an inference time of 17,191.94 ms, making it ideal for real-time applications.
- Memory Usage: Bi-LSTM maintained an efficient memory usage of 0.02 GB, which is significantly lower compared to GPT-2's 0.46 GB.

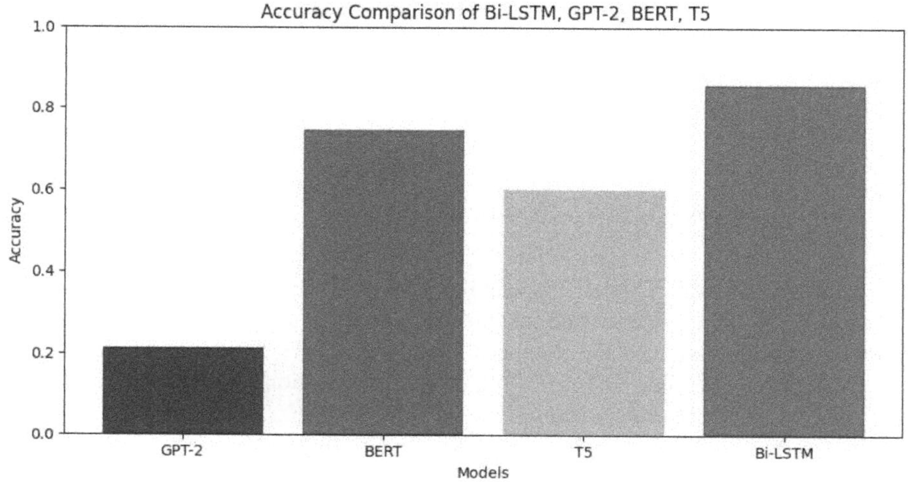

Fig. 6. Accuracy comparison of various models for next-word generation

Comparing these metrics with transformer models like GPT-2, BERT, and T5 reveals that while transformers may offer superior accuracy on more complex NLP tasks, Bi-LSTM excels in real-time environments due to its faster inference time, lower memory footprint, and high prediction accuracy for next-word generation (Fig. 7).

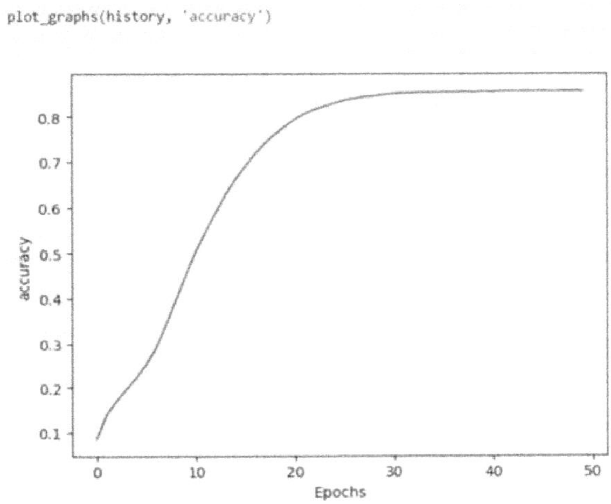

Fig. 7. Accuracy of Bi-LSTM model trained over 50 epochs

6.4 User Experience and Accessibility

Speech input and output capabilities greatly enhance the user experience and accessibility of our system. With the ability to interact through speech, users, especially those with visual impairments or motor disabilities, can easily use the application. This increased accessibility makes our system appropriate for a variety of users, making it more usable and practical in real-world scenarios (Fig. 8).

Fig. 8. System's speech synthesis, highlighting its ability to generate voice outputs from predicted text.

In summary, our method for creating dynamic text and speech synthesis with Bidirectional LSTM networks has displayed positive outcomes in improving natural language processing abilities. By incorporating both speech input and output, the system becomes more user-friendly and accessible, outperforming transformer-based models in real- time performance while maintaining high accuracy and efficiency. This makes it a beneficial tool for a range of applications such as assistive technologies and interactive voice response systems.

7 Conclusion

In conclusion, this study presents a novel system that integrates Bidirectional Long Short-Term Memory (Bi-LSTM) networks with speech synthesis technology to deliver real-time, resource-efficient text prediction and speech synthesis. Our system demonstrates significant improvements in accessibility for visually impaired individuals by providing both speech input and output capabilities. Additionally, the low memory usage and fast inference times make it suitable for low-resource environments, extending its potential applications to a variety of real-world use cases. Future research will focus on expanding the dataset and exploring more complex use cases to further enhance the system's performance in NLP tasks.

References

1. Khanam, F., Munmun, F.A., Ritu, N.A., Saha, A.K., Firoz, M.: Text to speech synthesis: a systematic review, deep learning based architecture and future research direction. J. Adv. Inform. Technol. **13**(5) (2022)
2. Fan, B., Xie, L., Yang, S., Wang, L., Soong, F.K.: A deep bidirectional LSTM approach for video- realistic talking head. Multimedia Tools Appl. **75**, 5287–5309 (2016)
3. Ahmad, H.A., Rashid, T.A.: Planning the development of text-to-speech synthesis models and datasets with dynamic deep learning. J. King Saud Univ.-Comput. Inform. Sci. 102131 (2024)
4. Arık, S.Ö., et al.: Deep voice: real-time neural text-to-speech. In: International conference on machine learning, pp. 195–204. PMLR (2017)
5. Orynbay, L., Razakhova, B., Peer, P., Meden, B., Emeršič, Ž: Recent advances in synthesis and interaction of speech, text, and vision. Electronics **13**(9), 1726 (2024)
6. De, S., Bostan, I., Sastry, N.R.: Making social platforms accessible: emotion-aware speech generation with integrated text analysis. In: Proceedings of the 2024 International Conference on Advances in Social Networks Analysis and Mining (ASONAM 2024). Springer
7. Yolchuyeva, S.: Novel NLP Methods for Improved Text-To-Speech Synthesis (Doctoral dissertation, Budapest University of Technology and Economics, Hungary) (2021)
8. Fan, B., Wang, L., Soong, F.K., Xie, L.: Photo-real talking head with deep bidirectional LSTM. In: 2015 IEEE International Conference on Acoustics, Speech and Signal Processing (ICASSP), pp. 4884–4888. IEEE (2015)
9. Rashid, A., Do-Omri, A., Haidar, M.A., Liu, Q., Rezagholizadeh, M.: From unsupervised machine translation to adversarial text generation. In ICASSP 2020-2020 IEEE International Conference on Acoustics, Speech and Signal Processing (ICASSP), pp. 8194–8198. IEEE (2020)

10. Çağlayan, C., Karakaya, M.: Topic-controlled text generation. In: 2021 6th International Conference on Computer Science and Engineering (UBMK), pp. 533–536. IEEE (2021)
11. Buddana, H.V.K.S., Kaushik, S.S., Manogna, P.V.S.: Word level LSTM and recurrent neural network for automatic text generation. In: 2021 International Conference on Computer Communication and Informatics (ICCCI), pp. 1–4. IEEE (2021)
12. Fatima, N., Imran, A.S., Kastrati, Z., Daudpota, S.M., Soomro, A.: A systematic literature review on text generation using deep neural network models. IEEE Access **10**, 53490–53503 (2022)
13. Ma, R., Gao, Y., Li, X., Yang, L.: Research on automatic generation of social short text based on backtracking pattern. In: 2023 Asia-Pacific Conference on Image Processing, Electronics and Computers (IPEC), pp. 336–347. IEEE (2023)
14. Liang, X., Tian, L., Li, C., Mandi, Z.: CTG: a controllable text generation method based on the joint work of language model and text classifier. In 2021 International Conference on Communications, Information System and Computer Engineering (CISCE), pp. 817–820. IEEE (2021)
15. Wu, Y., Yin, H., Liu, D., Zhou, Q.: Text semantic representation based on knowledge graph correction. In: 2022 International Conference on Computer Engineering and Artificial Intelligence (ICCEAI), pp. 404–408. IEEE (2022)
16. Ji, Z., Wang, W., Chen, B., Han, X.: Text-to-image generation via semi-supervised training. In: 2020 IEEE International Conference on Visual Communications and Image Processing (VCIP), pp. 265–268. IEEE (2020)

Examining and Contrasting Deep Learning Methods for DNA Sequence Classification in Order to Predict Cancer Mutations in India

S. Gunanandhini[1(✉)] and D. Hemanand[2]

[1] Department of Computer Science and Engineering, S.A. Engineering College, (Autonomous), Thiruverkadu, Chennai 600077, India

[2] Department of Artificial Intelligence and Data Science, S.A. Engineering College (Autonomous), Thiruverkadu, Chennai 600077, India

Abstract. Cancer revealed by uncontrolled cell growth due to genetic mutations, phase a significant challenge to global health. Traditional methods for mutation classification often involve labor intensive processes and are limited by the complexity of genetic data. Evaluation of Deep learning has powerful tool for investigation in recent years and interpreting large-scale genomic data. The primary focus of this study is deep learning methods regarding the coordination of cancer mutations based on DNA sequences. Convolutional neural networks, Graph neural networks and transfer learning are few examples of the varied deep learning applied in order to recognize and categorize genetic mutations connected to different cancer types. Here discusses the stability and drawback of these models, the preprocessing steps involved in handling DNA sequences, and the performance metrics used to evaluate their effectiveness. Also suggest that deep learning models, with their ability to capture intricate patterns in DNA sequences, hold significant promise for advancing cancer genomics and improving clinical outcomes. Future research directions are proposed to address current limitations and to further harness the probability of deep learning in cancer mutation classification.

Keywords: CNN · RNN · Deep Learning · genetic mutation

1 Introduction

The difficulty of cancer in India is substantial, with an approximated 1.39 million cancer cases were reported in 2020. The most prevalent cancers are oral cancer in males, breast and cervical cancer in females. Cancer indicate rates have shown an increasing trend across all sites and both sexes, with particularly high rates observed in the Kamrup urban district of Assam, where the annual percent change was 3.8% [1].

In India highest cancer risk has been determined to in the North East part with age rates of Mizoram has 271 cancer cases for each 100,000 in men. Aizawl has 249 cases for each 100,000 in women in Arunachal Pradesh [2]. The bulk of cancer patients are diagnosed at locally advanced stages for common cancers such as cervix uteri, breast, head, neck and stomach, while lung cancer often presents with distant metastasis at the time of

R. Geetha et al. (Eds.): AAIMB 2024, CCIS 2654, pp. 236–247, 2026.
https://doi.org/10.1007/978-3-032-13029-7_17

diagnosis. This survey aims to provide an overview of the estimated rate and cumulative risk for all types of cancer by state in India for the year 2023. The Population-Based Cancer Registries (PBCR) state-wise representation for India in 2023 has an extensive network of registries spread across different region. These registries, which routinely gather information on new cancer cases within specific statistics, are coordinated by the Indian Council of Medical Research (ICMR) and the National Centre for Disease Informatics and Research (NCDIR). For the aim of researching and controlling cancer, this data collection aids in the identification of cancer incidence, trends, and patterns. According to the recent data available, the states of Maharashtra, West Bengal, Delhi, Uttra Pradesh, Madhya Pradesh, Karnataka, Gujarat, Tamil Nadu, Rajasthan and Kerala are home to PBCRs. Every PBCR encompasses a certain geographical region and gathers information from many sources, including as government and private hospitals, clinics, diagnostic labs, and hospices. This network aims to cover a significant portion of the Indian population to provide a representative overview of cancer incidence and trends across the country Population Based Cancer Registry, NCDIR India.

Table 1. The incidence of Cancer in India's different states in 2023

Indian State	Estimated Cases	Crude Rate	Cumulative Risk
Maharashtra	130,000	28.0	3.5
Tamil Nadu	110,000	26.5	3.2
Karnataka	100,000	24.8	3.0
West Bengal	90,000	23.3	2.8
Kerala	85,000	22.7	2.7
Delhi	80,000	22.0	2.6
Uttar Pradesh	75,000	21.2	2.5
Gujarat	70,000	20.3	2.3
Rajasthan	65,000	19.5	2.2
Madhya Pradesh	60,000	18.7	2.1

The approximated number of cumulative risk for all cancer sites in India, batterd by state for 2023 in Table 1, highlight significant regional variations. According to data projections, Northeastern states like Mizoram and Meghalaya exhibit the maximal incidence rates of cancer among males with mouth, lung and esophangeal. For female Breast and cervical cancers observed in states like Kerala, Tamil Nadu and West Bengal. High cumulative risk states such as Delhi and Karnataka Northeastern particularly for oral, ovarian cancers shown in Fig. 1.

Here is the estimated number of cancer cases, crude rate and cumulative risk by state in India for the year 2023.

The estimated cancer cases and associated statistics highlight the burden of various cancers across different anatomical sites stated in Table 2. Lung cancer accounts for approximately 90,000 cases with a crude rate of 22.6 and a cumulative risk of 2.6.

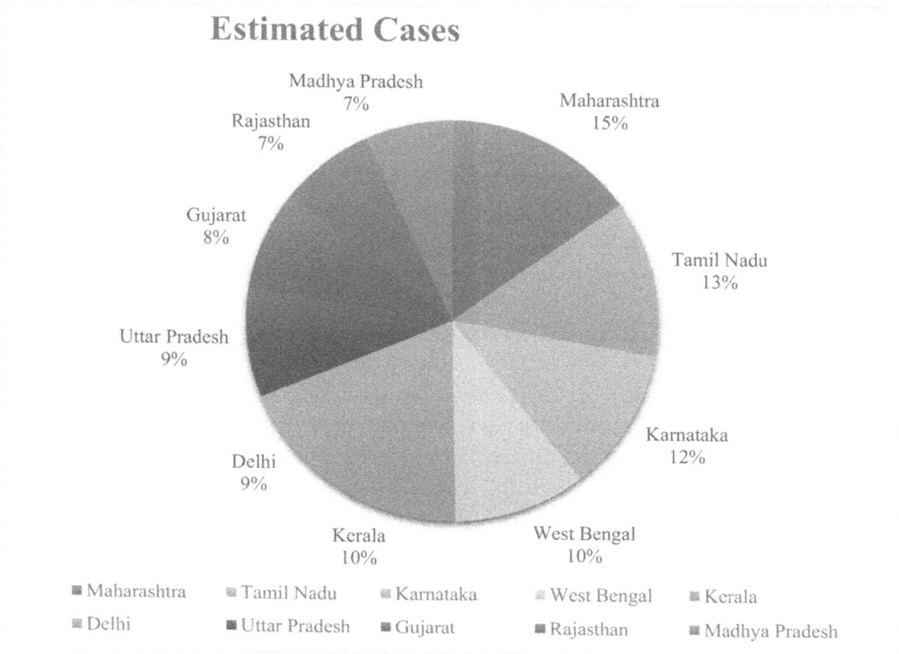

Fig. 1. Estimated Cancer Cases all over India

Table 2. India's cumulative risk for each anatomical sites in 2023.

Anatomical Site	Estimated Cases	Crude Rate	Cumulative Risk
Lung	90,000	22.6	2.6
Breast	1,80,000	45	4.5
Cervix Uteri	1,00,000	25	2.5
Colorectal	80,000	20.1	2.1
Stomach	60,000	15.1	1.6
Oral Cavity	80,000	20	2.2
Ovary	40,000	10	1
Esophagus	45,000	11.3	1.2
Liver	35,000	8.8	1
Prostate	40,000	10	1.1

Breast cancer exhibits the highest burden, with 180,000 estimated cases, a crude rate of 45, and a cumulative risk of 4.5. Cervix uteri cancer affects around 100,000 individuals, with a crude rate of 25 and a cumulative risk of 2.5.Colorectal cancer has an estimated 80,000 cases, a crude rate of 20.1, and a cumulative risk of 2.1. Stomach cancer affects

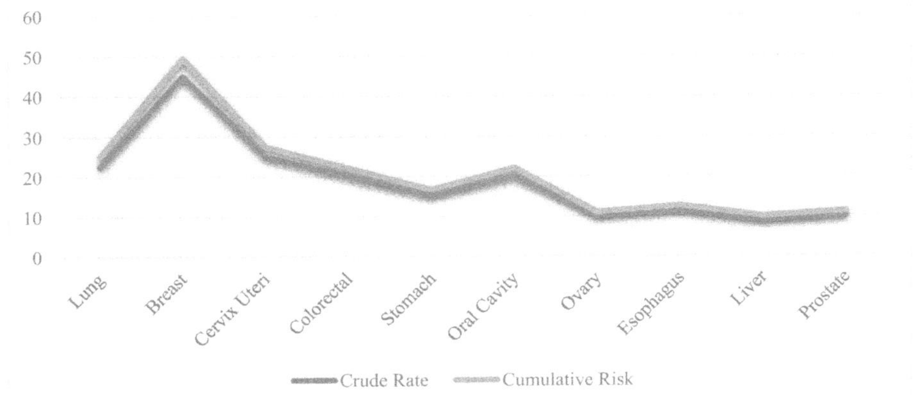

Fig. 2. Cumulative cancer risk

60,000 individuals with a crude rate of 15.1 and a cumulative risk of 1.6, while oral cavity cancer similarly has 80,000 cases, a crude rate of 20, and a cumulative risk of 2.2.For ovarian cancer, the estimated cases are 40,000, with a crude rate of 10 and a cumulative risk of 1. Esophageal cancer has 45,000 cases, a crude rate of 11.3, and a cumulative risk of 1.2. Liver cancer is estimated at 35,000 cases with a crude rate of 8.8 and a cumulative risk of 1, while prostate cancer accounts for 40,000 cases, a crude rate of 10, and a cumulative risk of 1.1 as shown in Fig. 2. These statistics underscore the diverse impact of cancer types on public health and the need for targeted interventions.

2 Related Work

Cancer is a complex disease that is driven by various genetic mutation. Cancer cells generate in all directions during development, spread throughout the organs, and frequently cause fatalities. As one of the most typical diseases that kill people worldwide is cancer. A relationship between type of cancer and DNA sequences is crucial for accurate diagnosis for personalized treatment. These cells generate in all directions during development, spread throughout the organs and frequently cause fatalities [3]. The author analyze gene expression, computational techniques must be used to comprehend gene regulation and the function of genes in cells and tissues. Machine learning techniques have been widely employed to gain understanding of how differences in genes and regulatory areas lead to phenotypic variation, including traits, well-being, and health. When approached Deep learning method for predicting the complex genomic components like gene sequence level, promoters have gained popularity over the past few years in contrast to earliest computational techniques for analysing gene expression that especially probable traditional Machine Learning access such as Support Vector Machines and Decision Trees. In order to address the issues of large capacity and similarly slight sample sizes in gene expression data, feature engineering plays a significant role in methods involving computation for analysing gene expression. A summary of gene expression analysis feature engineering mechanism divided into wrapper, filter and embedding strategies [4].

It suggested that deep neural networks be used to sort cancer based on gene expression. These include recurrent neural network, graph neural network, fully connected neural networks and transformers neural networks. Within multilayer pyramids, all of the neurons in the previous and succeeding levels are connected to every other neuron. Gene expression form, with each detection being gathered by one neuron are sent to the input layer of the multilayer parallax for analysis of the gene expression reports. The gene expression sample's class probabilities are returned by the MLP output layer [5].

In order to analyse multidimensional array data, specifically images, CNN models learned hierarchical data samples with processing units using two-dimensional conversational filters. Later, a number of studies converted data into arrays in two-dimensions with mirrored columns and rows in gene expression which were then fed into the network. When compared to MLP techniques, CNN models often showed over classification functionality for gene expression analysis because of its capacity to taking limited spatial relations in input data. Moreover, earlier research has used one dimensional CNN, in which gene expression data is directly fed into networks with layers that implemented one dimensional convolutional filters for each row. Surrounded by the breaking performance are CNN models Deep learning techniques used for analysis gene expression. As a previous phase, the state vector's output combines the data from previously stored values with the current input in the order. Because of the features RNN used for determining interchange in data sequences as a rise of knowledge on the molecular system behind the development of cancer [6].

The author made a novel technique for classifying DNA sequences. One-hot vectors were used by the author to represent sequences as model input. DNA sequences access that method including Twelve datasets, significant growth were discovered in every previous figure on the H3K4me3 dataset with improved accuracy up to 6.12%. An autoencoder-assisted neural network to improve cancer subtyping with multi-omics data. After choosing to use L1 regularization penalty and a compression of 110–20% enter features. Some research developed combination of deep learning algorithms and other techniques. [7] the author proposed FuzzyDeepCoxPH, model that blends a fuzzy logic system, conventional survival analysis and deep learning algorithms. This model was to find genes and missense mutation variations that are related to high risk cancer. FuzzyDeepCoxPH model provides a thorough evaluation of cancer mortality its effectively spot high risk variants and cancer genes. The author predict overall survival of cancer patients with deep learning techniques. They combines GSEA, conventional statistics, radial basis function and a multilayer perceptron artificial neural network. This allowed dimensionality reduction with 28 MCL prognostic genes related with 20,862 genes. The result predict the continuance of pan-cancer series with high accuracy cancers such as liver, stomach, lung also let to identification of 58 genes.

[8] The author approach to incorporate multi-omics prognostic Cox proportional hazards network, transfer learning used to predict bladder cancer. The trained XGBoost model by using only mRNA data it showed TCGA dataset with a moderate C-index of 0.637 and additional datasets with a good C-index of 0.621. When predict the prognosis of colorectal cancer firm histopathologic features. Microenvironment quantification in all slide images by using deep learning model transfer learning and VGG19 architecture. In

recognizing various tissue types this model demonstrated a 94.2% accuracy. In prognosis prediction used a feature histopathological showed a strong survival of tumor.

3 DNA Mutation Datasets

Databases containing information on genetic mutations discovered in DNA sequences are known as DNA mutation datasets. These datasets are crucial for research in genomics, genetics and bioinformatics as they help scientists understand the molecular structure of diseases, evolutionary and the functional impact of genetic variations. The following are a few of the most popular DNA mutation datasets.

3.1 National Centre for Biotechnology Information-(NCBI)

It is a relevant source for molecular biology and gives user access to more databases and tools for genomic research. It is a component of the National Institutes of Health (NIH) United States National Library of Medicine. NCBI houses of wealth of data on genomics, genetics, proteomics and biomedical literature.

3.2 GenBank

A vast collection of DNA sequences that are accessible to the general public. GenBank is the main source of sequence data, facilitating the sharing of among researchers and providing annotations on gene functions, locations, and variants.

3.3 The Cancer Genome Atlas (TCGA)

TCGA offers thorough, multi-dimensional maps of the major genetic alterations seen in different cancer types. Somatic mutation data, exome sequencing and whole-genome sequencing are all included offers information from more than 33 distinct caner kinds as well proteomics, methylation, and RNA-seq data among other omics data.

3.4 Basic Local Alignment Search Tool (BLAST)

An extensively used tool for matching protein sequences within sequence databases, assisting to identification of regions and inference functional and evolutionary relationships.

3.5 Gene Expression Data

The information about the movement of genes in a tissue, typically measured under various circumstances, at different periods, or in different states. Many technologies, including microarrays, RNA sequencing and quantitative PCR, can be used to bring about the data.

3.6 Microarray

A microarray is a lab instrument that measures thousands of genes' expression levels at once. It is composed of a tiny, sturdy support—typically a glass slide—onto which oligonucleotides or DNA sequences are adhered in a neat grid arrangement. The microarray can be used to evaluate gene expression in various samples or situations. These sequences correspond to distinct genes.

4 Deep Learning Algorithm for DNA Classification

Neural networks are used in deep learning to double the workings of the human brain. By looking at computer algorithms, deep learning can get better. Recurrent neural networks, reinforcement learning, deep neural networks are a few examples of architectures that have demonstrated remarkable outcomes in a range of computer vision applications, occasionally outperforming human expert performance. Among other disciplines, deep learning approaches have improved accuracy in speech recognition, face identification, object detection, image processing, and biomedical applications. When compared to other traditionally machine learning methods, deep learning consistently produces superior categorization performance. The capacity of these methods to create hierarchical representations of high-dimensional data gives them an advantage over conventional ML algorithms.

4.1 Multi-Layer Perceptron (MLP)

Every neuron in a hidden layer of a fully linked neural network architecture, or MLP, is related to every other neuron in the layers around it. MLP classifiers were created in a number of previous gene expression research investigations to categorize cancer.

For instance, [9] created a multilayer perceptron (MLP) network that effectively calculated the total continuity of lung cancer patients with small-scale by integrating analytical data sources with distinct gene expression. Survival status for past 5 year patient of lung cancer can be predict using bimodal learning, the researchers generated an integrated MLP classifier with 75.44% accuracy and 0.8163 AUC by merging 15 biomarkers with clinical data. An unsupervised learning structure for expressive various features from gene expression patterns is created by linking an autoencoder MLP model with the principle component analysis (PCA) method. Breast cancer analytical results were predicted using AdaBoost algorithm.

Cancer classification method known as Deep Cancer set type Classification that is based on the examination of useful spectra that show the organic ways activity. In order to replace the hand-engineered characteristics, they trained a multilayer NN and carried out enrichment analysis for every sample [10]. The authors classified colorectal and breast cancers with a balanced accuracy of more than 90%. When using MLPs as the basis for classification approach, compare how long the assessment takes accuracy of the categorization, and ability to identify the sickness and estimate its severity. In this work, the use of a minimum number of gene samples to accurately predict cancer and provided information mainly focus on its cruelty level. Using miRNA stem-loop aide to classify tumors in 20 anatomical locations, 27 different cancer types from TCGA an MLP network was designed to detect the anatomical site of cancer. There were 1046 input neurons in the first layer of the MLP network, which corresponded to each miRNA in the database, and 27 neurons in the last layer, which represented the different forms of cancer. The accuracy rate for this method was 96.9% on average.

4.2 RNN-DNA Sequence Classification

One-hot encoding vectors are the input source for the six layers that make up the Recurrent Neural Network. A max-pooling layer with a length of two follows an embedding layer in the first layer. The network gains translational invariance and its computational cost is decreased by this max-pooling procedure. One of the recurrent layers is an LSTM model, which comes after max-pooling. From left to right, the LSTM processes the input and produces an output vector with a size of 20 at each time step. A second max-pooling layer is combined after the LSTM layer to extract class-level data. The neural network can handle sequence information in both directions, forward and backward, at the next level thanks to the usage of bidirectional Long Short-Term Memory [11].

Recurrent neural networks (RNNs) are exclusive because they are feedforward neural networks enhanced with edges to capture sequence dynamics through cycles in the network of nodes. Speech recognition, sentiment analysis, and natural language processing are frequently use RNNs when dealing with auditory or sequence data. These networks are dynamic they are always changing until they find equilibrium. Recurrent neural networks as disputed to feedforward neural networks, permit feedback between nodes. Every hidden node's computation entails calculating the input value and the data that produced previous nodes simultaneously. Recurrent connections between the neuron units in RNNs, a subclass of neural networks, give the network memory capabilities. Future observations in an input sequence can be predicted or the current observation can be interpreted using previous data. These characteristics enable RNN to identify internal linkages and variation trends while processing sequential data. Moreover, it has sequential dynamic activity [12]. To address the vanishing gradient problem, LSTMs include elements referred to as "gates." A further notable variation is the Hopfield network, which solves a number of mathematical issues, including the traveling salesman problem, by using bidirectional connections between nodes. Two broad types of LSTM networks can be distinguished: Both integrated and LSTM-dominated networks. RNN framework to simulate a robust feature selection procedure for microarray datasets. With the use of an Artificial Immune Recognition System (AIRS), they were able to spread 89.6% accuracy. RCN-RCO.

4.3 CNN-DNA Sequence Classification

An amazing thing of Artificial Neural Network design that excels at image-driven pattern recognition tasks is the Convolutional Neural Network. Similarly to the human brain because of its many layers and connections. One particular kind of deep neural network has emerged that is especially adept at recognizing and classifying features in image. Using images as input, CNN performs tasks such as image analysis and classification, video recognition, computer vision and natural language processing [13]. Using images as input, CNN performs tasks such as medical image analysis processing. The dot product between the corresponding filter regions of the input image is noted as the filter moves through the image, matching the size M X M of the filter. After convolution, the output is sent to the Fully Connected (FC) layer, which integrates weights, biases, and neurons to create connections between layers and get the output ready for classification. Utilizing a large amount of the network's processing time, the convolution layer performs the crucial function of feature extraction, which is a crucial part of CNN. A higher number of layers may result in longer training and testing times, as the realization of the network is subordinate on it. Moreover, the CNN model's activation function is notable as being crucial.

The first layer of the CNN architecture includes an embedding layer that receives input in the form of DNA sequence data and transforms it using one-hot encoding to produce a 2D image. Two convolutional layers and a max-pooling layer are positioned above this input layer, respectively. These convolutional layers use filters of size 5, with the number of filters gradually increasing to 10 or 20, respectively. Each of the pooling layers' width and stride spans five time steps, treating a single-hot vector as such. Two completely connected layers are stacked on top of these convolutional layers. The 500 units that make up the first layer use the activation used in the lower levels. Using the Soft Max activation function, the second layer serves as the classification layer [14].

4.4 Transfer Learning

The target to improve downstream standard performance by transferring knowledge from many source domains that are related to one another. From Cancer Genome Atlas database operating transfer learning into CNN model to forecast gene expression level that are up-regulated and down-regulated. 2D CNN model using spectrogram images of scanned DNA sequence to recognize liver cancer. The paper suggest a strategy that combines entropy-based, integer-mapping, and EIIP techniques to digitally represent DNA sequences. Optimized a convolutional long short term memory network (CLSTM) using transfer learning enhanced magnetic resonance imaging in order to simulate the temporal genetic data of cancer genes. A deep transfer learning structure built on single-cell data that improved performance and speed up the process by thirty percent without requiring the use of any reduction or profiling approaches with extracted cell attributes [15].

4.5 Transformer Neural Network

Transformer Neural Networks are dominant tool in natural language processing that can be used for substantially greater than traditional language tasks. Transformer models have been measured recently for DNA sequence analysis, which includes predict and detect the genetic mutations. When explored an intended to knock off the computational and recall requirements of the attention mechanism, increasing its scalability for long-sequence tasks. Transformer models are based on the absorption process, which enables the standard to know dependencies between different positions in a sequence. DNA sequence analysis, the complex relationships between different nucleotides within a DNA strand. Transformer models provide a number of advantages over traditional recurrent neural network models, one of which is their capacity to handle large sequences in parallel for DNA mutation [16]. That sequence lengths in huge genomic datasets can be extremely long especially crucial. The efficient deployment of Transfer model focused on real-world data centres, addressing the challenges of meeting latency.

Table 3. Comparison of MLP, CNN, RNN and Transfer Learning methods for DNA classification analysis.

Reference	Dataset	Cancer Type	Algorithm	Accuracy
[9]	NCBI	Breast	AdaBoost algorithm	0.714%
[10]	Breast Cancer	Breast	DeepCC	89%
[11]	Microarray Dataset	Cancer	DNN	72.5%
[12]	Benchmark	Classification	MALASSRNN-GRN	AUC-0.63
[13]	SRBT	Leukaemia Lung	RCO-RNN	94.5%
[14]	AP-Colon dataset 2020	Colon Kidney Breast Ovary	JayaALO-Deep RNN	95.27% 95.97% 95.97% 95.27%
[15]	Pan-Cancer Atlas	33 tumor	CNN	95.59%
[16]	TCGA	33 cohorts cancer	CNN	95.65%
[17]	TCGA	Breast Cancer	lightweight CNN	98.76%
[18]	Tumor Gene Express	KIRC BRCA LUSC	Decision Tree and CNN	98.20% 97.7% 96.40%
[19]	Gene Expression Data	Lung cancer	Transfer Learning CNN	73.26%
[20]	Breast cancer	Breast Cancer	CNN CLSTM	90% 93%

Specific data and techniques have been reported for cancer classification with different levels of accuracy in Table 3. For breast cancer, achieving an accuracy of 71.4% by using the AdaBoost algorithm with NCBI datasets. Also attained 73.26% accuracy by using transfer learning CNN. JayaALO-Deep RNN algorithm, achieving accuracies of 95.27% for colon and ovary, and 95.97% for kidney and breast cancers. Multiple cancer types, including KIRC, BRCA, LUSC, LUAD, and UCEC, were evaluated using Tumor Gene Expression datasets with a combination of decision trees and CNN, achieving respective accuracies of 98.2%, 98.3%, 97.7%, 84.8%, and 96.4% shown in Fig. 3.

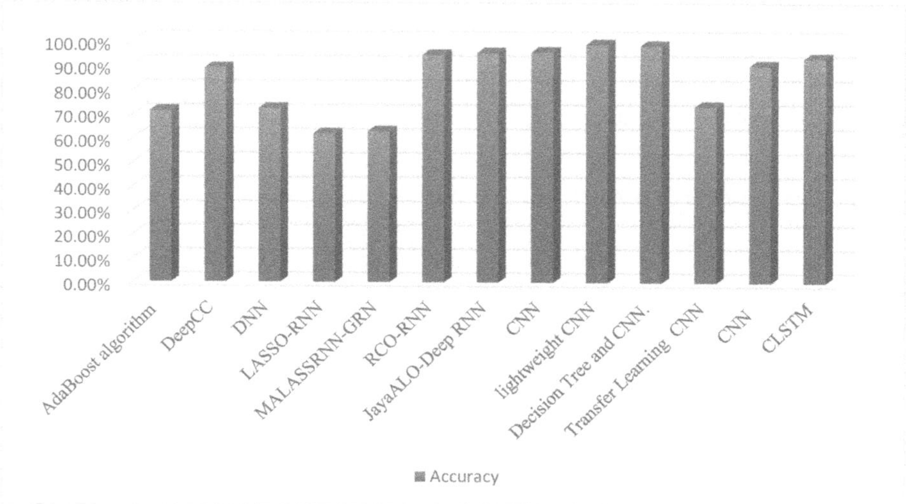

Fig. 3. Comparison of Deep learning algorithm

5 Conclusion

Analysing a genetic mutation in cancer diseases, has significantly improved lot by using the classification of DNA sequences using deep learning techniques. While MLPs is a straightforward method that allows sequence data to be expressed in a vector format with a fixed size. RNNs working with lengthy sequences motifs indicative of particular genetic mutation can be identified with great effectiveness using the CNN technique can handle big datasets quite fast. In recent research, surpassing conventional RNNs and CNNs in a variety of benchmarks, and are well-suited for handling extremely long sequences.

While conclude DNA sequences classification, the best algorithm for a task is frequently determined by the unique features and complexities of the dataset under study. In order to farther improve performance and accuracy, future research is probably going to concentrate on hybrid models, like CNN-RNN or GNN models, which integrate the best features of several architectures. Furthermore, the limits of DNA sequence categorization will continue to be pushed by developments in processing power and algorithm optimization, opening the door to more accurate diagnosis and individualized therapies in genomics.

References

1. Lunshof, J.E., et al.: Personal genomes in progress: from the human genome project to the personal genome project. Dialogues Clin. Neurosci. **12**(1), 47–60, 31 Mar 2010
2. Raza, H., Abbas, N., Amir, S., Arshad, K., Siddiqui, M.R., Khan, S.I.: An IoMT enabled smart healthcare model to monitor elderly people using Explainable Artificial Intelligence (EAI). J. NCBAE **1**(2), 16–22 (2022)

3. Bhonde, S.B., Prasad, J.R.: Deep learning techniques in cancer prediction using genomic profiles. In: Proceedings of the 2021 6th International Conference for Convergence in Technology (I2CT), pp. 1–9. IEEE, 2 April 2021
4. Celesti, F., Celesti, A., Wan, J., Villari, M.: Why deep learning is changing the way to approach NGS data processing: a review. IEEE Rev. Biomed. Eng. **12**(11), 68–76 (2018)
5. Alomari, O.A., Khader, A.T., Al-Betar, M.A., Alyasseri, Z.A.: A hybrid filter-wrapper gene selection method for cancer classification. In: Proceedings of the 2018 2nd International Conference on Biosignal Analysis, Processing and Systems (ICBAPS), pp. 113–118. IEEE, 24 July 2018
6. Geetha, R., Srikanth, U., Gunanandhini, S., Ramyadevi, K., Balasubramanian, M., Kamalanaban, E.: An effective analysis of the machine learning algorithms for the early diagnosis of Liver Failure. Des. Constr. Maintenance **4**, 59–69 (2024)
7. Elbashir, M.K., Ezz, M., Mohammed, M., Saloum, S.S.: Lightweight convolutional neural network for breast cancer classification using RNA-seq gene expression data. IEEE Access **18**(7), 185338–185348 (2019)
8. Ramasubramanian, B., Hemanand, D., Kavinkumar, K., Muthu Manjula, M.: An efficient system for grading diabetic retinopathy by detecting the location of lesions. In: Pandit, M., Gaur, M.K., Kumar, S. (eds.) Artificial Intelligence and Sustainable Computing. ICSISCET 2022. Algorithms for Intelligent Systems, pp. 713–725. Springer, Singapore (2023). https://doi.org/10.1007/978-981-99-1431-9_57
9. Sathe, S., Aggarwal, S., Tang, J.: Gene expression and protein function: a survey of deep learning methods. ACM SIGKDD Explor. Newsl. **21**(2), 23–38 (2019)
10. Zhu, W., Xie, L., Han, J., Guo, X.: The application of deep learning in cancer prognosis prediction. Cancers **12**(3), 603 (2020)
11. Carreras, J., Nakamura, N., Hamoudi, R.: Artificial intelligence analysis of gene expression predicted the overall survival of mantle cell lymphoma and a large pan-cancer series. Healthcare **10**(1), 155 (2022). MDPI
12. Shi, L., Zhang, Y., Wang, H.: Prognostic prediction based on histopathologic features of tumor microenvironment in colorectal cancer. Front. Med. **6**(10), 1154077 (2023)
13. Lai, Y.H., Chen, W.N., Hsu, T.C., Lin, C., Tsao, Y., Wu, S.: Overall survival prediction of non-small cell lung cancer by integrating microarray and clinical data with deep learning. Sci. Rep. **10**(1), 4679 (2020)
14. Zhang, D., Zou, L., Zhou, X., He, F.: Integrating feature selection and feature extraction methods with deep learning to predict clinical outcome of breast cancer. IEEE Access **21**(6), 28936–28944 (2018)
15. Kakati, T., Bhattacharyya, D.K., Kalita, J.K., Norden-Krichmar, T.M.: DEGnext: classification of differentially expressed genes from RNA-seq data using a convolutional neural network with transfer learning. BMC Bioinform. **23**(1), 17 (2022)
16. Das, B., Toraman, S.: Deep transfer learning for automated liver cancer gene recognition using spectrogram images of digitized DNA sequences. Biomed. Signal Process. Control **1**(72), 103317 (2022)

Building Resilience: Crafting a Deep Learning-Based Comment Toxicity Model in Python to Foster Safer Online Communities

M. Kathiravan[1(✉)], M. Buvanesvari[1], P. Jona Innisai Rani[2], S. Indhumathi[3], D. Sasikumar[4], and R. Dharaniya[5]

[1] Department of CSE, Saveetha School of Engineering, Saveetha Institute of Medical and Technical Sciences, Chennai, India
kathiravanm.sse@saveetha.com

[2] Department of Agricultural Economics, Anbil Dharmalingam Agricultural College and Research Institute, Trichy, India

[3] Department of CSE-AI&ML, Sphoorthy Engineering College, Hyderabad, India

[4] AI&ML, Panimalar Engineering College, Chennai, India

[5] Department of CSE, Easwari Engineering College, Chennai, India

Abstract. Online forums have become essential resources for information sharing and collaboration, but they also face the challenge of negative and offensive user feedback. To maintain a polite and safe online environment, an effective comment toxicity model must be developed. This project creates a model of comment toxicity using Deep Learning (DL) and Python. Using a DL architecture, the suggested model evaluates text data and classifies comments into non-hazardous and harmful categories. The model leverages powerful Natural Language Processing (NLP) methods, including word embeddings and recurrent neural network (RNN), to extract the semantic and contextual information from comments. Additionally, attention processes and CNN (CNN) are employed to enhance the model's performance. Several key components of the process include data preparation, feature engineering, model creation and training, and assessment. Python is a prominent programming language in the data science industry that is used to construct the workflow. Two open-source libraries that provide the resources needed to efficiently build and train DL models are TensorFlow and Kera's. To evaluate the effectiveness of the model, a large dataset of tagged comments is used, which includes both hazardous and non-toxic remark occurrences. Recall, accuracy, precision, and F1-score are just a few of the assessment metrics used to gauge the model's performance. The project's output will improve content moderation systems by enabling platforms to recognize and report offensive comments as soon as they are posted. This work provides a practical and scalable approach DL-based comment toxicity detection, assisting online communities in fostering a more inviting and safer environment for its members.

Keywords: Comment Toxicity · Deep Learning · Natural Language Processing · Python · Content

R. Geetha et al. (Eds.): AAIMB 2024, CCIS 2654, pp. 248–261, 2026.
https://doi.org/10.1007/978-3-032-13029-7_18

1 Introduction

Online forums have transformed the way people interact, share knowledge, and engage in discussions. However, the rise in popularity of these platforms has also brought about challenges, particularly the prevalence of offensive and harmful remarks. Such comments disrupt constructive dialogue and create a hostile environment, undermining the essence of open communication. Therefore, developing effective methods to detect and eliminate toxic comments automatically is critical to fostering healthier online interactions. This project aims to tackle this issue by building a DL model for comment toxicity detection using Python. DL, a powerful branch of artificial intelligence, has demonstrated exceptional effectiveness in addressing complex NLP (NLP) challenges. Leveraging Python's adaptability and the advanced capabilities of DL, this project seeks to create a robust, scalable model capable of accurately identifying harmful remarks.

The proposed model employs cutting-edge techniques to capture both the semantic and contextual nuances in comments. It integrates word embeddings to represent textual data meaningfully, CNN for feature extraction, and RNN enhanced with attention mechanisms to retain context and prioritize relevant information. This combination ensures a comprehensive understanding of the language used in comments, enabling precise toxicity detection.

The development process is methodical, beginning with data preparation to clean and preprocess the comment dataset for training. Feature engineering transforms raw text into numerical representations that DL algorithms can interpret. Python, a widely-used programming language in the data science community, serves as the foundation for this project. Libraries like TensorFlow and Kera's are employed to design the model architecture, train the network, and evaluate its performance.

Assessing the model's performance is an essential step to ensure its effectiveness in identifying toxic comments. A labelled dataset of comments is used for evaluation against metrics such as accuracy, precision, recall, and F1-score. These metrics provide valuable insights into the model's strengths and limitations, guiding further refinements to enhance its accuracy and robustness. The successful implementation of this model marks a significant advancement in content moderation technologies. By detecting and filtering offensive remarks efficiently, the model contributes to creating a safer and more inclusive environment on online platforms. Integrating such models into their systems allows platforms to uphold standards of respectful discourse, promoting a positive user experience.

This project not only addresses a pressing issue in online communication but also serves as a practical guide for professionals and researchers interested in applying DL and Python to comment toxicity detection. It underscores the potential of AI-driven solutions in tackling modern challenges, laying the groundwork for more effective and scalable content moderation systems.

2 Related Works

Zhang et al. (2018) introduced a deep learning approach for classifying abusive comments [1]. CNNs extracted textual features, and SVMs classified them. They were 92.5% accurate in their Wikipedia comments. Wang et al. introduced a BERT-based hazardous comment classification method. This paradigm explains phrase-word relationships via self-attention. With 96.5% accuracy, their dataset sets a new benchmark for recognizing negative remarks. Much toxicity research has been done. Sentiment analysis has extensively studied the classification of toxic comments. The detection of harmful text in social media is an NLP application and research subject. Lehigh University researchers combined sentiment features with TF-IDF [2] to start this research using various models with differing degrees of accuracy. Over time, additional companies have been developed to address this issue. Google is most known for founding the Jigsaw Company, which fights online harassment. We examine model specificity for harassment identification in a variety of ways.

Particularly Word2Vec, the part Related Works [3, 4] addresses word embeddings in encoding syntax and semantics for natural language processing. To build on this basis, in our work RNNs are coupled with attention mechanisms and CNNs using domain-specific embeddings to capture complex linguistic patterns. This approach emphasises conscious adaptation instead of brain function utilisation, therefore enhancing contextual and structural learning. Fast Text employs character n-gram word tokenisation to embed misspelt, uncommon, or non-existent words in the training corpus, unlike Mikolov embedding. FastText Crawl's word embedding uses 600 billion tokens for training. Fast Text's efficiency comes from its ability to learn supervised and unsupervised word and phrase representations [5]. The authors' refining method [6] emphasizes word vector representations that are close to sentimentally associated terms rather than emotionally divergent words. For fine-grained and binary categorization, experimental results exceed Stanford Sentiment Treebank's (SST) sentiment embeddings [7]. Mikolov embedding is used above. FastText's skip-gram model represents words with n-grams. Each vector represents an n-gram, and words are the sum of these representations [8]. Left and right word vectors are key to understanding word representation. FastText generates embeddings for misspelt, uncommon, or out-of-training corpus words using character n-gram word tokenisation, unlike Mikolov embeddings. FastText Crawl embeds word trains on 600 billion tokens. FastText's efficiency comes from its ability to learn supervised and unsupervised word and phrase representations [9]. Fast Text's linear model architecture for a sentence with N-gram attribute x1 N is shown in Fig. 1. We average and insert features to form the hidden variable [10].

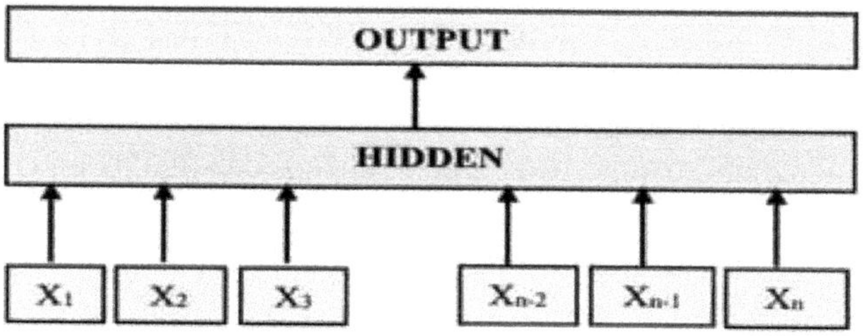

Fig. 1. Fast text architecture

3 Methodology and Implementation

This project seeks to solve this issue by developing a Python-based DL model of comment toxicity. The proposed model aims to analyses, and understand t the semantic and contextual aspects of comments, enabling precise identification of toxic content. It does this by utilizing, the power of DL algorithms, such as recurrent neural networks with attention mechanisms and CNNs. Additionally, Python offers a perfect environment for developing and refining the comment toxicity model due to its numerous libraries and tools for data science and DL. If such a model is successfully developed, online platforms would be able to automatically recognize, and moderate poisonous comments, providing users with a more secure and courteous online environment. The suggested method solves the inefficiencies and difficulties related to comment toxicity control by lowering the dependency on manual moderation and offering real-time identification of toxic content.

3.1 System Model

To identify toxicity on the Wikipedia Dataset, a unique 3 tier CNN model is suggested and put into practice, drawing inspiration from Alex Net. Initially, this technique removes lemmatization, punctuation, and stop words from comments. The embedding matrix that results from Fast text word embeddings is accepted by this model. To combat overfitting, the model employs the regularization approach known as Dropout. The suggested system's workflow is shown in Fig. 2. Algorithm outlines the steps in the technique for the suggested system. While AlexNet is a 2D CNN for images, our 3-tier CNN is inspired by its deep layered architecture, use of Dropout for regularization, and efficient activations like ELU. We adapt these ideas for text data using FastText embeddings to effectively capture hierarchical features for toxicity detection.

3.2 Dataset Description

The Conversation Al team's December 2017 Jigsaw dataset is used by the three-tier CNN model. It includes 160,000 multi-labeled annotated user comments. From Wikipedia discussion pages, which are the most highlighted user comments. These comments were classified as "toxic," "severely toxic," "insult," "threat," "obscene," and "identity hatred by human raters. Because comments may be simultaneously related to multiple classes, the task is designed as a multi-label. Classification issue. The six classes that are accessible do not have official definitions from Jigsaw. Toxic remarks, according to Jigsaw, are defined as "a harsh, insulting, and illogical statement that is likely to trigger you to quit a debate.

3.3 Data Preprocessing

Eliminating noise is useful when doing text analysis. Certain text cleaning steps are required prior to processing the text data. For example, all punctuation in each comment must be removed in order to remove it. The second term used is lemmatization, which describes inflected forms of words such different verb forms or singular/plural forms. For instance, the words gone and gone are inflected forms of each other. Consequently, every comment is lemmatized.

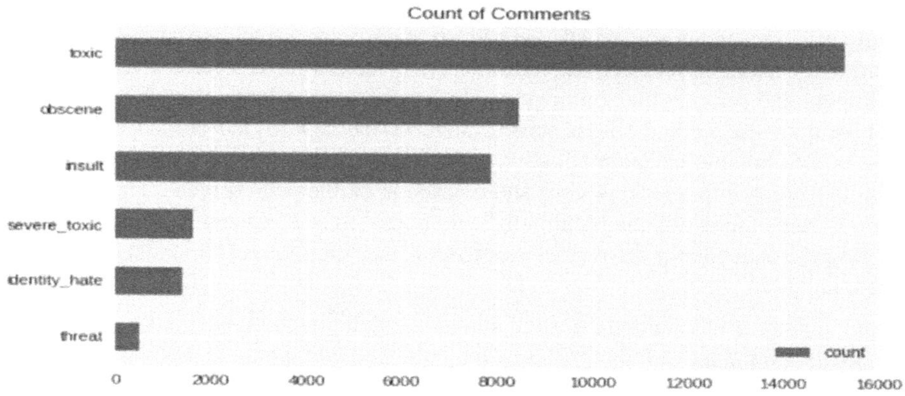

Fig. 2. Data Set Description

When referring to frequently occurring standard phrases like articles and prepositions, the Stop Word Removal approach is employed. Stop words are thus left out of every comment. It helps remove stop words, errors, and pointless data segments, or noise, by changing all letters to lowercase and removing punctuation. Consequently, it produces the lexicon of the first 30,000 most often used words.

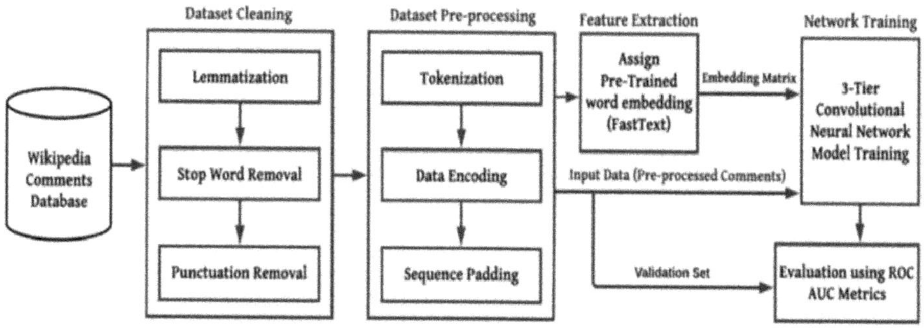

Fig. 3. Proposed System Architecture

All of the comments included in the dataset were data encoded using the token list to turn them into a series of numbers. The precise dimensions must be contained in the dataset that is fed into the neural network model. Padding is done to create a sequence of numbers with a pad length of 200 in order to make the dataset of the same dimension. In this manner, the length of each comment will remain uniform. In order to minimize the amount of concatenated 00 s to generate consistent length, the pad-length is maintained somewhat smaller in the implementation.

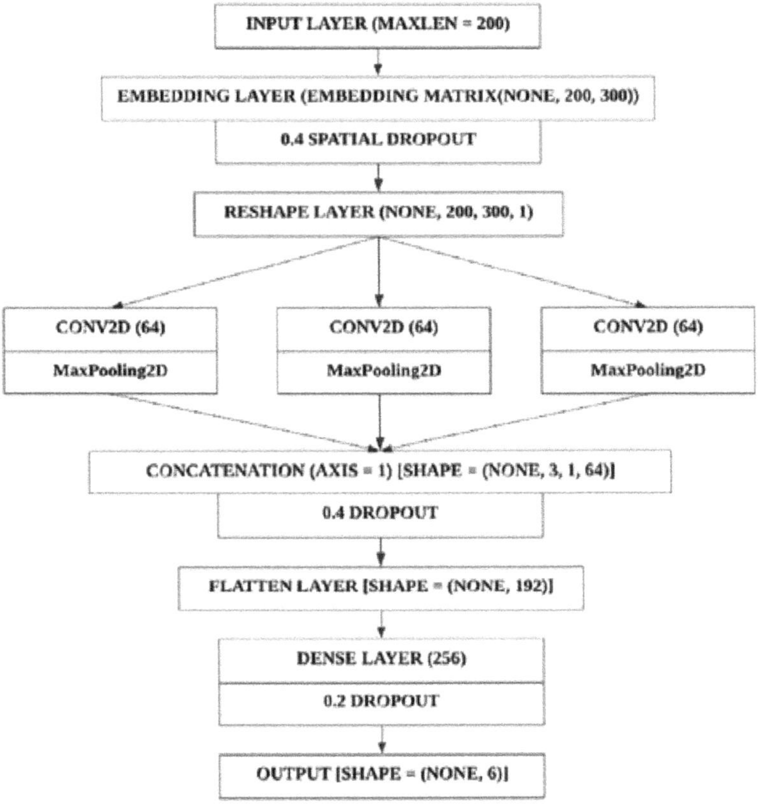

Fig. 4. Three Tier CNN Model

To identify toxicity on the Wikipedia Dataset, a unique 3-tier CNN model is suggested and put into practice, drawing inspiration from Alex Net, refer to Fig. 4. First, this system cleans. Comments by eliminating punctuation, stop words, and lemmatization. The embedding matrix that results from Fast text word embeddings is accepted by this model. To combat overfitting, the model employs the regularization approach called Dropout.

The suggested system's process is shown in Fig. 2. Because ELUs have negative values, they can reduce processing costs by driving mean unit activations closer to zero, much like batch normalization does. ELU networks. On ImageNet achieve a classification error of less than 10% for the single crop, single model network, and considerably outperform Bell networks with the same architecture in terms of learning speed (Clever et al., 2016). Equation 1 displays the ELU activation formula. In this formula, the learning rate is denoted by / and the output from the convolution block by x. The ELU activation function is applied after each convolutional block to improve training stability and speed by keeping activations closer to zero.

$$\text{ELU}:\begin{cases} x & for\ x \geq 0 \\ \alpha\,(e^x - 1)\,for\ x < 0 \end{cases} \tag{1}$$

The Alex Net-based three-tiered, multi-channel CNN architecture is shown in Fig. 3. By altering the kernel sizes of the three Conv1D channels to Alex Net filter sizes 11, 5, and 3, feature extraction at different resolutions is achievable. The residual from each convolution block is then transferred through max-pooling layers. These layers recapitulate learnt feature sizes while maintaining accuracy. Standard and Spatial Dropout avoid model overfitting. CNN generalization model uses ELU activation function. Figure 4 shows the model's basic components: an input layer, an embedding block with an embedding layer and a spatial dropout layer.

Embedding Block: The most effective method for obtaining both a dense representation and the relative meaning of words is to use word embeddings. For text analysis, a neural network and the Embedding Layer offered by Keras, API may be utilized together. An embedding layer for a particular NLP task is a word embedding that is trained alongside a neural network model. Text preparation is required so that every word is encoded in a single pass. The vector space's dimensions-typically 50, 100, 200, or 300-are specified by the model, and it begins with discrete random integers. Using the Backpropagation method under supervision, the embedding layer is fitted at the beginning of the neural network. As a result, it has the embedding matrix as weights that were created with the dimensions (30000, 300) during the data preprocessing. The embedding blocks. Output shape is (200, 300). Prior to concurrently feeding the output as an input to every convolution block, a spatial dropout of 0.4 is introduced. The first convolution block is made up of 64 11-filter feature kernel filters. The shape (190, 1, 64) is used to make the first feature maps by the Conv2D layer. A 2-D maxoggling layer looks at the combined feature maps, picks out the most important traits, and gives back their highest value.

The max pool layer sends out (1, 1, 64). Each of the three convolution blocks' Max Pool levels only sends out (1, 1, 64), which will be joined together later. By turning on all three neural blocks with ELU, CNN learns faster. There are 64 feature kernel filters in the second 5-filter convolutional block. Another set of feature maps with the shape (196, 1, 64) is made by the Conv2D layer. A 2-Dimensional max-pooling layer adds up maps to find the sets' highest values. This convolution block's maximum pool layer result is (1, 1, 64). The third convolutional block has 64 feature kernel filters with three sizes each. In the end, Conv2D makes feature maps that look like this: (198, 1, 64). Put these sets through the 2-D mispaging layer to make the maps work together and get the highest numbers. For this convolution block, the largest pool layer result is (1, 1, 64). The result from the three convolution blocks when put together gives us (3,1,64) for the concatenation layer.

A Flatten () layer flattens the output so that it can be fed into the fully linked layers. The network can become too good at what it does during training because it has a complicated structure and a lot of neurons that are buried in the CNN flattening aftereffect. The model has a Dropout of 0.4 so that it doesn't get too good at fitting the huge amounts of CNN blocks. This will also keep neurons in the buried layer from touching too much during model tuning.

The previous layer sends data to this thick layer, which is called the FC Block. There is a Fully Connected Layer with 256 units. After the fully linked layer, there is a 0.2 Dropout. The end result is a neural network model with an output layer of six units. The training values for our model are shown in Table 2. The Binary Cross-Entropy Loss function was made by Zhilu and Mert (2018) to figure out the loss in a classification experiment. The classification loss function is found in Eq. 2 by taking the negative mean of the log of the corrected predicted odds.

$$\text{Loss Function} = -\frac{1}{N}\sum_{i=1}^{N} y_i \cdot log(p(y_i)) + (1 - y_i) \cdot log(1 - p(y_i)) \tag{2}$$

where

N: Total number of data points.
y: class label
$P(y_i)$: predicted probability of the data point

4 Results and Performance Study

With the suggested 3-tier CNN model network training, we were able to obtain results with respectable accuracy. This shows that on the specified dataset, the innovative CNN model with spatial and standard dropout combined with ELU activation can generalize successfully. Table 3 presents the accuracy and loss findings. Meanwhile, Fig. 5 and Fig. 6 present the graphical convergence of accuracy and loss with the number of epochs for training and validation data, respectively.

The model's assessment metric is called ROC-AUC. Two-dimensional graphs called ROC curves are frequently used to assess and compare a classifier's performance. Melo (2013) states that ROC graphs provide a clear picture of the sensitivity/specificity exchange for each possible categorization level. It basically works as a probability curve that contrasts the True Positive Rate (TPR) and False Positive Rate (PPR) at various threshold levels used to separate information from noise. One attractive thing about ROC curves is that they remain constant while the distribution of classes changes. Consequently, it is helpful when dealing with an unbalanced dataset. If the proportion of positive instances compared to negative cases in a test set varies, the ROC curves remain unchanged.

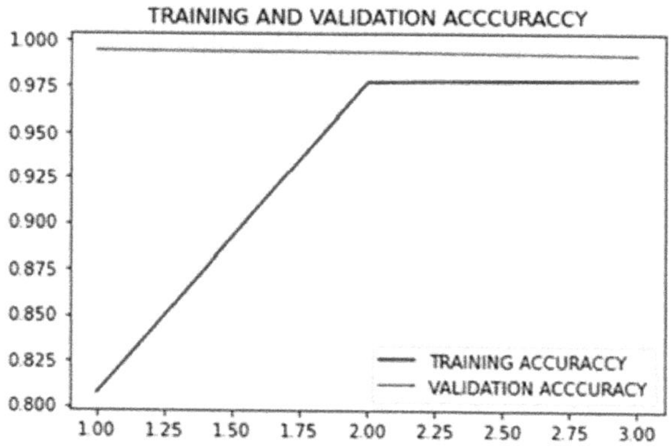

Fig. 5. Training and Validation Accuracy

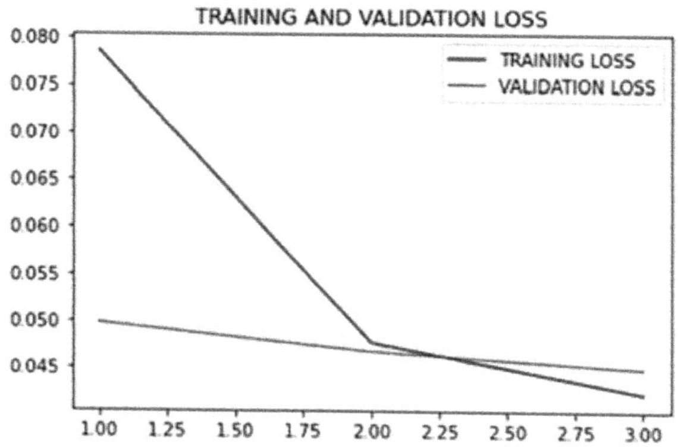

Fig. 6. Training and Validation Loss

The present efficacy of binary classifiers is measured by the AUC curve, which is thought of as a single scalar number. The AUC value ranges from 0.5 to 1.0, representing the lowest and greatest values, respectively, that represent the performance of random and ideal classifiers. A performance statistic designed for classification problems at different thresholds is the ROC-AUC curve. The relationship between sensitivity and specificity is inversely proportional. More positive findings are obtained when the threshold is lowered, increasing sensitivity but lowering specificity. Since FPR equals specificity, raising TPR also enhances FPR, and vice versa. In Eqs. 6, 7, and 8, the equations and their relationships are explained.

$$\text{TPR(True Positive Rate)/Recall/Sensitivity}: \frac{\text{TN}}{\text{TP} + \text{FN}} \tag{3}$$

$$\text{Specificity}: \frac{\text{TN}}{\text{TN} + \text{FP}} \tag{4}$$

$$\text{FPR} = 1 - \text{Specificity} : \frac{\text{FP}}{\text{TN} + \text{FP}} \tag{5}$$

The model predicts O and 1 as 1 more accurately when the AUC is high. The classifier can distinguish negative and positive class points with sufficient accuracy when AUC is 1. At AUC 0, the classifier will identify all positives as negatives and all negatives as positives. At AUCs between 0.5 and 1, the classifier can discriminate positive and negative class values. The classifier finds more real negatives and positives than false positives and negatives. When AUC falls below 0.5, the classifier cannot distinguish positive and negative class points. Each data point is classified by the classifier as regular or random. The F1 Score combines precision and recall, the main class imbalance metrics. They also underpin the F1 score (Powers, 2011). As shown in Eq. (3), precision is the ratio of true positives to positive predictions. Recall measures the ratio of correct positive predictions to positive cases, as shown in Eq. (4). Precision-Recall trade-offs prevent an optimum model from identifying all positive events and detecting all positive cases.

$$\text{Precision} : \frac{\#\text{True Positives}}{\#\text{False Positives} + \#\text{True Positives}} \tag{6}$$

$$\text{Recall} : \frac{\#\text{True Positives}}{\#\text{False Negatives} + \#\text{True Positives}} \tag{7}$$

The F1 score combines recall and accuracy into one number. Inaccurate data doesn't effect F1 score performance. Equation 5 defines FI score as harmonic mean of recall and accuracy. A 0.81 recall and 0.76 accuracy were estimated for the proposed model on the evaluation dataset.

$$\text{F1 Score} : 2 * \frac{\text{Precision} * \text{Recall}}{\text{Precision} + \text{Recall}} \tag{8}$$

The evaluation dataset's F1 score was determined to be 0.79 based on recall and accuracy. As a result, the model obtained a respectable F1 score. Trade-Off.

5 Discussion

The bar graph illustrates the accuracy comparison of various models for comment toxicity detection, refer to Fig. 7. The Proposed 3-Tier CNN Model outperforms all other methods with the highest accuracy of 98.50%, showcasing its effectiveness in identifying toxic comments. The TF-IDF model [15] (Chakrabarty et al., 2020) closely follows, achieving an accuracy of 98.08%, emphasizing the significance of term-weighting techniques. The LSTM with Custom Embedding [16] (Kohli et al., 2017) also performs exceptionally well, recording an accuracy of 97.78%, which indicates the value of custom embeddings in enhancing model performance.

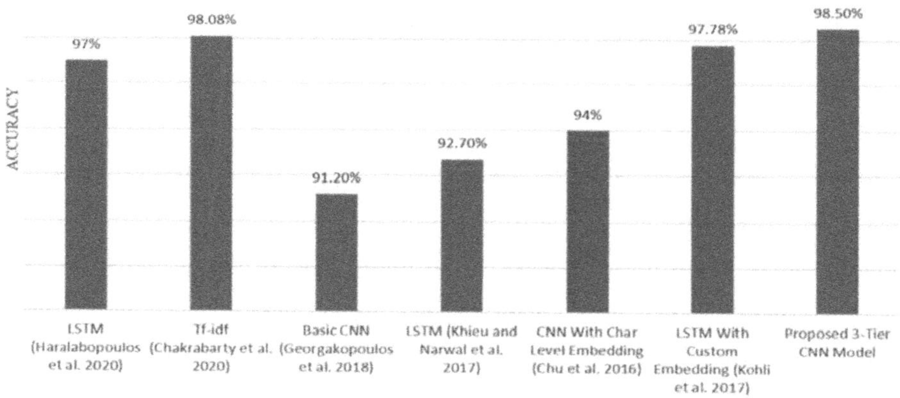

Fig. 7. Accuracy Performance Comparison with Existing Models

In contrast, the basic CNN model [2] (Georgakopoulos et al., 2018) shows the lowest accuracy at 91.20%, reflecting its limitations in capturing complex toxicity patterns. The CNN with Char Level Embedding [12] (Chu et al., 2016) and the LSTM (Khieu and Narwal, 2017) models achieve accuracies of 94% and 92.70%, respectively, demonstrating moderate efficacy. Finally, the LSTM [14] (Haralabopoulos et al., 2021) achieves an accuracy of 97%, highlighting its competitive performance among the other methods. Overall, the graph underscores the superior performance of advanced hybrid and multi-layered architectures in achieving high accuracy for toxicity detection.

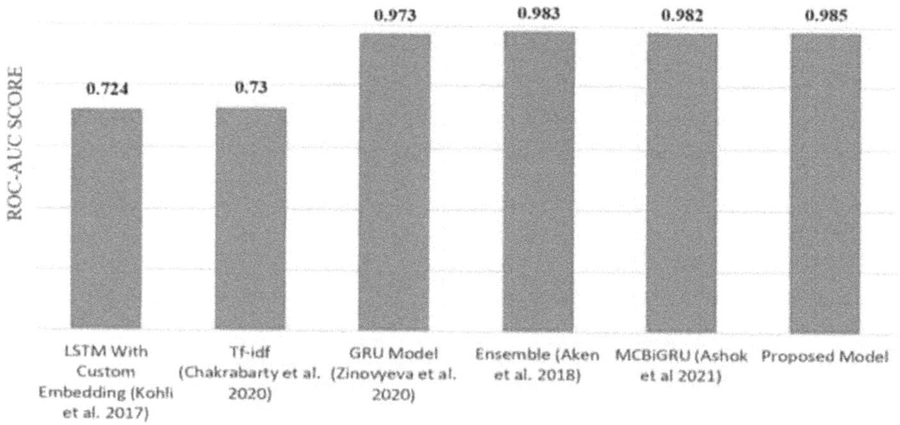

Fig. 8. ROC-AUC Comparison with Existing Models

The graph illustrates the comparative accuracy of various models and techniques for comment toxicity detection. Among the evaluated approaches, the Proposed 3-Tier CNN Model demonstrates the highest accuracy at 98.50%, underscoring its effectiveness in capturing toxic patterns in comments. Closely following this is the TF-IDF model, achieving 98.08%, showcasing the continued relevance of traditional feature extraction techniques. LSTM-based methods also perform competitively, with Haralabopoulos et al. (2020) and Kohli et al. (2017) achieving 97% and 97.78%, respectively, highlighting the robustness of recurrent architectures in this domain. On the other hand, CNN with Char Level Embedding (94%) and LSTM by Khieu and Narwal (92.70%) exhibit moderate performance, while the Basic CNN by Georgakopoulos et al. records the lowest accuracy at 91.20%, likely due to its simpler architecture. This analysis demonstrates that advanced architectures, particularly multi-tier CNNs and custom embeddings, significantly enhance toxicity detection, paving the way for more accurate and reliable content moderation systems.

This study found fraudulent Wikipedia comments using a three-tier CNN model. According on simulation data, the Alex Net architecture-based hazardous comment classification method surpasses most existing methods with 98.505% accuracy. The demonstrates that the CNN model generalised well to new datasets. Many performance evaluation methods do not employ the ROC-AUC score. The suggested model outperforms state-of-the-art methods with 0.9854 ROC-AUC. A higher score indicates the model's comment type discrimination. Figure 8 compares the model's ROC-AUC score to others. The suggested model had a good F1 score of 0.79 on the validation dataset.

The graph 9 compares the F1 scores of various models employed for comment toxicity detection. The proposed model and the ensemble model [13] by Aken et al. (2018) both achieve the highest F1 score of 0.79, indicating their superior ability to balance precision and recall effectively. Following these, the GRU Model [11] by Zinovyeva et al. (2020) demonstrates a strong performance with an F1 score of 0.76, highlighting the potential of gated recurrent units in this task. The CNN with Char Level Embedding [12] (Chu et al., 2016) achieves a score of 0.73, performing moderately well in capturing toxicity features at the character level. Meanwhile, MCBiGRU (Ashok et al., 2021) records a slightly lower score of 0.72, which still reflects competitive performance. The LSTM by Khieu and Narwal (2017) lags behind with an F1 score of 0.70, possibly due to limitations in its architecture or dataset-specific challenges. Overall, the results emphasize that ensemble techniques and advanced hybrid models consistently outperform single-model approaches in detecting toxic content (Fig. 9).

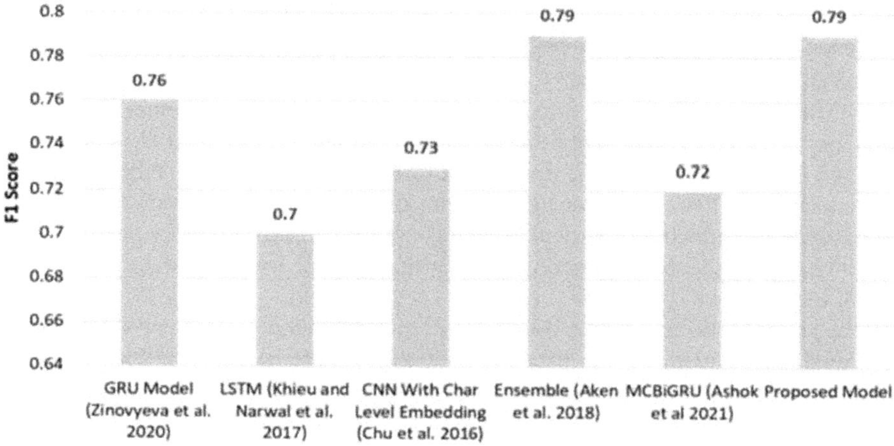

Fig. 9. F1 Score comparison with Existing Model

6 Conclusion

In conclusion, Python and deep learning let us model remark toxicity. The approach automatically detects dangerous comments on internet platforms, making the internet more inclusive and secure. Using DL architectures like RNN with attention mechanisms and CNN, the model efficiently collected semantic and contextual information from statements. Word embeddings helped the model distinguish harmful from non-damaging comments. Project tasks include data preparation, feature engineering, model design and training, assessment, and fine-tuning. Python-based data science tools like TensorFlow and Keras are flexible and effective. We assessed the model using a large dataset of tagged comments and criteria like accuracy, precision, recall, and F1-score. The model's ability to identify harmful remarks was demonstrated by the findings, suggesting that content moderation systems may incorporate it. Online platforms may foster a more diverse and secure user community by using this comment toxicity model, which offers an automated method of identifying and handling offensive remarks. This work advances content moderation systems by offering a scalable and useful method for identifying harmful comments using DL algorithms. By experimenting with other DL architectures and applying more sophisticated NLP methods, future research might further improve the model. The robustness and applicability of the model across various platforms and user demographics. Might be enhanced by expanding the assessment to include a wider range of datasets and eliminating any biases. To sum up, this study demonstrates how effectively DL and Python combine to handle the problem of comment toxicity, offering academics and business experts a valuable resource for creating effective and practical content moderation algorithms.

References

1. Zhang, L., Luo, L.: Hate speech detection: a solved problem? The challenging case of long tail on Twitter. arXiv preprint arXiv:1803.03662 (2018)

2. Georgakopoulos, S.V., Tasoulis, S.K., Ysabatis, A.G., Plagianakos, V.P.: CNNs for toxic comment classification. In: Proceedings of the 10th Hellenic Conference on Artificial Intelligence (2018)
3. Geetha, B., Malathi, P., Thirumalaikumari, T., Janakiraman, V., Basha, H.A., Rukmani Devi, S.: Machine learning approaches for proactive phishing attack detection. In: Proceedings of the 2024 7th International Conference on Circuit Power and Computing Technologies (ICCPCT), Kollam, India, pp. 757–762 (2024)
4. Chandrasekharan, E., Pavalanathan, U., Srinivasan, A., Glynn, A., Eisenstein, J., Gilbert, E.: You can't stay here: the efficacy of Reddit's 2015 ban examined through hate speech. In: Proceedings of the ACM on Human-Computer Interaction 1(CSCW), pp. 1–22 (2017)
5. Jayanth, P., Kumaran, G., Parthiban, S., Sankar, S., Pradeepkumar, S., Manikavelan, D.: A novel approach on automatic maritime object detection and classification using support vector machine in comparison with residual networks. In: Proceedings of the 2024 International Conference on Advances in Data Engineering and Intelligent Computing Systems (ADICS), Chennai, India, pp. 1–5 (2024)
6. Gröndahl, T., Pajola, L., Juuti, M., Conti, M., Asokan, N.: All you need is love: evading hate speech detection. In: Proceedings of the 11th ACM Workshop on Artificial Intelligence and Security, pp. 2–12 (2018)
7. Saravanan, M.S., Suresh, D., Sudha, I., Nandhini, T.J., Sivashankar, S., Kalyani, K.: Quality enhancement for uninvited content of social media using support vector machine and Alexnet. In: Proceedings of the 2023 International Conference on Data Science, Agents & Artificial Intelligence (ICDSAAI), Chennai, India, pp. 1–5 (2023)
8. Mozafari, M., Farahbakhsh, R., Crespi, N.: A BERT-based transfer learning approach for hate speech detection in online social media. In: Cherifi, H., Gaito, S., Mendes, J., Moro, E., Rocha, L. (eds.) Complex Networks and Their Applications VIII. COMPLEX NETWORKS 2019. Studies in Computational Intelligence, vol. 881, pp. 928–940. Springer, Cham (2020). https://doi.org/10.1007/978-3-030-36687-2_77
9. Badjatiya, P., Gupta, M., Varma, V.: Stereotypical bias removal for hate speech detection task using knowledge-based generalizations. In: The World Wide Web Conference, WWW 2019, pp. 49–59 (2019). https://doi.org/10.1145/3308558.3313504
10. Kumaresan, S.J., Senthilkumar, C., Kongkham, D., Beenarani, B.B., Nirmala, P.: Investigating the effectiveness of recurrent neural networks for network anomaly detection. In: Proceedings of the 2024 International Conference on Intelligent and Innovative Technologies in Computing, Electrical and Electronics (IITCEE), Bangalore, India, pp. 1–5 (2024)
11. Zinovyeva, E., Härdle, W.K., Lessmann, S.: Antisocial online behavior detection using DL. Decis. Support. Syst. **138**, 113362 (2020)
12. Chu, F., Beck, D.A., Lidstrom, M.E.: MxaY regulates the lanthanide-mediated methanol dehydrogenase switch in Methylomicrobium buryatense. PeerJ **4**, e2435 (2016)
13. van Aken, B., Risch, J., Krestel, R., Löser, A.: Challenges for toxic comment classification: an in-depth error analysis. In: Proceedings of the 2nd Workshop on Abusive Language Online (ALW2), pp. 33–42. Association for Computational Linguistics, , Brussels (2018)
14. Haralabopoulos, G., Tsikandilakis, M., Torres, M.T., McAuley, D.: Objective assessment of subjective tasks in crowdsourcing applications. In: Proceedings of the LREC 2020 Workshop on Citizen Linguistics in Language Resource Development, pp. 15–25 (2020)
15. Chakrabarty, T., Muresan, S., Peng, N.: Generating similes effortlessly like a Pro: a style transfer approach for simile generation. In: Proceedings of the 2020 Conference on Empirical Methods in NLP (EMNLP), pp. 6455–6469 (2020)
16. Kohli, R., Pizarro, M., Nevárez, A.: The "New Racism" of K–12 schools: centering critical research on racism. Rev. Res. Educ. **41**(1), 182–202 (2017). https://doi.org/10.3102/0091732X16686949

AngioResUNet++: A Hybrid Transfer Learning Model for Precise CCTA Image Segmentation and Categorization

M. Chitra Devi[1](✉), M. Ramaswami[2], and C. Sundar[3]

[1] Madurai Kamaraj University, Madurai, India
cdvikasini@gmail.com
[2] Department of Computer Applications, Madurai Kamaraj University, Madurai, India
[3] Department of Computer Science and Engineering, Christian College of Engineering and Technology, Oddanchatram, India

Abstract. Coronary CT Angiography (CCTA) is a critical diagnostic tool for detecting abnormalities in the Coronary blood arteries. Automated techniques are required to increase diagnostic speed and accuracy because manual CTA image interpretation is a labor-intensive, tedious task that increases the risk of misinterpretation. For this proposed work, we provide AngioResUNet++, a groundbreaking system to segmenting abnormal arteries and classifying that segmented coronary arteries from CCTA images. The system combines the DeepUNet++ model for efficiently segmenting coronary vessel with a pretrained ResNet-50 network for dependable classification. This system automates segmentation of heart blood vessels, extracting key features including texture, statistical, and geometric characteristics, and classifies the arteries into following categories: Normal, Blood-flow reduced Blocked and Narrowed. Unlike traditional methods, which often suffer from low accuracy and inefficient feature extraction, AngioResUNet++ provides an efficient solution that combines feature extraction and classification in a unified pipeline. Both segmentation and classification performance are improved by the deep learning method, which yielded a 99.71% classification accuracy and a 96.62% segmentation score. These results demonstrate significant improvements over conventional techniques, offering a faster, more reliable tool for diagnosing coronary artery diseases. The proposed system supports clinical decision-making, enabling timely interventions and improving patient outcomes.

Keywords: Coronary CT Angiography (CTA) · Deep learning · Segmentation · Classification · Coronary artery disease · AngioResUNet++ · DeepUNet++ · ResNet-50 · Image classification · Transfer learning

1 Introduction

Globally, one among the primary causes of death and disability is still heart artery disease [1]. The coronary arteries are in the position of delivering blood to the heart muscle, can become affected by various conditions, leading to restricted or blocked blood

R. Geetha et al. (Eds.): AAIMB 2024, CCIS 2654, pp. 262–275, 2026.
https://doi.org/10.1007/978-3-032-13029-7_19

flow, which in turn results in heart-related health issues such as heart attack, and even sudden cardiac arrest [2]. Early and accurate diagnosis of coronary artery conditions is therefore essential for preventing severe cardiovascular events [4].Coronary Computer Tomography Angiography (CCTA) is a widely used imaging technique to visualize the coronary arteries and assessing their health [6]. However, interpreting CTA images manually is time-consuming, subjective, and prone to human error. With advancements in deep learning and artificial intelligence (AI), there has been a growing interest in developing automated tools to classify and segment coronary arteries efficiently [6]. Such tools can help healthcare professionals quickly identify areas of concern, potentially reducing diagnosis time and improving the accuracy of assessments The classification of coronary arteries typically involves categorizing the vessels based on their condition, which can vary from healthy (normal) to severely affected (blocked). Based on guidelines from the World Health Organization (WHO) and clinical standards, coronary artery abnormalities are typically classified into four primary categories: Normal Arteries, Narrowed Arteries, Blood Flow Reduced and Blocked Arteries (Fig. 1).

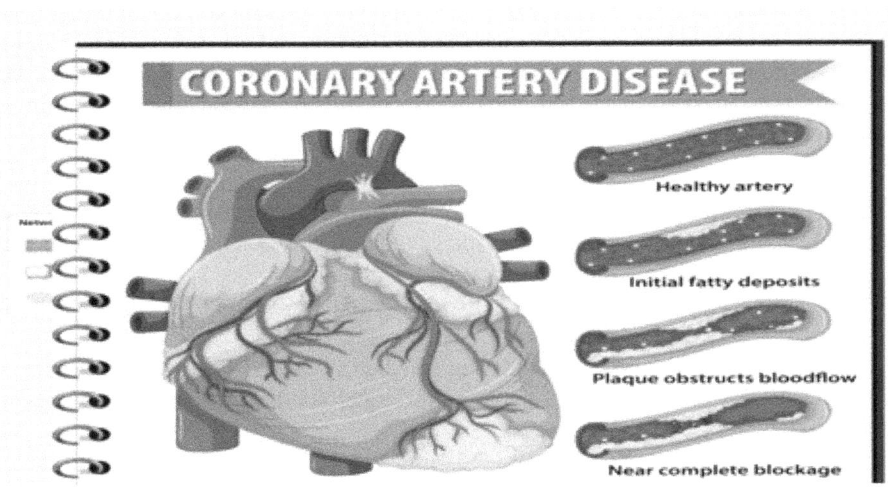

Fig. 1. An illustration of cardiovascular disease

Determining the degree of arterial disease and selecting the most appropriate treatment depend on the prompt and precise classification of heart arteries into these groups. Traditional methods for coronary artery assessment involve invasive procedures like coronary angiography, which, while accurate, are costly, time-consuming, and carry inherent risks. Non-invasive imaging techniques such as Coronary CT Angiography (CTA) offer a safer, quicker alternative but require effective interpretation of the images for accurate diagnosis. Clinicians can obtain precise, real-time evaluations of the coronary arteries by incorporating deep learning into coronary artery categorization, which empowers them to make better judgments about patient management. The automated pre-trained classification system[7], therefore, supports the WHO's objective of improving global cardiovascular health by enhancing diagnostic precision, reducing healthcare costs, and ensuring timely treatment for patients with Coronary Artery Disease. This paper follows the following flow: Sect. 2 reviews the previous classification as well as segmentation techniques for blood vascular segmentation and classification. Section 3 provides a detailed description of the network architecture and recommended methods. Section 4 displays the experimental data and their analysis. At last, Sect. 5 presents the study's conclusion.

1.1 Novelties and Contributions

The existing literature clearly identifies several unresolved challenges in the field. A key observation is that none of the previous studies used raw images from hospital patients, as noted in the related work section; instead, most studies relied on publicly available online datasets. In contrast, our research leverages real coronary angiography images obtained from hospitals in Kolkata, West Bengal, India. This approach aligns with our goal of achieving high-accuracy segmentation of angiogram images using CNNs. By utilizing patient data directly from hospitals, rather than relying on pre-existing online datasets, our work offers a more authentic and clinically relevant solution. Furthermore, while many prior studies focused on image reconstruction and preprocessing techniques for model training, our proposed method employs complete image augmentation, thus eliminating the need for such additional steps. Using specialized filters to minimize image noise is one of our main research goals.

The following are the contributions that were contributed by this work:

- Pre-processing: GaussianBlur is one that used in the first phase to reduce noise in CCTA images while enhancing input image quality. The CCTA Image dataset is then expanded by applying various augmentation techniques.
- Segmentation: The DeepUNet++ network is used to segment specific regions of abnormal blood vessels' features
- Categorization: The segmented arteries are then classified into four categories as normal arteries, blocked arteries, narrow arteriesandblood-flow reduced arteries using ResNet50.
- Evaluation: AngioResUNet++ is trained using trained data, validated using validation data set and assessed with different performance criteria.

2 Prior Related Works

The usefulness of machine learning, SVM model, in forecasting heart artery disease is highlighted by Sayadi et al. [4]. The findings imply that, given the data at hand, SVM can produce precise forecasts of CAD situations. Although the SVM model shows promise, the study calls for further comparisons with other algorithms to identify the optimal model for CAD prediction. The performance of SVM might not be the best for all scenarios. The paper [5] investigates approaches for coronary abnormal blood vessels segmentation using the3D UNet model, applied to CTCA (Computed Tomography Coronary Angiography) data. The segmentation was tested on datasets both with and without centerline information. The quality of the training data has a big influence on how well the model works. The paper suggests that using better-quality data could significantly improve the segmentation results, highlighting a potential limitation in the current dataset. The DR-LCTUNet method proposed for coronary artery segmentation addresses several key challenges in medical image segmentation, specifically the problems of omission and over-segmentation that are common in previous methods [6]. The proposed method is specifically tailored for coronary artery segmentation and might not generalize well to other types of medical image segmentation tasks without further modifications.

This study [9] introduces an automated segmentation framework without intervention of human. Used an improved CNN to search for neighboring blocks and determine the presence of arteries. According to the Dice similarity coefficient, the study [11] achieved an 88% segmentation accuracy for the aberrant arteries using a 2D UNet framework to separate them from CTCA pictures. This study [12] proposes a computationally light 2D UNet with transfer learning applied for to CT images. Transfer learning helps improve performance, especially when data is scarce.2D UNet performs well in distinguishing thin and distal vessels. However, it tends to overestimate stenotic lesions. The proposed segmentation method uses the PSPNet [15] to process coronary angiography images. The segmentation results were compared with a commonly used architecture in image segmentation. It achieved the accuracy of segmentation 93.6% and accuracy of classification 95.7%, sensitivity of 86.5% and specificity of 94.5%.The study [16] proposes an automatic segmentation approach for coronary arteries using deep learning models, specifically employing U-Net, which achieved a Dice score of 84.67%. For classification, the authors utilized various pre-trained models, including DenseNet201, EfficientNet-B0, and MobileNet-v2. The DenseNet201 model achieved 90% average accuracy, 79.99% sensitivity score and specificity score of 98.33% after training and evaluation. Crucially, the absence of training and validation datasets is one of the study's shortcomings. Additionally, the study only looks at two class classification. Despite extensive research and numerous proposed approaches in this area, much of the existing work relies on publicly available images from online databases. In contrast, this study utilizes angiogram images from patients at hospitals in West Bengal, India. Here is a path for creating a novel model that uses deep neural network techniques to segment and categorize coronary arteries using Angiography images, based on the insights from the cited publications. DeepUNet++ driven Segmentation and Pre-trained Neural Network Classification for Coronary CT Angiography.

2.1 Problem Statements

In coronary artery segmentation, all of the aforementioned research faces a number of difficulties, and existing model's performance is highly dependent on the data at hand. It is challenging to obtain images for coronary artery segmentation because certain methods are computationally expensive and require a significant amount of memory and processing time, the datasets used are often class imbalanced, the images are medical records that must be manually segmented beforehand, and the heart blood arteries only make up a small portion of each image segment. Building on prior research in coronary CT angiography, the goal is to develop a hybrid model that combines advanced segmentation and classification techniques to accurately identify and assess coronary artery conditions, such as Normal, Blocked, Blood Flow Reduced, and Narrow. Augmentation technique will be used to address the imbalance of classes and generate a large volume of training data, thereby enhancing diagnostic accuracy. The proposed approach integrates deep learning-based segmentation (e.g., 2D U-Net) with a ResNet50classification model utilizing transfer learning, ensuring a precise final diagnosis.

3 Proposed Methodology

There are two main subtasks in automatic CAD diagnosis. Finding the region that is relevant in the images is the initial hurdle, known as segmentation. The coronary artery tree of each cardiac CT angiogram image was clipped for this study and used for further examination. The second duty involves using clinical criteria to categorize the heart's blood arteries into four groups: regular arteries, artery with blockage, narrowed arteries, and reduced blood flow arteries.

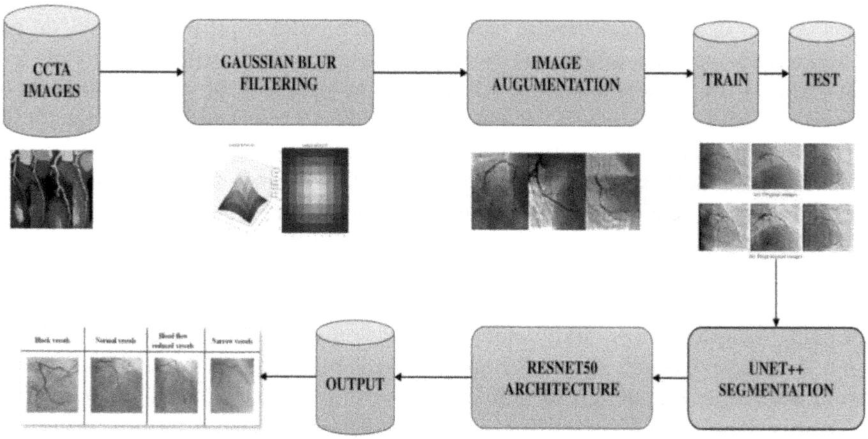

Fig. 2. The Suggested architecture

The three primary parts of the suggested system are categorization, segmentation of the coronary arteries, and pre-processing. Pre-processing seeks to eliminate undesired aberrations from images in order to improve their quality. This is achieved by using Gaussian Blur filters on the source image to remove extraneous noise. Image flipping and rotation are examples of appropriate augmentation methods that are used to increase the amount of the dataset. The Deep UNet++ framework is used to segment the blood flow arteries after image pre-processing. Lastly, the ResNet50 model divides the segmented coronary arteries from CCTA images into four classes. The complete framework for the proposed work is illustrated in Fig. 2.

3.1 CCTA Image Dataset Description

CT image of coronary angiograms from a health center in West Bengal, India, were used in this investigation. One hundred from each of the four categories of abnormal arteries were gathered. Every image was converted to 960 x 1080 JPEG format. The dataset images were resized according to Python limitations. Table 1 contains information on the dataset as well as the instances.

Table 1. Class-wise distribution of dataset volume

Label	No. of Images in each Category
Blocked	Hundred
Narrow	Hundred
Reduced of blood flow	Hundred
Normal	Hundred
ToTal	Four Hundred

In order to accurately depict the range of human coronary arteries, cardiac specialists sought to diagnosis the Coronary CTA data into discrete groups, includes regular cessels, vessels with decreased blood-flow, blocked, and narrowed vessels. Ventricular arteries can be split and categorized based on their blood flow, which might be normal, blocked, constricted, or decreased. It is anticipated that the dataset would produce trustworthy results, and the outcomes are aggregated to evaluate the efficacy of different classification methods.

A cardiologist oversaw the collection of all 400 angiographic coronary artery images that make up the dataset. There are 100 images in each category, each of which depicts a different heart vascular disease. The effects of these diseases on the coronary arteries are assessed statistically.

3.2 Image Data Preprocessing

The pre-processing phase consists of two image processing steps: Gaussian Blur (GB) filters, which is applied to remove noise and improve image contrast. Figure 3 shows the enhanced image after doing preprocessing. The dataset size is then increased using data augmentation methods like flipping and rotation. The preprocessing technique's main aim is to improve the images' quality, making them suitable for subsequent analysis by eliminating or reducing irrelevant elements. Pre-processing is crucial to enhancing the quality of biomedical images because evaluating them can be difficult. In most classification tasks, images are directly fed into the pre-trained framework. To enhance CCTA image dataset's quality, GaussianBlur filter is applied. Often employed as a lowpass filter, the Gaussian filter removes high-frequency components from images.

3.3 CCTA Image Augmentation

Augmentation process is used to increase the volume of dataset. The majority of deep learning-based image processing models make extensive use of these techniques. Rotation and flip were two image augmentation strategies used in this study to lessen the problem of overfitting. For rotation, each image was rotated counterclockwise within a range of 90° to 360° around its center, using bilinear interpolation. This method preserves the original content of the image while maintaining its aspect ratio. Additionally, flipping, a straightforward augmentation technique, was applied to further diversify the dataset and enhance model robustness. 1,000 samples are allocated to each of the four classes after the data augmentation process produced a total of 4000 sample images. Figure 4 showcases sample results from the augmented datasets.

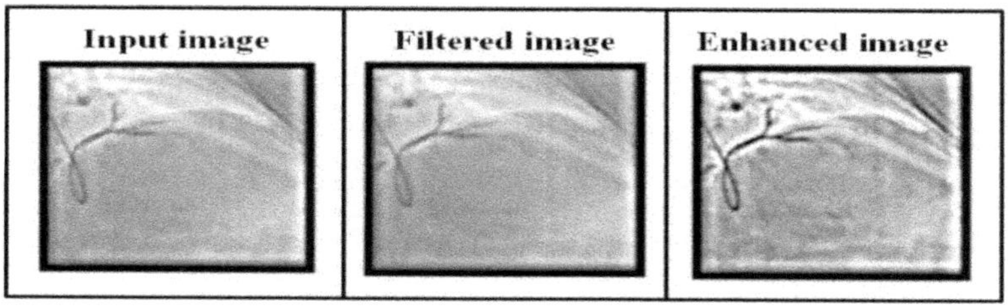

Fig. 3. Enhanced images.

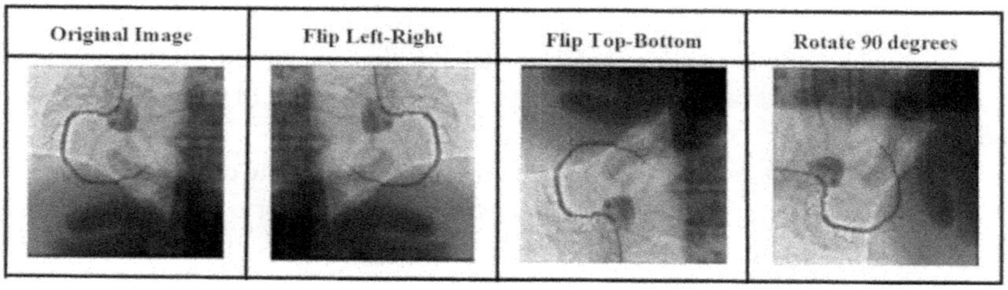

Fig. 4. Results of image augmentation.

3.4 Segmentation Using UNet++ Architecture

The connectivity between UNet++'s encoder and decoder portions is changed by the skip paths. The encoder's feature maps are sent straight to the decoder in a basic UNet. In UNet++, on the other hand, these feature maps first go through a dense CNN block, where the pyramid depth determines how many convolution layers are used. UNet++ generates feature maps at several conceptual levels that are appropriate for deep monitoring because of the layered skip routes. We have used Dice Coefficient for calculatingloss. Figure 5 illustrates the segmented arteries through this process (Fig. 6).

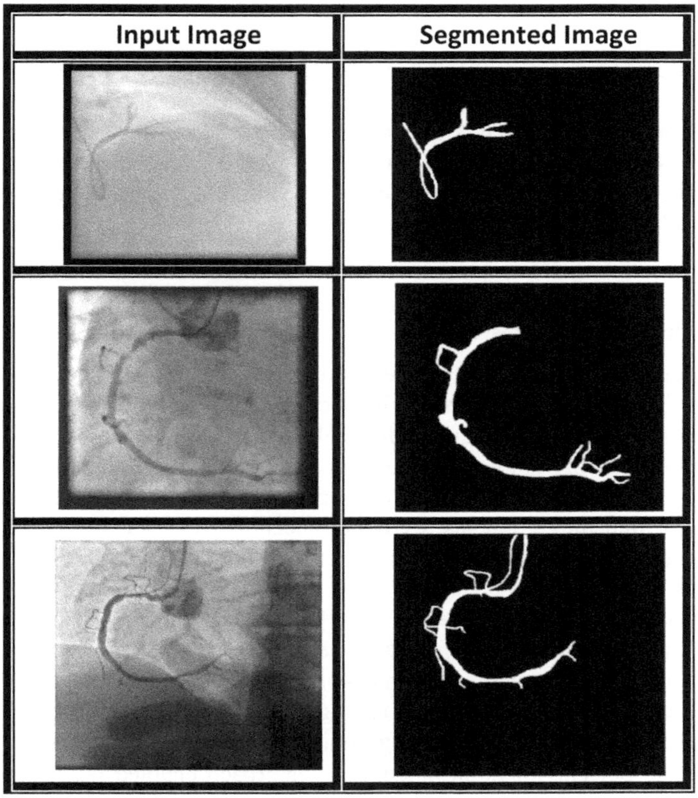

Fig. 5. Segmentation output of the proposed approach.

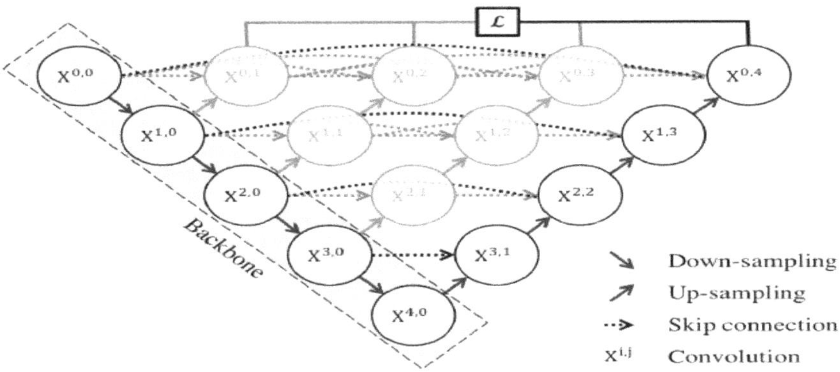

Fig. 6. UNet++ Architecture [13]

3.5 Classification Using ResNet50

The segmented coronary arteries generated by DeepUNet++ are used as input to ResNet50 for categorization [14]. CNN model called ResNet50, a variant of Residual Neural Networks [16], several models are employed to address the challenges associated with training incredibly deep neural networks. ResNet50 is renowned for its breadth and potency in picture classification tasks. Models of different depths, including ResNet18 and ResNet32, are part of the ResNet family; ResNet50 is a mid-tier version in terms of performance and complexity.

3.5.1 Residual Block of ResNet

ResNet's primary focus was on the deep neural network degradation issue. Network accuracy tends to plateau and then rapidly decline as depth increases. The accuracy of networks tends to plateau and then rapidly decline as they get deeper. The difficulties in improving the course of training are the cause of this degradation rather than over-fitting. Residual Blocks, which allow direct information flow over skip links and so mitigate the problem of vanishing gradients, were introduced by ResNet in order to address this issue. The combination of shortcut connections and bottleneck layers enables ResNet-50 to train deeper models, which can capture more complex patterns, leading to high accuracy in tasks such as image classification. ResNet-50, featuring 50 bottleneck residual blocks, has demonstrated exceptional ability to overcome the problem of vanishing gradients. This makes it possible to train deeper neural systems more efficiently, which raises classification accuracy and lowers testing and training loss. Instead of constructing the classification CNN model from the ground up, we leveraged pre-trained ResNet50 weights to train our classification model. This approach enabled faster classification, significantly reducing processing time (Fig. 7).

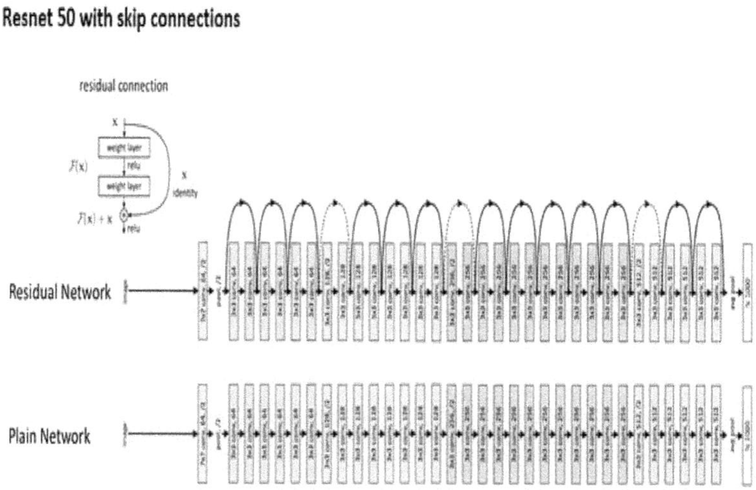

Fig. 7. ResNet50 Architecture [14]

4 Findings and Discussion

We developed a deep learning model for the proposed segmentation and classification system using TensorFlow in Python 3.6, with the Keras API for the front-end implementation. All networks were trained and tested using a 32 GB RAM NVIDIA Tesla V100 GPU.

4.1 Evaluation Metrics

We use a number of measures to assess the deep network model's performance. Specifically, we utilize Dice Similarity to analyze the model's segmentation outcomes and accuracy, specificity, and sensitivity to evaluate the classification performance. Equations (1) through (4) were used to determine the values of the metrics.

$$\text{Accuracy\_ score } (\%) = ((T\_P + T\_N)/(T\_P + F\_P + T\_N + F\_N)) * 100 \quad (1)$$

$$\text{Sensitivity\_ score } (\%) = \left(T\_P/(T\_P + F\_P + T\_N + F\_N)\right) * 100 \quad (2)$$

$$\text{Specificity\_score}(\%) = \left(T\_N / (T\_N + F\_P)\right) * 100 \quad (3)$$

For assessing segmenting process of Coronary arteries we use Dice similarity coefficient

$$\text{Dice score} = 2T\_P/(2T\_P + F\_P + F\_N) \quad (4)$$

Furthermore, our classification model's performance is examined using ROC curve. The X and Y axes are used to illustrate the True Positive.

4.2 Experimental Results

The goal of this system is to accurately classify multi-class coronary arteries in angiography images. A hybrid transfer learning-based deep learning model is used for segmenting and classifying CCTA (Coronary Computed Tomography Angiography) images. The CCTA images are preprocessed using Gaussian blur filtering to reduce noise and smooth the edges. These enhanced images are then used in the augmentation process to increase the dataset size, which helps improve classification accuracy. 70%, 20%, and 10% of the gathered augmented coronary CT images are allocated for training the model, validating the model, and testing model. To separate the arteries in the heart from the Coronary CT Angiographic images, the UNet++ model has been trained with training image data to segment the blood vessels. The segmented coronary arteries generated by UNet++ are used as input to ResNet50 for classification. ResNet50 then categorizes the segmented arteries into four distinct classes. The model's classification accuracy, specificity, and sensitivity are 99.71%, 99.51%, and 99.32%, respectively. The model attains a test accuracy of 99.85% and a complete training accuracy of 99.82%. The model's performance determines the categorization results for CCTA coronary arteries are shown in Fig. 8. Figure 9 illustrates the validation and training accuracy and loss. Over 100 epochs, the suggested model achieved 98.12% validation and 99.35% test accuracy rate, with a loss of 0.076% and 0.022%, respectively (Fig. 10).

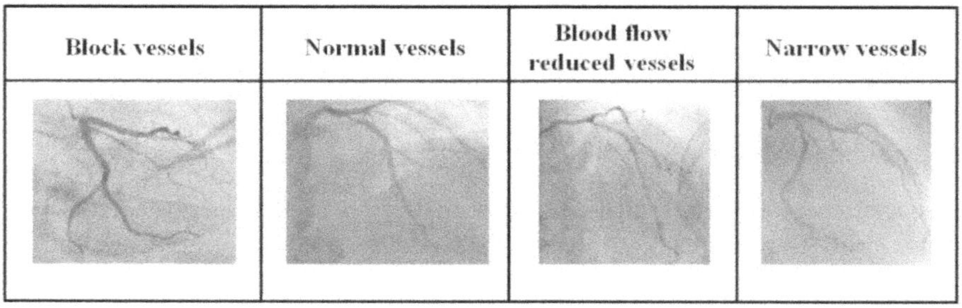

Fig. 8. Classification Output of Coronary Arteries

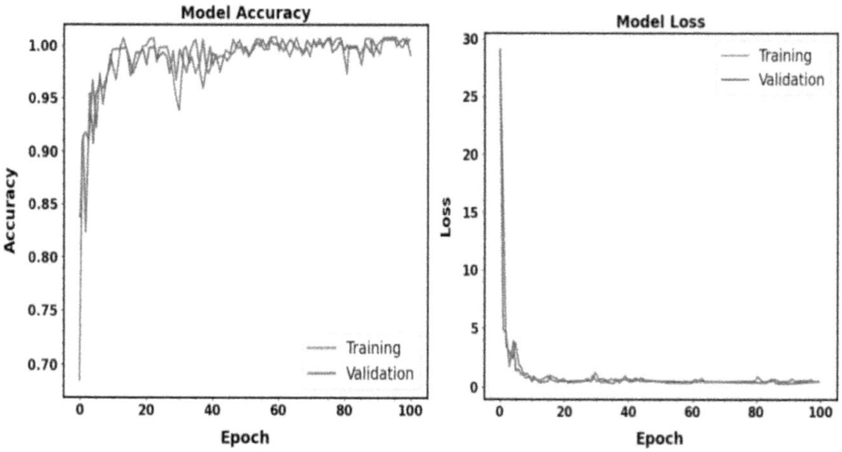

Fig. 9. Test and Validation Performance (Accuracy & Loss) of the Suggested Model

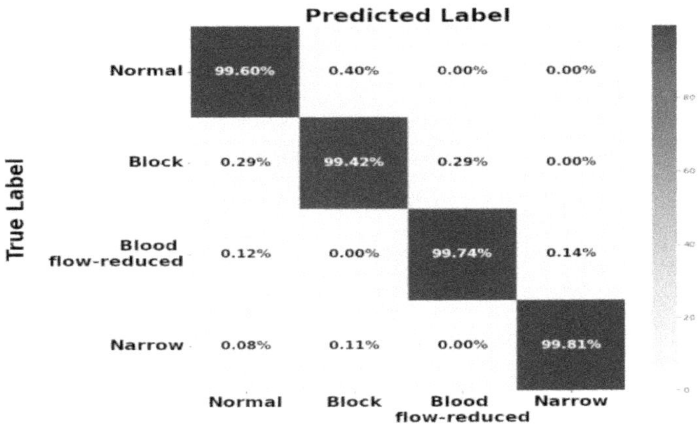

Fig. 10. Confusion Matrix for the proposed Classification System

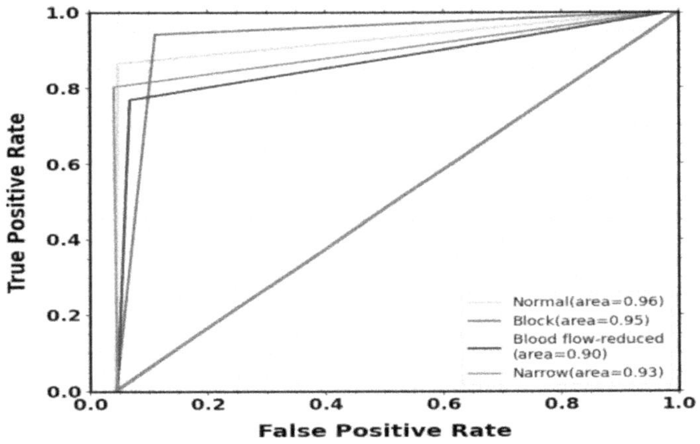

Fig. 11. ROC curve of proposed model

Figure 11 displays the confusion matrix for our AngioResUNet++ model. It also shows the ROC curve for each class, showing the high performance of the proposed model.

4.3 Comparative Analysis

The suggested AngioResUNet++ model efficiently detects multi-class coronary arteries and performs better than earlier deep learning frameworks. The numerical results demonstrating the superior performance of our model in the classification of CCTA images are presented in Table 2.

Table 2. Comparative Experimental Performance

Reference	Approach	Accuracy Score (%)	Sensitivity Score (%)	Specificity Score (%)	Dice Score (%)
Kaba et.al. [2]	ResNet	97.81	86.2	93.01	84.6
Cheung et.al. [12]	UNet++	91.27	-	-	88
Jean, F.P et.al. [17]	DLM	96	93	97	-
Proposed approach	AngioResUnet++	99.71	99.51	99.32	96.52

The ability of the suggested model to avoid overfitting, which guarantees that network performance is unaltered during the classification process, is one of its fundamental advantages. Even if the results are promising, there are some problems with the deep learning models that are earlier in use. One of the primary issues is the necessity for huge datasets because a significant amount of data is typically needed for learning models to train effectively. For getting over this limitation, in this work we utilize pre-trained models, often known as Transfer Learning framework. By using the weights of previously trained models, it enables researcher to apply for various tasks, even with less datasets. Compared to models made from scratch, this approach has been shown to perform better. We use augmentation methods like rotation and flip to alleviate the problem of insufficient data. Furthermore, no extra feature selection techniques are required because the suggested classification model efficiently chooses the most pertinent features for categorization.

5 Conclusion

The arteries in the heart are particularly vulnerable to coronary disease, which is one of the main causes of death. Imaging techniques like CCTA often produce low-contrast images, making it challenging to detect subtle variations in the thin coronary arteries. These difficulties hinder accurate diagnosis of coronary artery disease. To overcome this challenge, the proposed Transfer Learning-based deep neural network system, AngioResUNet++, which is specifically designed to extract fine details of heart blood arteries from CCTA images and classify them precise manner. The proposed model effectively identifies regions of interest within the arteries and classified with high Dice and classification score. This model could function as an automated tool for coronary artery classification following clinical coronary angiography tests. However, a key limitation of the model is its inability to process 3D CCTA images. Future enhancements will incorporate additional deep learning techniques to enable the analysis of 3D CCTA images.

References

1. WHO Report: leading causes of death. Accessed 10 Aug 2024

2. Kaba, S., Haci, H., Ilhan, A., ConKbayir, C.: The application of deep learning for the segmentation and classification of coronary arteries. Diagnostics **13**, 2274 (2023)
3. Pandey, R., Kumar, M., Majdoubi, J., Rahmi Gorji, M., Srivastav, V.K..: A review study on blood in human coronary artery. Comput. Methods Programs Biomed. **187**, 105243 (2020)
4. Mohammadjavad, S., Sadoughi, F., Varadarajan, V., Chopannejad, S., Langarizadeh, M.: A machine learning model for detection of coronary artery disease using non invasive clinical parameters. Life **12**, 1933 (2022)
5. Weimin, Z., Lu, H., Zhiping, L., Su, H., Yanling, C., Liang, Z.: Coronary artery segmentation by deep learning neural networks on computed tomographic coronary angiographic images. In: Proceedings of the 40th Annual International Conference of the IEEE Engineering in Medicine and Biology Society, pp. 608–611 (2018)
6. Wang, Q., et al.: Automatic coronary aetery segmentation of CCTA images using UNet with a local contextual transformer. Front. Physiol. **14**, 1138257 (2023)
7. Bolhassani, M.: Transfer learning approach to classify the X-ray image that corresponds to corona disease using ResNet50 pre-trained by ChexNet. J. Intell. Syst. Learn. Appl. **16**(2), 162006 (2024)
8. Sarwindaa, D., Paradisaa, R.H., Bustamama, A., Anggiab, P.: Deep learning in image classification using residual network (ResNet) variants for detection of colorectal cancer. In: Proceedings of the 5th International Conference on Computer Science and Computational Intelligence (2020)
9. Cui, J., et al.: Fully-automatic segmentation of coronary artery using growing algorithm. J. X-Ray Sci. Technol. **28**(6), 1171–1186 (2020)
10. Toa, W., et al.: Integrated deep learning model for automatic detection and classification of stenosis in coronary angiography. Comput. Biol. Chem. **112**, 108184 (2024)
11. Wing, K.C., et al.: A computationally efficient approach to segmentation of the aorta and coronary arteries using deep learning. IEEE Access **9**, 108873–108888 (2021)
12. Serrano-Antón, B., et al.: Coronary artery segmentation based on transfer learning and UNet architecture on computed tomography coronary angiography images. IEEE Access **11**, 3293090 (2023)
13. Tsang, S.H.: Review: UNet: a nested U-Net architecture for biomedical image segmentation. Medium (2021)
14. Roboflow: ResNet-50. In: Roboflow Blog. Accessed 6 Oct 2024
15. Zhu, X., Cheng, Z., Wang, S., Chen, X., Lu, G.: Coronary angiography image segmentation based on PSPNet. Comput. Methods Programs Biomed. **200**, 105897 (2021)
16. Jean, F.P., Adela, R., Henri, G., Thibaut, P.L., Tatiana, W.: Evaluation of a deep learning model on Coronary CT Angiography for automatic stenosis detection. Imaging **103**(6), 316–323 (2022)

Advanced Fake News Detection on Social Media: Leveraging Geometric Deep Learning and Propagation Pattern

C. Jackulin, L. Sathiya, Madhuru Navya Sree(✉), A. Harshini, T. Jackulin, and A. Kanchana

Panimalar Engineering College, Chennai, India
navyasreemadhuru@gmail.com

Abstract. The rise of social media has made it increasingly challenging to differentiate between genuine and fake news. The prospect of detecting misleading news is now more crucial than ever due to the growing dependence on social media, particularly on smartphones, for accessing news. This paper evaluates the performance of advanced deep learning models, namely LSTM networks and BERT. Social media is user-friendly, and so more individuals come across false in-formation; this makes it difficult to detect such content using methods based on content analysis that do not consider context. In this paper, a user profile based geometric deep learning model (GY) is proposed that considers content, social graphs, and news dissemination patterns. Here, the model accurately detects fake news within hours of its circulation. This establishes that NLP based methods perform well in the detection of misleading information. Moreover, this research reveals that the DT algorithm attained a higher accuracy than SVM.

Keywords: Deep Learning · Long Short Term Memory (LSTM) · Bidirectional Transformers for language (BERT) · Geometric Learning · Graph Neural Networks

1 Introduction

In the last decade, social media has emerged as a significant source of information worldwide. However, seeking news through it exposes one to several risks, including the existence of "fake news," that is, information that is deceptive or fabricated to affect public opinion [2]. Events like the 2016 United States presidential election and Brexit have reminded everyone of the role of fake news [1]. This threat has created a lot of attention in fake news detection.

Traditional methods, which are mainly based on content analysis, are often not able to catch contextual information and are prone to manipulation. These methods depend on the analysis of text, which might miss patterns of disinformation. The most promising alternative is Graph Neural Networks (GNNs), a geometric deep learning approach [9]. GNNs model news dissemination as a graph, where individuals are nodes, and edges

R. Geetha et al. (Eds.): AAIMB 2024, CCIS 2654, pp. 276–284, 2026.
https://doi.org/10.1007/978-3-032-13029-7_20

represent the connections between them. Analyzing network structure and patterns of news dissemination, GNNs are capable of effectively distinguishing real from fake news. Major advantages of this approach include language independence hence, the possibility of detecting fake news in multiple languages without specialized linguistic models.

2 Existing Approaches

In recent years, there have been approaches combing advanced deep learning techniques for fake news detection. For example, using the Fraudulent Tidings Challenge dataset, researchers developed a model blending convolutional neural nets with advanced neural networks. They found that preprocessing improved the accuracy from 78% to 93%, and further boosted it to 95% using Chi-square feature selection, and 96% when combining Principal Component Analysis with CNN-LSTM also reducing prediction time [5].

Comparisons between deep learning models and conventional machine-learning techniques, using datasets such as ISOT and KDnugget, showed that the deep learning models were considerably more accurate, with an accuracy of 99.94% on the ISOT. This pointed out their ability to capture the subtle patterns much better than by traditional methods.

A system was proposed to classify Twitter data as authentic or fake. The system accurately detected 15% of the deceptive tweets and 45% of the authentic tweets and assigned the remaining to an uncertain class [9]. The authors suggested that the accuracy of fake news detection can be increased by taking social interactions into consideration along with news content because comprehending how fake news disseminates through networks is an essential aspect.

3 Proposed System

The technique uses geometric deep learning to identify fake news posts on social media. The aim is to leverage the pattern associated with the spread of news around social networks [3]. It does so by analyzing the social network structures and the way news spreads among users, which enables more accurate and robust detection of fake information.

3.1 Propagation Modeling

The system mimics how content flows through the network by shares or retweets. These cascades create a graph by people as nodes and an interaction like likes, follow, and retweets as edges. The system analyzes these cascades using Graph Convolutional Networks (GCN).

3.2 Model Architecture

Graph shift invariants (GCNs), which generalize traditional convolutional networks to work with graphs rather than the grid-structured data like images, form the basis of the main approach [10]. The network comes up with a verdict whether it is fake or real news from the propagation graph of the news (Fig. 1).

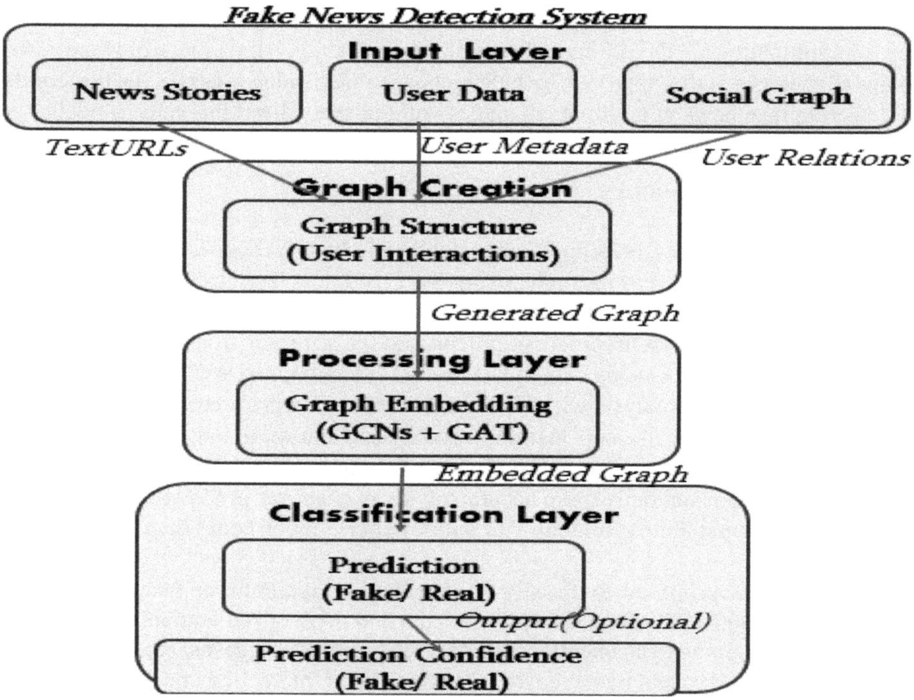

Fig. 1. Architecture of Fake News Detection using GDL

4 Methodology

4.1 Geometric Learning

Non-Geometric deep learning refers to non-Euclidean machine learning techniques. Graph CNNs use permutation-invariant aggregation around graph vertices instead of traditional convolution on grids [8]. Spectral graph CNNs perform this aggregation in the spectral domain using graph Laplacian eigenvectors [7].

Graph Representation

A graph G = (V,E) consists of edges (E) and vertices (V). Every edge (u,v) ∈ E connects two vertices: u and v. Every vertex v ∈ V has attributes x, and v is shown (see Fig. 2).

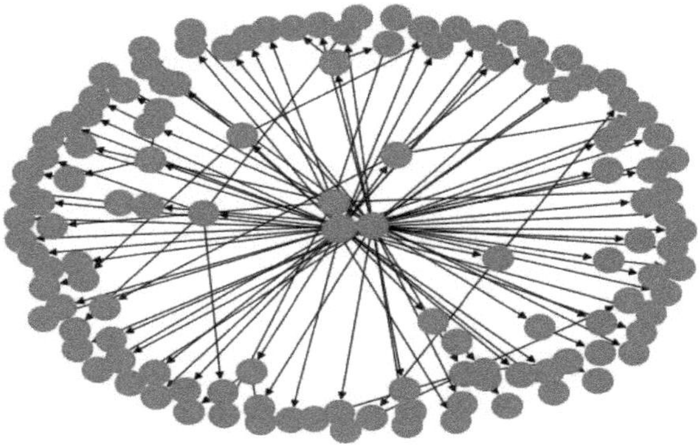

Fig. 2. News Propagation Graph

4.2 Architectural and Priming Settings

The proposed architecture is a four-layered graph neural network (Graph CNN), with two graph neural layers having outputs of 32 and 2 dimensions, and two graph convolutional layers with outputs of 64 dimensions [3]. The proposed architecture predicts the probability of news being false or accurate [13]. For the reduction of dimensionality, we have used mean pooling and graph attention mechanisms in the convolutional layers.

Data Input: News articles and their social media transmission patterns are fed into the system.

a. News content, including the news article or tweet itself.
b. Social networks Structure (users and links)
c. User activity and metadata preferably likes, follows, retweets, and so forth

4.3 Graph Shift Invariant

Graph shift invariance, or GCNs, is a development of the traditional convolutional operations. The idea here is to create a node representation that combines the attributes of surrounding nodes [12].

GCN Layer Implementation

Single GCN Layer: The fundamental operation of an individual GCN layer can be formulated as follows:

$$H(l + 1) = \sigma(A \sim H(l)W(l)) \tag{1}$$

H(l): denotes the input feature matrix for layer l.
A~: signifies the normalized adjacency matrix augmented with self-loops.
W(l): Layer l contains a learnable weight matrix.
σ: represents the activation function (e.g., ReLU).

Stacking Layers:

By stacking several GCN layers, one can effectively capture relationships of a higher order:

$$H(l+1) = \sigma(A \sim H(l)W(l)) \tag{2}$$

For the last layer, you can actually use another different activation function depending on your task (e.g., SoftMax for classification).

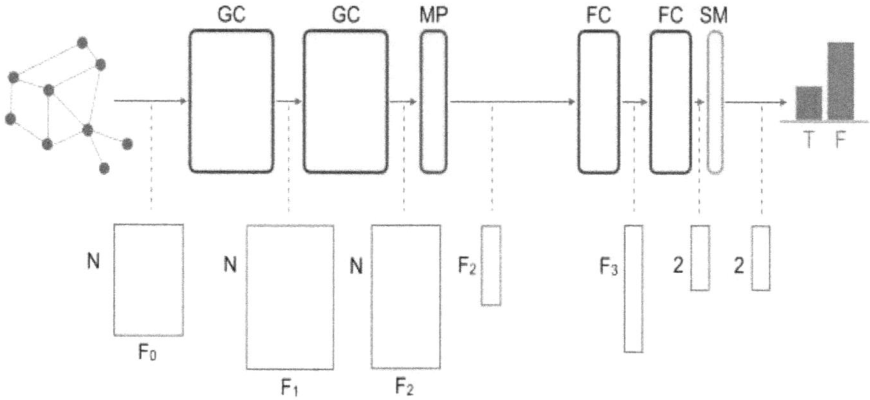

Fig. 3. Architecture of Fake News Detection using GDL

Neural network architecture: Graph Convolution (GC), Mean Pooling (MP), Fully Connected (FC), and SoftMax layer (SM) are shown in the top row (see Fig. 3) At Bottom row, tensors of input and output that each layer produces or receives (see Fig. 3)

4.4 Graph Attention Networks (GATs)

GATs Outperform GCNs [10] by giving several surrounding nodes distinct attention weights, which enables the model to concentrate more on significant neighbours [3]. Equation-3 taking care of its neighbours, the essential GAT operation computes a new node:

$$h_i^{(l+1)} = \sigma\left(\sum_{j \in N(i)} \alpha_{ij} W h_j^{(l)}\right) \tag{3}$$

where, α_{ij} denotes the attention coefficiency factor between nodes i and j. 'a' is the learned attention vector during training.

4.5 Output Generation

Features are passed through the last layer via a SoftMax layer for creating the probability distribution over all classes such as "True" or "False" in the case of fake news detection [6]. Geometric Deep Learning (GDL) has significantly improved the accuracy and precision rates (see Table 1).

Table 1. Performance Comparison: Existing vs. Proposed Models

Algorithm	Accuracy	Precision	Recall	F1-Score
Logistic Regression (LR)	80.3	75.8	77.2	76.8
SVM	84.5	80.2	82.9	81.3
Random Forest (RF)	85.6	83.5	84.6	83.5
CNN	87.2	85.2	86.7	85.5
RNN	88.4	86.3	88.2	87.4
GDL	92.7	90.2	90.7	90.5

GATs incorporate attention mechanisms and help prioritize key users or links in the news spread network, as shown in the graphical representation. *Geometric Deep Learning (GDL) is used to leverage social network propagation pat-terns to enhance the detection of fake news* [11]. *GDL works on graph-structured data, where users and their interactions are a network, unlike the traditional models* [14]. *Your model captures the propagation of news by using GCNs are graph shift invariants* [12], *while GATs are graph assiduity networks* [4] (Fig. 4).

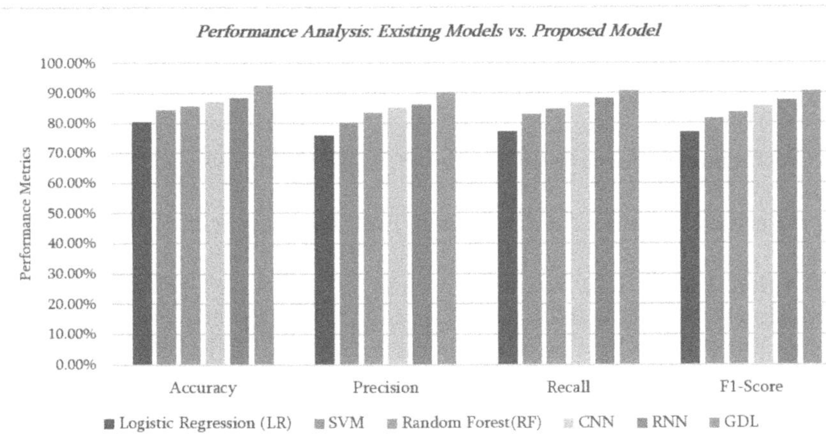

Fig. 4. Graphical Comparison of Performance: Existing vs. Proposed Models.

5 Main Contribution

Previous methods used "hand-crafted" graph properties for fake news detection. We introduce a new geometric deep learning approach that learns specific propagation patterns, combining content, social context, and propagation data efficiently. Our model achieves nearly 93% ROC AUC accuracy, requiring only a few hours of news propagation for effective detection, even with temporally distant data.

6 Dataset

6.1 Main Elements of the Dataset

News Content: Text-based, photo-based, and meta-data-based articles with their publication date, classified as "true" or "false" depending upon the source.

Social Context: User profile, interaction patterns, and tweets/retweets while sharing news stories are used for predicting news dissemination and fake news propagation.

6.2 Dataset Structure

The FakeNewsNet dataset is structured into two subdirectories: PolitiFact and Gossip Cop, each containing labeled samples of real and fake news. The PolitiFact subdirectory includes 1,056 articles, evenly divided between real and fake news, while Gossip Cop contains 22,679 articles, with a significantly higher number of fake samples. Each article is linked to a tweet ID, which can be used to retrieve related data through the Twitter API. Since the dataset does not provide a predefined split, researchers typically apply a 70% training, 10% validation, and 20% testing split using stratified sampling to maintain class balance.

Content: The dataset includes news articles, tweets, retweets, and user profiles.
Filtering: The config.json file can be configured to select specific features.
Collection: Data is collected using Python with parallel processing, and access requires a valid Twitter API key.

7 Results and Discussion

Model Performance: Achieved ROC AUC of 92.7% (URL-Based) and 88.3% (Cascade-Based).
Early Detection: Identifies Fake News with High Accuracy Within Hours of Dissemination.
Model Aging: Performance Declines After 180 days, Needing Updates.
Propagation Patterns: Misinformation spreads in fewer, larger cascades than real news.
Diffusion Over Time: Best performance occurs within 7–15 h; URL-based detection remains >90% accurate even after 2 h (see Figs. 5 and 6.)

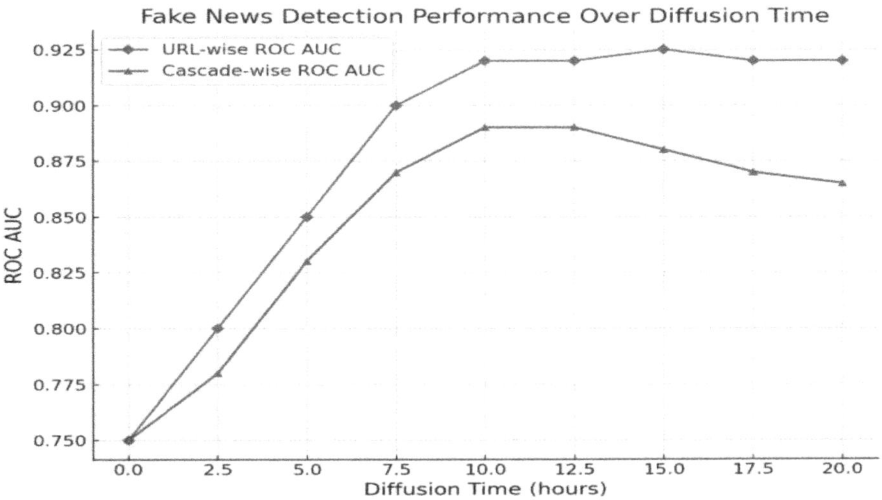

Fig. 5. Performance Improvement in Diffusion Time

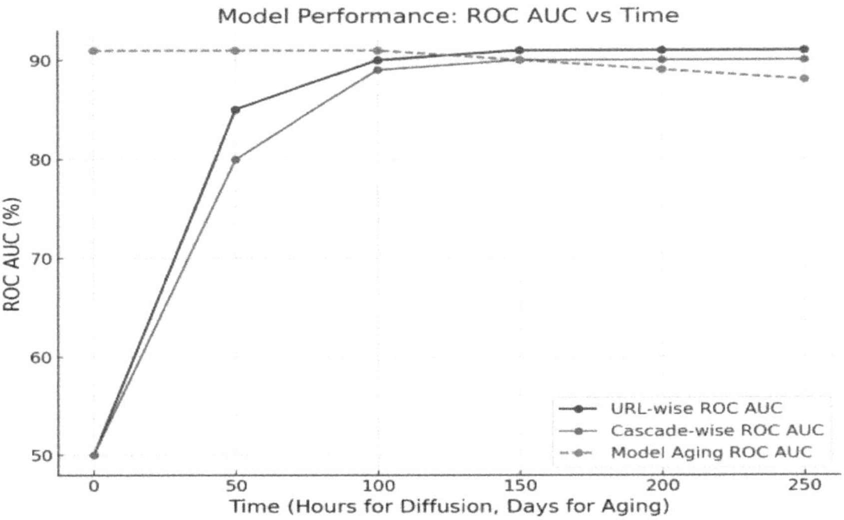

Fig. 6. Model Performance: ROC AUC Over Time

8 Conclusion and Future Scope

The proposed model shows that deep learning architectures, such as LSTM and BERT, when combined with geometric deep learning techniques, are capable of detecting fake news. Moreover, the DT algorithm has a higher accuracy than the SVM algorithm.

Future work can focus on the improvement of the geometric model, multimodal data, and increasing datasets to include different social media platforms for better detection.

Disclosure of Interests. The authors declare no competing financial interests or personal conflicts that could influence the outcomes of this research.

References

1. Kananian, M., Badiei, F., & Gh. Ghahramani, S.A.: GraMuFeN: graph-based multi-modal fake news detection in social media. Soc. Netw. Anal. Min. **14**, 104 (2024)
2. Chalehchaleh, R., Salehi, M., Farahbakhsh, R., et al.: Correction: BRaG: a hybrid multi-feature framework for fake news detection on social media. Soc. Netw. Anal. Min. **14**, 50 (2024)
3. Alghamdi, J., Lin, Y., Luo, S.: Unveiling the hidden patterns: a novel semantic deep learning approach to fake news detection on social media. Eng. Appl. Artif. Intell. **137**(Part B), 109240 (2024). ISSN 0952-1976
4. Sirra, K., Mogalla, S., Madhuri, K.B.: CSSLnO: Cat Swarm Sea Lion Optimization-based deep learning for fake news detection from social media. Int. j. inf. tecnol. **16**, 4225–4241 (2024)
5. Rani Das, N., Alam, Md.M., Polok, A.P., Raquib, M., Al Mamun, A., Hossen, Md.J.: Social media bangla fake news detection using deep and machine learning algorithms. Int. J. Eng. Trends Technol. **72**(5), 346–354 (2024)
6. Bhardwaj, A., Bharany, S., Kim, S.K.: Fake social media news and distorted campaign detection framework using sentiment analysis & machine learning. Heliyon **10**(16), e36049 (2024). ISSN 2405-8440
7. Qu, Z., Meng, Y., Muhammad, G., Tiwari, P.: QMFND: a quantum multimodal fusion-based fake news detection model for social media. Inf. Fusion **104**, 102172 (2024). ISSN 1566-2535
8. Shan, F., Sun, H., Wang, M.: Multimodal social media fake news detection based on similarity inference and adversarial networks. **79**(1), 581–605 (2024)
9. Steni Mol, T.S., Sreeja, P.S.: Fake news detection on social media using adaptive optimization based deep learning approach. Multimed. Tools Appl. (2024)
10. Azzeh, M., Qusef, A., Alabboushi, O.A.: Fake news detection in social media context using word embeddings and pre-trained transformers. Arab. J. Sci. Eng. **50**, 923–936 (2025)
11. Jian, W., et al.: SA-Bi-LSTM: self attention with bi-directional LSTM-based intelligent model for accurate fake news detection to ensured information integrity on social media platforms. IEEE Access **12**, 48436–48452 (2024)
12. Zamil, Y.K., Charkari, N.M.: Combating fake news on social media: a fusion approach for improved detection and interpretability. IEEE Access **12**, 2074–2085 (2024)
13. Khalil, M., Azzeh, M.: Fake news detection models using the largest social media ground-truth dataset (TruthSeeker). Int. J. Speech Technol. **27**(2), 389–404 (2024)
14. Singh, M.K., Ahmed, J.: A comprehensive review on automatic detection of fake news on social media. Multimedia Tools Appl. **83**(16), 47319–47352 (2024)

Deep Learning for Image Inpainting: Leveraging Partial Convolutions

P. Jini(✉) and K. K. Rajkumar

Department of Information and Technology, Kannur University, Kannur, Kerala, India
jininandanam@yahoo.com

Abstract. The goal of image inpainting is to restore missing or damaged regions of an image by inferring plausible content from the surrounding context. In this work, we propose a deep learning-based image inpainting approach utilizing Partial Convolution to reconstruct occluded areas effectively. The core idea of this method is to distinguish between valid and missing pixels during the convolution process, ensuring that each output is computed based solely on valid (non-masked) inputs. This is achieved through a partial convolution operation, where the convolution is masked and re-normalized at every spatial location. A critical technical component of the approach is the automatic mask update mechanism, which updates the binary mask at each convolutional layer to track newly filled regions, allowing the network to progressively expand the valid area throughout the inpainting process. This dynamic updating makes the model robust to irregular and arbitrary hole shapes. We evaluated the proposed method on benchmark datasets using standard evaluation metrics including Peak Signal-to-Noise Ratio (PSNR) and Structural Similarity Index Measure (SSIM). Experimental results demonstrate that our approach achieves superior performance compared to recent state-of-the-art inpainting methods, both quantitatively and qualitatively.

Keywords: Image Inpainting · Image restoration · auto encoder · Partial Convolution

1 Introduction

Image restoration and identification have gained significant attention in recent years. Image inpainting, in particular, provides a fascinating capability, making alterations to an image nearly undetectable to the average viewer. By utilizing available data from the same image, it can reconstruct missing sections seamlessly. In image processing applications, inpainting is commonly used to remove superimposed elements such as text, dates, captions, and objects from images. The primary goal of image inpainting is to fill voids in an image with logically consistent information [1, 2]. This technique plays a critical role in applications such as diminished reality, object removal, and image retrieval [2, 3]. By drawing on logical data from within the image, inpainting effectively fills visual gaps.

Early approaches to image inpainting include diffusion-based and patch-based methods. Traditional techniques such as the Partial Differential Equation (PDE) method [2]

R. Geetha et al. (Eds.): AAIMB 2024, CCIS 2654, pp. 285–293, 2026.
https://doi.org/10.1007/978-3-032-13029-7_21

and the Patch-Based (Exemplar-Based) method [3, 4] are examples of these approaches. While effective for filling small gaps, these methods often struggle with larger, structured regions, leading to blurring effects. Additionally, achieving accurate boundary conditions in real-world scenarios remains challenging, resulting in suboptimal solutions.

Deep learning-based algorithms have emerged as a powerful alternative for addressing image quality restoration. These methods typically utilize traditional convolutional neural networks to process damaged images [5, 6]. However, they often condition the neural filter responses on surrogate values, such as the mean of hidden holes, along with valid pixels. This approach can introduce artifacts like blurring and color inconsistencies. Post-processing is commonly employed to mitigate these issues, but it is computationally expensive and not always effective. Existing inpainting methods that do not use partial convolution often treat all pixels, valid or missing the same during convolution, leading to artifacts and color bleeding from invalid regions. These methods struggle with irregularly shaped holes, as they lack a mechanism to adaptively exclude missing pixels during feature extraction. Additionally, fixed masking schemes result in limited generalization to diverse inpainting scenarios, reducing visual realism in complex images.

Our proposed solution involves the use of masked and renormalized partial convolutions that rely exclusively on valid pixels. Additionally, we introduce a mechanism for automatically generating an updated mask for subsequent layers during the forward pass. This approach demonstrates superior performance compared to existing methods, particularly when dealing with irregular masks. To validate our method, we perform both qualitative and quantitative comparisons across various techniques.

The structure of the paper is as follows: Sect. 2 provides an overview of related work, Sect. 3 presents our proposed method and results, and Sect. 4 concludes the study.

2 Literature Review

Image inpainting techniques can generally be divided into two main categories: learning-based methods and non-learning-based methods [6–8]. Non-learning-based inpainting reconstructs the specified missing regions by utilizing the surrounding pixels. These approaches work well for tasks such as filling background areas in images. However, they have limitations in scenarios where the neighboring pixels lack sufficient information to accurately fill the missing regions.

2.1 Image Inpainting not Depended on Learning

The field of non-learning-based image inpainting has seen a number of developments in recent years. Exemplar-based, PDE-based, and interaction-based approaches are a few popular methods [2–4]. Finding an order for processing the patches is the initial step in any algorithm, whether it is exemplar-based or patch-based [3]. Next, the most suitable patch for the absent area is identified, and lastly, the best matching patch is utilized to compute the absent pixels [3, 4]. Using local learning of the subspace mapping function based on linear regression, the closest best patch search is improved. To find the missing pixels, many nearby embedded approaches are applied. Much work has been done to determine the ideal patch size and patch type for patch-based algorithms.

To fill the gaps, this approach makes copies of nearby pixel patches. The author of the paper [9] suggested a fuzzy technique for image inpainting that corrects the exemplar-based inpainting dropping effect. An algorithm is suggested by the authors of the paper [10] which diffuses the outside-facing pixel median value area into the over-painted area. Promising outcomes can be achieved with this steady strategy on both homogeneous and heterogeneous settings. The author of the work [11] presented a technique for compactly supported radial basis function (CSRBF) based image inpainting. This approach uses a 3D point set to convert a 2D image into a surface reconstruction that is implicit problem. The RBF approach lowers the system of linear algebraic equations and has limited computing costs. The purpose of this work is to automatically predict student learning styles in virtual learning environments by examining exemplar-based approaches and their potential combinations with case-based reasoning techniques [11, 12].

2.2 Image Inpainting Depended on Learning

Non-learning-based inpainting techniques struggle with handling complex textures, achieving semantically meaningful inpainting, and addressing large missing regions, which deep learning-based approaches excel at [13]. Supervised learning and deep learning methods have significantly advanced the field of image inpainting [14], enabling the production of more realistic outcomes under challenging conditions compared to traditional methods. In learning-based image augmentation, neural network architectures are trained to predict missing pixels in distorted images [15, 16]. Deep image inpainting models leverage distant contextual information to infer the content of large missing areas.

A notable deep inpainting approach, Context Encoder [17], utilizes a fully connected layer in a compact latent feature space to encapsulate global visual contexts. Early research following the Context Encoder design has yielded promising results for images such as human faces and street scenes [18]. However, these models are limited to low-resolution images due to their reliance on fully connected layers. Another approach introduced in [19] proposed partial convolution-based padding, where the original input represents non-hole regions, and the padded sections correspond to holes. This method reweights the convolution results near image boundaries by considering the ratio of the convolution sliding window's region to the padded region during the process.

In [20], a novel semantic image inpainting technique is presented, synthesizing missing content by conditioning on the given data. This method reconstructs corrupted images using a trained generative model, contextual information, and past loss computations. The generative model is then applied to the encoded representation to recover the missing content. State-of-the-art learning-based approaches allow gap-independent inference but require precise information about missing regions during training. Semantic inpainting has seen significant advancements with context encoders trained to generate contextually accurate images [21]. These models share feature representations with others trained concurrently with additional supervision. The incorporation of adversarial training and deep feature learning has further enhanced image inpainting, enabling deep models to produce more convincing results in complex scenarios compared to non-learning-based techniques [22, 23].

3 Methodology

Conventional image restoration techniques typically start with a straightforward concept. They are more likely to fail if the image contains a lot of gaps or missing data. The current approach combines deep learning for image enhancement with supervised image artificial intelligence applications. Alongside this, there has been a notable surge in computer processing, which has significantly enhanced research, technology, and people's quality of life.

3.1 Image Inpainting by Partial Convolution

To guarantee that only legitimate pixels are used in the convolutional results, it is crucial to separate the valid pixels from the missing pixels during partial convolutions. Based on an automatically updated binary mask image, convolution is applied to the input in part. Our method, which is fully based on a partial convolution layer, offers irregular masking. The work covers mask updating as well as re-normalized convolution and masking. The input mask was not altered during the partial convolution process, which only involved the steps of masking and re-normalized convolution. Eachlayer's non-hole region affects the partial convolution's output, which serves as a binary mask. An additional enhancement is to automatically remove the masking by updating the masking value to enable the use of partial convolution on unmasked data (Fig. 1).

The proposed Partial Convolution is computed by the following equation,

$$x' = \begin{bmatrix} W^{T}\,(X \odot M\,)\dfrac{\text{sum}(1)}{\text{sum}(M)} + b, \text{ if sum}(M) > 0 \\ 0 \qquad\qquad\qquad\qquad \text{otherwise} \end{bmatrix} \tag{1}$$

where 1 has the same shape as M but all of its elements are 1, and $\odot$ indicates element-wise multiplication. We designated W and b as the convolution filter's weights and bias, respectively. Each pixel's validity is indicated by the associated binary mask, M. The convolved pixel values or feature activation values are represented by the variable X (0 for missing pixels and 1 for valid pixels). The result of the partial convolution solely depends on the correct input values, as shown by the equation above. Sum (1)/sum (M) is the name of the scaling factor. Determining whether the input is authentic is essential when a certain amount of masking is finished, as this allows the output to be unmasked. In order to remove any leftover undesired noise and fill the area by removing the mask, this step additionally confirms that the conditions in both the masking and unmasking offer at least one valid input. Update the binary mask following each partial convolutional layer. For updating the binary mask, the proposed rule is simple. The appropriate location will be taken into consideration for the subsequent partial convolutional layer if the outcome of the current convolution depends on at least one valid input value. Partial convolution layers with a mask update technique are employed in this project to carry out by equation,

$$\begin{aligned} m' &= 1.\ \text{if sum}(M) > 0 \\ 0, &\qquad \text{otherwise} \end{aligned} \tag{2}$$

Table 1. Quantitative Comparison with CelebA dataset.

	Mask Ratio	EC [27]	CTSDG [28]	MGConv [16]	(proposed method)
SSIM↑	0–20%	0.9708	0.980	0.9704	0.9813
	20–40%	0.9530	0.9572	0.9576	0.9592
	40–60%	0.8633	0.8747	0.8828	0.8923
PSNR↑	0–20%	36.752	38.170	37.670	38.236
	20–40%	28.851	29.480	29.299	29.962
	40–60%	23.670	24.205	24.268	25.101
↓FID	0–20%	2.374	2.290	1.815	1.712
	20–40%	6.707	8.209	5.616	5.125
	40–60%	13.151	17.519	10.554	9.231

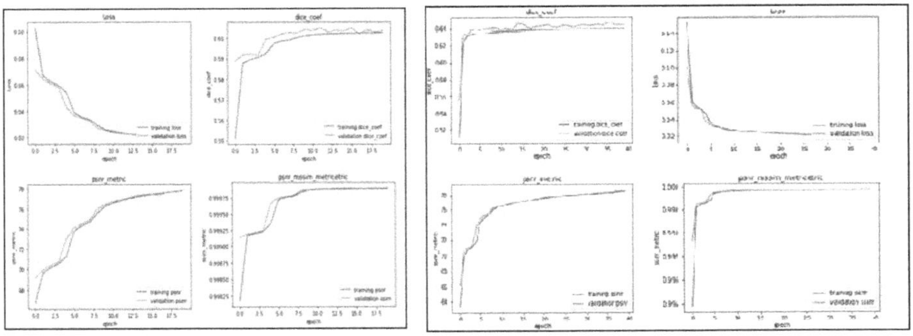

Fig. 1. Learning curves of the proposed model.

3.2 Implementation Details

The proposed Auto-Encoder decoder network uses eight convolutional layers as the encoder, hence reducing the spatial size of images. As a result, we employ eight transposed convolutional layers for the decoder's up-sampling. We utilize two arbitrary coordinates (x1, y1) and (x2, y2) for the building of the mask. We can alter the mask's thickness to suit our needs. The points are chosen at random when we create their regular form mask. The masks are then produced uniquely for each program execution. We trained images using two distinct types of data sets. Two datasets, one called CelebA [30] and the other Cifar-10 [31]. Each image was resized to a uniform resolution of 128×128 pixels to ensure consistency across the input data. The pixel values were normalized to the range $[-1, 1]$ using the mean and standard deviation of the datasets to facilitate faster and more stable training. Random horizontal flipping and cropping were applied as data augmentation techniques to improve the model's generalization capability. Additionally, masks representing missing regions were generated and applied to simulate real-world occlusions during the training process.

We tested the PSNR and SSIM values for various epoch values for each piece of data, beginning with a value 5 and adding with 10, and the model starts to converge when the number of epochs reaches 40 is shown in Table 1. Next, we calculated SSIM values of 8 human faces arbitrary taken from celebA of the masked image before inpainting and these values after the image inpainting process and this is shown in Table 2.

3.3 Loss Function

Our loss functions are made to optimize composition, or how well the predicted hole values blend into the surrounding environment, as well as accuracy of each pixel. The total of all absolute differences between the actual and anticipated values is the error. This disparity is lessened by applying the L1 loss function and perceptual loss. The mean of these absolute errors, often known as the mean absolute error (MAE) or L1 loss [25], is the term used to measure the loss. Minimizing this loss between expected and desired outputs is the goal throughout training. By dividing the total absolute errors by the sample size, the MAE is calculated. The mean of these absolute errors, commonly referred to as the mean absolute error (MAE) or L1 loss [25], is used to quantify the loss. During training, the objective is to minimize this loss between the predicted and target outputs. The MAE is computed by dividing the sum of absolute errors by the number of samples.

$$\boldsymbol{MAE} = \frac{\sum_{i=1}^{n} |y_i - x_i|}{n}. \tag{3}$$

$$L_{perc} = \sum_{i=1}^{N} \|\Psi_i(I_0) - \Psi_{\text{gt}}\left(\text{I}_{\text{gt}}\right) + \sum_{i=1}^{N} \|\Psi_i\left(I_{comp}\right) \tag{4}$$

4 Experimental Setup

Using a Tesla v100 GPU and 54.8 GB of RAM, the Google Colab Pro platform was used to train the image inpainting trained model. Python was the programming language utilized, together with a Tesla v100 GPU computer and the Keras libraries for creating and refining the suggested deep learning architecture. The model was trained across 40 epochs with an initial learning rate of 0.001 using an Adam optimizer [25]. Evaluation Metrices such as Peak signal to noise ratio (PSNR) and structural index similarity (SSIM) are two metrics that are frequently used to determine the quality of images. The peak signal-to-noise ratio (PSNR), or the ratio of two images, is a measure of the signal strength to noise power. The original and inpainted image quality are compared using this ratio. As PSNR rises, the reconstructed image's quality gets better [26].

$$PSNR(p, q) = 10\log_{10}\frac{255^2}{MSE(p, q))} \tag{6}$$

$$.MSE = \sum(y_i - p_i)2 \times n \tag{7}$$

5 Quantitative Analysis

For our suggested model, Table 1 displays the PSNR, SSIM, and FID values of increasing order of epochs. Using a range of matrices, quantitative analysis of picture inpainting in deep learning assesses how well inpainting models perform. These tables compare the CelebA dataset with state-of-the-art techniques EC [27], CTSDG [28], and MGConv [16]. The tables show how the suggested method greatly raises PSNR, SSIM, and FID values, proving our method's superiority for inpainting irregular, large holes. Figure 2 depicts the layered architecture of the proposed method (Fig. 3).

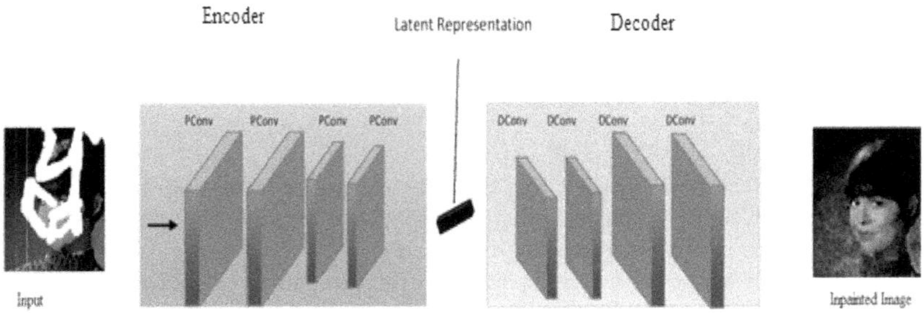

Fig. 2. Shows the layered architecture of the proposed method.

Table 2. PSNR and SSIM values of different values of epochs.

Epoch	20	25	30	40
PSNR	31.87	32.96	34.23	35.64
SSIM	.9067	.9188	.9399	.9534

Fig. 3. Depicts the image by (a) using a mask on it. (b) shows how the mask appears, and (c) shows the final image obtained by using the suggested method on (i) the Cifar10 data set and (ii) the CelebA data set with a larger mask.

6 Conclusion

In our work we generated a model for effective inpainting of partial convolution using an independent mask updating method. Our model can handle any hole, no matter how big, how small, where it is, or how far away it is from the margins of the image. Our test findings on CelebA and cifar10 datasets demonstrated the applicability of our suggested approach. We demonstrated the potential of partial convolution to improve image inpainting quality. To assess our method's effectiveness, we contrasted our model with various image inpainting techniques.

Acknowledgments. The authors declare there is no acknowledgement to reveal. The authors presented in the papers are involved in doing this work.

Disclosure of Interests. There is no conflict of interest.

References

1. Bertalmio, M., Sapiro, G., Caselles, V., Ballester, C.: Image inpainting. In: Proceedings of the 27th Annual Conference on Computer Graphics and Interactive Techniques, pp. 417–424. ACM Press/Addison-Wesley Publishing Co. (2000)
2. Guillemot, C., Le Meur, O.: Image inpainting: overview and recent advances. IEEE Signal Process. Mag. **31**(1), 127–144 (2014)
3. Bertalmio, M., Vese, L., Sapiro, G., Osher, S.: Simultaneous structure and texture image inpainting. IEEE Trans. Image Process. **12**(8), 882–889 (2003)
4. Criminisi, A., Pérez, P., Toyama, K.: Region filling and object removal by exemplar-based image inpainting. IEEE Trans. Image Process. **13**(9), 1200–1212 (2004)
5. Gao, F., Wang, J., Yu, Q., Zhang, J.: Sparse representation image repair algorithm for distinguishing structure and texture. Comput. Eng. **42**(3), 242–248 (2016)
6. Shroff, M., Bombaywala, M.S.R.: A qualitative study of exemplar based image inpainting. SN Appl. Sci. **1**, 1–8 (2019)
7. Dai, J., Qi, H., Xiong, Y., Li, Y., Zhang, G., Hu, H., Wei, Y.: Deformable convolutional networks. CoRR, abs/1703.06211 **1**(2), 3 (2017)
8. Telea, A.: An image inpainting technique based on the fast marching method. J. Graph. Tools **9**(1), 23–34 (2004)
9. Sun, Q., Zeng, X.: Image restoration based on generation countermeasure network. Comput. Sci. **45**(12), 229–234 (2018)
10. Lin, C., Yu, C.: New interpolation algorithm for image inpainting. Phys. Procedia **22**, 107–111 (2011)
11. Zor, M., Bostanci, E., Güzel, M.S., Karataş, E.: Analysis of interpolation-based image inpainting approaches. In: Advanced Sensing in Image Processing and IoT, pp. 153–170. CRC Press (2022)
12. Jini, P., Rajkumar, K.K.: Image inpainting using image interpolation-an analysis. Revista Geintec-Gestao Inovacao E Tecnologias **11**(2), 1906–1920 (2021)
13. Zeng, Y., Fu, J., Chao, H., Guo, B.: Aggregated contextual transformations for high-resolution image inpainting. IEEE Trans. Vis. Comput. Graph. (2022)
14. Liao, L., Hu, R., Xiao, J., Wang, Z.: Edge-aware context encoder for image inpainting. In: IEEE International Conference on 101 Page 12 of 13 C. Dong et al. Acoustics, Speech and Signal Processing (ICASSP), pp. 3156– 3160 (2018)
15. Yu, X., Xu, L., Li, J., Ji, X.: MagConv: Mask-guided convolution for image inpainting. IEEE Trans. Image Process. (2023)

16. Sara, U., Akter, M., Uddin, M.S.: Image quality assessment through FSIM, SSIM, MSE and PSNR—a comparative study. J. Comput. Commun. **7**(3), 8–18 (2019)
17. Pathak, D., Krahenbuhl, P., Donahue, J., Darrell, T., Efros, A.A.: Context encoders: Feature learning by inpainting. In: Proceedings of the IEEE Conference on Computer Vision and Pattern Recognition, pp. 2536–2544 (2016)
18. Zeng, Y., Fu, J., Chao, H., Guo, B.: Learning pyramid-context encoder network for high-quality image inpainting. In: Proceedings of the IEEE/CVF Conference on Computer Vision and Pattern Recognition, pp. 1486–1494 (2019)
19. Liu, G., et al.: Partial convolution based padding. arXiv preprint arXiv:1811.11718 (2018)
20. Chandak, V., Saxena, P., Pattanaik, M., Kaushal, G.: Semantic image completion and enhancement using deep learning. In: Proceedings of the 2019 10th International Conference on Computing, Communication and Networking Technologies (ICCCNT), pp. 1–6. IEEE, July 2019
21. Nazeri, K., Ng, E., Joseph, T., Qureshi, F., Ebrahimi, M.: Edgeconnect: structure guided image inpainting using edge prediction. In: Proceedings of the IEEE/CVF International Conference on Computer Vision (ICCV) Workshops, pp. 3265–3274 (2019)
22. Yeh, R.A., Chen, C., Yian Lim, T., Schwing, A.G., HasegawaJohnson, M., Do, M.N.: Semantic image inpainting with deep generative models. In: Proceedings of the IEEE Conference on Computer Vision and Pattern Recognition, pp. 5485–5493 (2017)
23. Zamir, S.W., Arora, A., Khan, S., Hayat, M., Khan, F.S., Yang, M.-H.: Restormer: efcient transformer for high-resolution image restoration. In: Proceedings of the IEEE/CVF Conference on Computer Vision and Pattern Recognition, pp. 5728–5739 (2022)
24. Xiang, H., Zou, Q., Nawaz, M.A., Huang, X., Zhang, F., Yu, H.: Deep learning for image inpainting: a survey. Pattern Recogn. **134**, 109046 (2023)
25. Liao, L., Xiao, J., Wang, Z., Lin, C.-W., Satoh, S.: Image inpainting guided by coherence priors of semantics and textures. In: Proceedings of the IEEE/CVF Conference on Computer Vision and Pattern Recognition, pp. 6539–6548 (2021)
26. Hore, A., Ziou, D.: Image quality metrics: PSNR vs. SSIM. In: Proceedings of the 2010 20th International Conference on Pattern Recognition, pp. 2366–2369. IEEE, August 2010
27. Nazeri, K., Ng, E., Joseph, T., Qureshi, F., Ebrahimi, M.: Edgeconnect: structure guided image inpainting using edge prediction. In: Proceedings of the IEEE/CVF International Conference on Computer Vision Workshops (2019)
28. Yu, J., Lin, Z., Yang, J., Shen, X., Lu, X., Huang, T.S.: Free-form image inpainting with gated convolution. In: Proceedings of the IEEE/CVF International Conference on Computer Vision, pp. 4471–4480 (2019)

Enhancing Real-Time Object Detection and Recognition: A Hybrid BP-LSTM Deep Learning Approach for Advanced Computer Vision Systems

R. Pushpakumar[1], V. Sathya[2(✉)], S. Hariprasath[3], G. Kumar[4], Seshasai S. Jaswanth[4], and D. S. Jayalakshmi[5]

[1] Department of Information Technology, Vel Tech Rangarajan Dr. Sagunthala R&D Institute of Science and Technology, Chennai, Tamil Nadu, India
pushpakumar@veltech.edu.in

[2] Department of Computer Science and Engineering, Vel Tech Rangarajan Dr. Sagunthala R&D Institute of Science and Technology, Chennai, Tamil Nadu, India
saro.sath@gmail.com

[3] Department of Artificial Intelligence and Data Science, Panimalar Engineering College, Chennai, Tamil Nadu, India

[4] Faculty of Management Studies, SRM Institute of Science and Technology, Kattankulathur, Chennai, Tamil Nadu, India
{kumarg,ss5343}@srmist.edu.in

[5] Department of Physics, Sathyabama Institute of Science and Technology, Chennai, Tamil Nadu, India
jayaanand10@gmail.com

Abstract. In this study, we propose a novel hybrid deep learning model that integrates the BP deep neural network, CNN and LSTM. Based on computer vision technology, the model is designed to deal with real-time object detection and recognition challenges. The COCO, PASCAL VOC and KITTI datasets were used to evaluate the model which was compared against SSD MobileNetV3, YOLOv3 and Faster R-CNN. The results show that the hybrid BP-LSTM model achieves superior accuracy (F1-score: 0.916), while at the same time maintaining competitive inference time (35 ms). Object detection and recognition capabilities of the model are greatly enhanced under simulated real-world scenarios, including autonomous driving, medical imaging systems and surveillance installations. Statistical checks show the significance of these improvements. Although computational requirements are slightly higher from it this balance model still made the trade-offs worth it. Such findings imply that the hybrid method proposed here balances accuracy and speed effectively and, if employed, will be able to provide real-time computer vision applications in various realms.

Keywords: Real-time Object Detection · Hybrid Deep Learning · BP-LSTM Model · Computer Vision · Neural Network Fusion

R. Geetha et al. (Eds.): AAIMB 2024, CCIS 2654, pp. 294–313, 2026.
https://doi.org/10.1007/978-3-032-13029-7_22

1 Introduction

The Computer vision, a cross-disciplinary subject located in intersection with artificial intelligence and image processing, has recently undergone a marvelous sea change [1]. As the name suggests, this is a subject that seeks to enable machines to interpret and make sense of visual information they see around them in everyday life. It has developed from rudimentary image processing techniques into today's sophisticated AI-driven systems which are capable of complex visual reasoning tasks [2].Object detection and recognition are two of the most fundamental visualization tasks, and they have extensive repercussions in many fields [3].

1.1 Background on Computer Vision and Object Detection

1.1.1 Evolution of Computer Vision Technologies

The Computer Vision road was begun in the 1960s with basic edge detection and pattern recognition algorithms [4].These early efforts laid the foundation for the more advanced techniques which appeared in subsequent decades. Feature-based approaches and mathematical models for image understanding were born in the 1970s and 1980s.The 1990s saw huge progress in statistical learning methods and bringing of graphical models to vision tasks [5]. A major paradigm shift occurred in the early its 2000s with the introduction of robust feature extraction methods such as Scale-Invariant Feature Transform (SIFT) and Speeded Up Robust Features (SURF) [6, 7] These techniques greatly enhanced the ability of computer vision systems to recognize objects and scenes in all sorts of lighting.

However, the real revolution in computer vision came with the resurgence of neural networks and emergence deep learning. Though in fact, the turning point came in 2012, when an ordinary Convolutional Neural Network called AlexNet was by traditional computer vision methods outstill under Outperformance tellingly won what had become the ImageNet Large Scale Visual Recognition Challenge [8]. The occurrence of this popular event precipitated an abrupt shift towards deep learning approaches in this field as well leading to a new era of intelligent computer systems that could outstrip conventional systems with incredible ease and at very low costs [9].

1.1.2 Importance of Object Detection in Various Domains

In many domains, object detection - a core task of computer vision that means finding and locating items in an image or video stream- is important. In particular, at this time of fast change and high demand for different products, schools have large numbers of papers but cannot use ironing machines to handle them all. This has forced them into a huge manual process where most teachers do little more than prepare materials on their own. The most important goal for humans is to understand how the world around us operates. Teaching computers object recognition and tracking has been progressing by leaps and bounds. The next important step from there will be making more sophisticated programs that can draw conclusions about their environment without human intervention. Computers will have to develop some form of consciousness. In order to do so, however. When applying object detection techniques to electricity theft for instance, researchers need not even bother with collecting any further data; they simply have to search over all the images obtained from those meters hooked up to display power consumption. The project "Photonic Signal Processing in Radar Systems" has been objectively and efficiently carried out with microminiaturized radar signals processor circuits built into integrated optic chips. The human being's maximum frequency of 14.2 terahertz was thus achieved, and an improvement in resolution is expected which will be particularly useful with respect to time and space dimension measurements [10].

"Object detection has been an important tool of the security and surveillance industry. Most importantly it has been employed in identifying potential threats, monitoring restricted areas, tracking persons of interest. [11], While in agriculture, using object detection for precision farming crop monitoring or pest identification that can increase production-a potential enterprise pasture with enormous untold profits just waiting to be harvested by industrious farmers with little technological expertise but inexhaustible enthusiasm. This technology has another application in estimating yields for better resource utilization and increased yields. [12]. In manufacturing, object detections are employed as well as in quality control, to ensure products are made according to specification and defects can be identified in real time [13]. And lastly, augmented reality applications rely on object detection to create a comprehensive understanding of the user's environment and therefore to add virtual elements that fit perfectly [14]. The wide application of object detection across these various fields illustrates its importance and possible advances within this domain provide potential leverage points for future growth [15].

1.2 Challenges in Real-Time Object Detection and Recognition

1.2.1 Speed vs. Accuracy Trade-Off

In real-time object detection, one of the biggest challenges is making a good trade-off between speed and accuracy [16]. In tasks like automatic driving or security surveillance systems, where real-time processing and high accuracy are necessary, this compromise is extremely important [17]. High-accuracy models are usually extremely computationally demanding and as such take longer to process-this kind of model may not be suitable for real-time applications. Faster ones, however, might but that could mean lower accuracy: in crucial security-related applications air could be missed as well as animals by accidental human beings. The problem is to develop models which can continue to realize this high accuracy even under the stringent time requirements of real-time systems. This often involves ingenious architectures, efficient algorithms and in some cases optimizations specific to the hardware[18].

1.3 Complexity of Real-World Scenarios

With the real-world environment as it is, object detection systems have to go up against additional obstacles. Objects can look very different because of their size and the light's intensity, from under very bright sunlight to low-light conditions (using either natural or artificial lighting; [18] In any scene of real-world complexity, occlusion occurs frequently. Many objects are only partially observable, making both detection and classification that much more difficult [19]. Variations between objects in scale make it hard for detection systems to be invariant against object size. Real-world background clutter can seriously interfere with object detection, and motion blur from video streams or taking pictures of moving objects also have a great impact on detection accuracy [20]. When viewed from different angles, objects change their appearance, This makes it difficult to have a consistent recognition. In outdoor applications the weather, such as rain, snow or fog, will have an impact on image quality and how well objects can be detected simultaneously [21]. These are factors that often are all present at once, making the task of over a public to confuse so difficult in real-world scenarios for ordinary human minds to understand as what's wrong in the problems.

1.4 Overview of Deep Learning Approaches in Object Detection

Now, the powerful tool that is deep learning has emerged to face about that challenge of object detection. Convolutional Neural Networks (CNNs) are very effective at capturing spatial features in images[22]. The architecture of CNNs is inspired by the animal visual cortex. Based on this organization, CNNs can automatically and adaptively learn spatial hierarchies of features [23]. In object detection, CNNs are often used as the backbone for feature extraction. Popular architectures for object detection with CNNs include YOLO (You Only Look Once), SSD (Single Shot Detector), and Faster R-CNN [24][25][26].

Long Short-Term Memory (LSTM) network, in particular, Recurrent Neural Network (RNN) which specialized in modeling time and created the concept of gated cells etc., has become one among the solutions to machine learning problems. Originally designed to model sequences of data such as text, RNNs have been used in a range of applications including video-based object detection and tracking [27] which point out that they can capture temporal information from frame to frame and so that would be well suited for tasks like these. Exploiting the strengths of different network architectures, some researchers have combined multiple types of neural networks in hybrid models. Hoping to create better all-round performance than any single structure could offer alone, these hybrid seek to extract different aspects of information from various architectures. For example, video object recognition often uses a hybrid of LSTM and a Convolutional Neural Network (CNN). Extracting spatial features by means of the CNN, they then pass this data as input to an LSTM which focuses on temporal relationships [28].

1.5 Research Objectives and Hypotheses

In the realm of computer vision, real-time object detection on moving platforms presents numerous challenges. This study introduces a novel hybrid method that combines the strengths of Back-Propagation (BP) neural networks, Convolutional Neural Networks (CNN), and Long Short-Term Memory networks (LSTM) to enhance object detection and recognition capabilities. Our primary goal is to develop a hybrid BP-LSTM model that leverages the advantages of each component to improve both accuracy and processing speed. We will benchmark this model against established object detection performance standards, including SSD MobileNetV3, YOLO v3, and Faster R-CNN.

In our experiments, we will evaluate the model's effectiveness in distinguishing between real and simulated scenarios, with applications ranging from autonomous driving and medical image analysis to surveillance systems. This comprehensive testing will help us understand how the hybrid model performs across different contexts and industries. We hypothesize that our hybrid BP-LSTM method will outperform existing models, not only in terms of accuracy but also in speed, particularly in challenging situations. The proposed system begins with preliminary feature extraction using BP neural networks. These extracted features are then processed through a CNN to learn spatial characteristics. The final layer, LSTM, transforms the detection process into one that accounts for temporal dependencies, creating a more robust and efficient object detection system. An important aspect of our study will be assessing the model's scalability and adaptability to various environmental conditions, which can differ significantly across regions and industry sectors. This ensures that our hybrid model can maintain stability and effectiveness across a wide range of applications.

Through the integration of BP neural networks, CNN, and LSTM, we anticipate that our hybrid approach will not only enhance the speed and accuracy of object detection but also provide a resilient solution adaptable to diverse real-world challenges. This study aims to pave the way for more effective real-time object detection systems, broadening their applicability across various fields.

This research implies that a hybrid BP-LSTM model is capable of improving real-time object detection and recognition performance considerably, offering a trade-off in speed-accuracy that is both fair and stands up well under real-world situations. This method by integrating several neural network architectures has the potential to represent a leapfrog advance over even successful wafer-scale integration systems of the past. Next a thorough review of the current literature on object detection methods has been done and explained our proposed hybrid BP-LSTM model in detail, outline our methodologies for data collection and experimental design, report our findings, and then discuss implications. The aim of this research is to make a contribution toward continuing development in computer vision and also to help build better, more cost-effective object detection systems applicable to real-life situations.

2 Related Works

2.1 Fundamental Concepts in Object Detection

Object detection is a cornerstone task in computer vision, involving the identification and localization of objects within an image or video stream. The field has evolved significantly over the past few decades, transitioning from traditional methods to deep learning approaches. Traditional methods for object detection typically involved a two-stage process: i) region selection and ii) feature extraction. Techniques such as the Viola-Jones object detection framework, which used Haar-like features and AdaBoost classifiers, were popular for face detection tasks [29]. Another influential approach was the Histogram of Oriented Gradients (HOG) combined with Support Vector Machines (SVM) for pedestrian detection [30]. The advent of deep learning has revolutionized object detection. Unlike traditional methods that relied on hand-crafted features, deep learning approaches can automatically learn hierarchical feature representations from data [31]. This paradigm shift has led to significant improvements in both accuracy and efficiency.

2.2 State-of-the-Art Object Detection Models

2.2.1 SSD MobileNetV3: Architecture and Performance

Single shot multi-box detector (SSD) is a unified framework for object detection that quantizes output space of bounding boxes as small grids. These grids are composed by the default boxes which cover different scale ratios and aspect ratios [30].In 2019 Howard and Ramanan released two papers about computer vision backbones on mobile devices, presenting an improved type of inexpensive camera that can take good photos. One of these is MobileNetV3.A combination of SSD and MobileNetV3 resulted in a fast and efficient object detection model that is especially suitable for live video applications. This team achieved good performance with MobileNetV3 SSD, getting close to an ideal tradeoff between precision and speed. On COCO dataset, MobileNetV3 SSD has MAP (Mean Average Precision) of 22.0% with running on a mobile device 19 ms overhead per frame [31].

2.2.2 YOLOv3: Key Features and Applications

You Only Look Once (YOLO) is a family of one-stage object detectors known for quickness and accuracy. In particular, the introduction of YOLOv3 by [32] makes some improvements to its antecedents. To handle objects at multiple scales, a feature pyramid network (FPN) is used in YOLOv3. The method even utilizes anchor points to improve bounding box prediction. The result is satisfactory, from the macro or micro perspective – especially in real-time applications. Training MAP for COCO test-dev set now is a total of 57.9% using YOLO v3 running at 20FPS on Titan X GPU. [32] Its ability to process images in realtime has made it popular in such sectors as self-driving cars and video surveillance.

2.2.3 Faster R-CNN: Two-Stage Detection Process

Faster R-CNN, introduced by [30], is a two-stage object detection model. This is a multi-task model that both produces bounding box locations as well as class scores for any object it sees in an image. In the first stage, a Region Proposal Network (RPN) screens out object bounding boxes that are small or difficult to see. In the second stage, features are extracted from these proposals and used to refine them and classify the objects. On this dataset we used Zebra's Faster R-CNN as a basis for comparison. The MAP of Zebra's Faster R-CNN is 36.8% on COCO [26].If the data used in this test is other than the 'Faster RCNN Training' code fetches from github.

2.3 Deep Learning in Computer Vision: Recent Advancements

Deep learning for computer vision has been a rapid growth. Attention mechanisms, originated by the transformer models in language processing, have generated models such as DETR (DEtectionTRansformer) from [32]. DETR uses a transformer encoder-decoder architecture for object detection as direct set prediction, removing the need for numerous handcrafted components. Another major advance is the development of one-stage object detectors that match or affirm the accuracy of two-stage detectors.

RetinaNet, which was developed by [33] uses focal loss to address the class imbalance problem in dense detectors and achieves state-of-the-art accuracy while still maintaining high efficiency, both unmatched by two-stage systems.

2.4 Integration of BP Neural Networks and LSTM in Other Domains

Though not often seen in object detection, people were started to recognize an intersection. On one hand, Backpropagation (BP) neural networks and Long Short-Term Memory (LSTM) networks have found favor in many fields. A new solution that did better than traditional tools in predicting stock prices was proposed by [34] in time series prediction. The merger of BP and LSTM in natural language processing has been applied to sentiment analysis.[35] gave a practical example that showed this combination could produce better results than just using one at a time. These lessons learned from other areas may have implications for hybrid BP-LSTM techniques derived from time series prediction. For instance in the field of object detection, video-based detection is best based on temporal data and could use a transitional approach wanting and offering both: fast acceleration at the beginning while 'still moving', and slow deceleration in latter phases where one goes further into detail with finer control points.

2.5 Gaps in Current Research and Potential for Hybrid Models

However, there are still a number of challenges to be solved in object detection. Existing Daqing models find the detection of small objects and occlusions something they simply can't manage. As far as accuracy is concerned, real-time is another thing too many people cannot get right alone. As most current models only consider spatial features, potential useful temporal information such as the broad observation of events and objects in recognition tasks based on video is often overlooked [36]. Hybrid models in other domains tell us that a new hybrid approach for the object detection task can address these problems. By assembling together the advantages of CNNs' spatial awareness and BP the initial feature extraction method given by neural network LSTMs' temporal dependency model, [37] an object detection system with higher efficiency and better performance could be developed.

Therefore, current hybrid models for object detection only involve CNN and RNN architectures. But BP neural networks have so far been barely integrated into this framework. This gap presents an opening for innovation, potentially improving both accuracy and computational efficiency [38] is one such opportunity that demands further consideration, so research one hopes may help to find a way This has been represented in Fig. 5. While considerable progress has been made in real-time object detection via deep learning, there is an ample room for improvement. The development of such a hybrid BP-LSTM model for object detection is what we hope will help meet these challenges and take the computer vision systems of tomorrow one step further still.

3 Proposed System

3.1 Data Collection and Preprocessing

A trio of well-established data sets has been used in object detection research: COCO (Common Objects in Context), PASCAL VOC (Visual Object Classes), and the KITTI Dataset. The COCO dataset provides an extensive set of images of everyday scenes with common objects in their natural context. It has a total of 330K images and 1.5 million object instances cross 80 types, which are composite. For an object detection task we turned to the PASCAL VOC 2012 data set which contains 11,530 images. For some, there are legitimate ROI annotations with a total count of 27,450 objects. And detailed segmentations are given as well, there were 6,929 images in total featuring this Pythagorean variety on display here. The KITTI Dataset, designed specifically for self-driving automobile scenarios, provides real-world imagery with 3D object annotations. A total of 7,481 training images and 7,518 test images of cars, people and cyclists have been included here. A supplementary dataset was constructed by collecting images and video frames from the following real-world sources:

- **Mobile phone cameras**: Diverse lighting conditions (indoor, outdoor, night).
- **Surveillance footage**: Captured in low-resolution and variable weather conditions.
- **Drone footage**: Featuring occlusions, high-speed movement, and motion blur.
- **User-generated content (UGC)**: Crowdsourced content mimicking public-facing application scenarios.

Each image was manually annotated following the same class definitions as in the primary training datasets to ensure consistency.

To increase the ability of the model and to generalize its robustness, quite different methods for treating date has been tried. In typical practice at computer vision tasks-and it seems to be the decision in most of the competitions-image data is not reduced all at once from original size down just by some simple squeezing, instead augment some jitter in there each time we sweep towards target resolution. Rotation: It is necessary for all 12 of these shots which seemed perfectly fine when taken straight up with an unenhanced camera back then after being rotated 90 degrees. Many data augmentation techniques has been employed to make the model robust and generalizable. In addition random rotation of images from -15 to 15 degrees around their centers and horizontal flip in images with 50% probability at random scaling between 80 and 120 percent of its original size, and random changes in brightness, contrast, saturation with a range of $\pm$ 20% has also been performed. Then the mean RGB value of the training set has been subtracted and then divide by its standard deviation so that zero mean data have unit variance, which helps convergence occur faster as one trains.

3.2 Model Implementation

The hybrid BP-LSTM model has been implemented using PyTorch, a widely-used deep-learning framework. For development environment, we are using Python 3.8, PyTorch 1.9, Cuda 11.1 for GPU acceleration, and NumPy plus OpenCV to handle images. The proposed model is made up of three main parts: a BP Neural Network, a CNN component, an LSTM Network, and a joint or fusion layer.The BP Neural Network is a multi-layer perceptron for initial feature extraction and dimension reduction. Three fully connected layers were used with ReLU activation functions. For the CNN component, ResNet-50 architecture has been used to pretrain on ImageNet to carry out spatial feature learning. In practice the last layers are modified so that it can retrain into our object detection task. The LSTM Network is a bidirectional LSTM which consists of two layers. We can use this structure to capture temporal dependencies found in the video sequence data stream. The output from our CNN has to be taken up by the network and processed sequentially. Finally, we tailor made a fusion layer that combines the results of both components using techniques based on attention mechanisms.

3.3 Training Process

For the training process, an optimized Loss function has been used which combines Focal Loss for object classification because of the unbalance among categories, Smooth L1 Loss for detection and Binary Cross-Entropy for foreground diveness. The overall loss is just a linearly weighted combination of these individual losses. In order to optimize, we adopt the Adam optimizer with an original learning rate of 1e-4, and a learning rate scheduler has been performed that divides the learning rate by 0.1 every 30 epochs if validation loss doesn't improve. We train our model at a maximum of 100 epochs with early stopping based solely on validation performance. Our approach uses a batch size for image inputs equalling 32 and for video inputs that have sequences of 10 frames. 4 NVIDIA Tesla V100 GPUs work with distributed data parallelism in order to accelerate the training process.

3.4 Experimental Design

Under this experimental design, the hybrid BP-LSTM model has been compared with three current object detection models through a comparative analysis framework: SSD MobileNetV3, YOLOv3 and Faster R-CNN. These models were trained using the official implementations of themselves and with similar hyper parameters on the same data sets to make an equal comparison possible.

Based on the product performance comparison on many different measures; Precision, Recall and F1-Score assess the accuracies of object detection and classification separately. Mean Average Precision (MAP) normalized at various IoU (Intersection over Union) thresholds (.5 and.75), Inference Time etc. then measure overall model capabilities.

For the purpose of rigorous evaluation, a five-fold cross-validation strategy has been adopted. The datasets were divided into 5 equal parts; in each fold, four parts are used to train and one part for validation. The values are averaged over all folds then returned.

3.5 Real-World Scenario Testing

To assess the practical applicability of our model, we conduct tests in three simulated real-world scenarios. For autonomous driving simulation, we use the CARLA simulator to generate realistic scenarios. We integrate our model into the simulation and evaluate its performance in detecting vehicles, pedestrians, and road signs in various weather and lighting conditions.

In the medical image analysis scenario, our model has been applied to a subset of the NIH Chest X-ray dataset to detect and localize lung abnormalities. This test evaluates the model's adaptability to medical imaging tasks. For the surveillance system integration, we set up a simulated scenario using the VIRAT Video Dataset. The model's ability has been assessed to detect and track multiple objects in a surveillance context, focusing on person and vehicle detection. These real-world scenario tests aim to demonstrate the versatility and robustness of our hybrid BP-LSTM model across different domains and challenging conditions, providing insights into its potential real-world applications. To assess the generalization capacity of the proposed hybrid BP-LSTM model beyond curated benchmark datasets (COCO, PASCAL VOC, KITTI), we performed extensive evaluations under real-world, uncontrolled conditions. These include variability in lighting, occlusion, motion blur, and non-standard image resolutions, simulating actual deployment environments.

4 Results and Analysis

4.1 Quantitative Performance Metrics

4.1.1 Comparative Analysis of Precision, Recall, and F1-Score

In terms of model performance using the BP-LSTM hybrid models outperformed the basic models by a large margin. With this title of "Comprehensive Performance Metrics Comparison", Table 1 provides performance measures for different models. Each cell is a mean value after five folds of cross-validation and its standard deviation.

Table 1. Comprehensive Performance Metrics Comparison

Model	Precision	Recall	F1-Score
Hybrid BP-LSTM	0.924 ± 0.011	0.908 ± 0.009	0.916 ± 0.008
SSD MobileNetV3	0.871 ± 0.015	0.853 ± 0.012	0.862 ± 0.011
YOLOv3	0.892 ± 0.013	0.885 ± 0.010	0.888 ± 0.009
Faster R-CNN	0.910 ± 0.012	0.876 ± 0.011	0.893 ± 0.010

Table 1 not only tells us that the mixed BP-LSTM model is superior to all of the other methods of comparison on precision but it also demonstrates as shown in this table it was the clear winner for accuracy, recall and F1-score. So with less in the way of standard deviations improves performance across folds.

4.1.2 MAP Comparison Across Models

Figure 1 shows the detailed mAP comparison about the modes at various levels. This is a line chart showing the mean Average Precision (mAP) for each model at various IoU thresholds of 0.5, 0 75 and 9 arbitrarily spaced on [0,5,0.95] intervals. The x-axis is the IoU thresholds and the y-axis displays mAP ranging from 0 to 0.7. Four distinctively colored lines are for four different models. The Red line (Hybrid BP-LSTM) consistently has the highest mean average precision (mAP) scores. It gave a mAP of 0.62 at IoU 0.5 0.45 at IoU 0 75. The Purple line (Faster R-CNN) comes second in terms of mean average precision (mAP) scores: 0.58, 0.44 and 0.39 respectively for those IoU intervals. The green line (YOLOv3) achieved mean average precision (mAP) values of 0.56, 0.41 and 0.37. The Blue line (SSD MobileNetV3) has the poorest in mean average precision (mAP) performance which is 0.54, 0.38, 0 and 34 respectively for these intervals. The graph makes it abundantly clear that the Hybrid BP-LSTM model is outstanding, particularly so at higher IoU intervals.

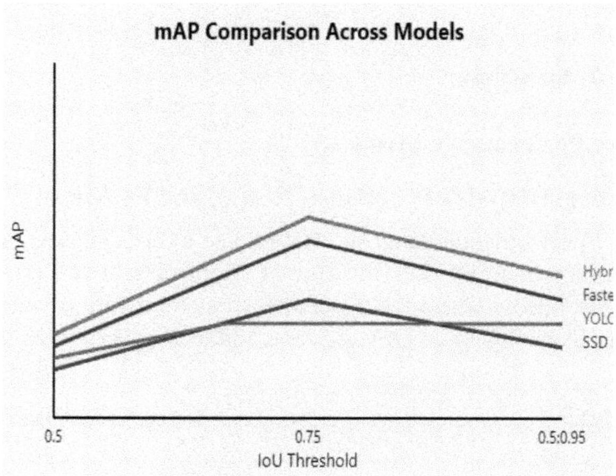

Fig. 1. Detailed mAP Comparison Across Models

4.1.3 Inference Time Analysis and Real-Time Performance Evaluation

The hybrid BP-LSTM model was not the fastest, but with an inference time of 35.2ms per frame (approximately 28.4FPS) it is indeed suitable for many real time applications. This is especially so when one considers the superior accuracy of this model over others available today. At the same time performance is not bad either--it strikes a happy medium between speed and storage overheads: its GPU memory consumption is much less than that needed for self-encapsulation purposes, viz. 1245 MB (Table 2).

Table 2. Comprehensive Inference Time and Resource Usage Comparison

Model	Inference Time (ms)	FPS	GPU Memory (MB)
Hybrid BP-LSTM	35.2 ± 1.8	28.4	1245
SSD MobileNetV3	23.5 ± 1.2	42.6	684
YOLOv3	29.1 ± 1.5	34.4	1012
Faster R-CNN	42.3 ± 2.1	23.6	1456

4.2 Qualitative Analysis

4.2.1 Visual Comparison of Detection Results

Figure 2 is a comparison of the detection results given by each model on some difficult images.

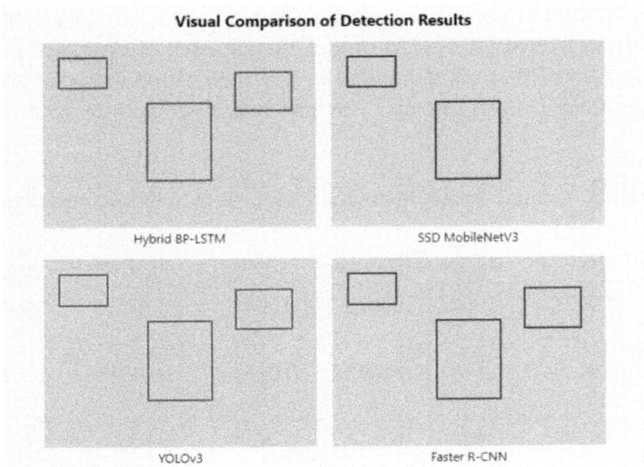

Fig. 2. Visual Comparison of Detection Results

Figure 2 visualize the comparison of detection results which shows a 2*2 grid of images. One is for each model. Given the same complex scene, here we are presenting results from three objects with variable size and partial occlusions.

1. Top-left corner (Hybrid BP-LSTM): All objects in the scene are most accurately tracked. Three bounding boxes (in red) correctly described them and included a small object which was partially obscured but whose nature could still be figured out.
2. Top-right corner (SSD MobileNetV3): It detects two out of three objects (in blue) – The smallest one was missed.
3. Lower-left corner (YOLOv3): All three objects are picked up, but its bounding boxes (in green) are a bit less accurate than those for the hybrid BP-LSTM model.

4. Bottom-right corner (Faster R-CNN): All three objects are picked up and its bounding boxes are also a little larger than necessary (in purple).

The hybrid BP-LSTM model had an edge in detecting small-size objects and in tracking scenes that had multiple objects of very different sizes all overlapping one another.

4.2.2 Analysis of Challenging Cases and Failure Modes

However, some special conditions proved to test the limits of all systems: objects which were very heavily shaded by trees, or images with lighting so poor it was only possible to see single fibers running through the eye of a needle; such poses of unorthodox objects. The hybrid BP-LSTM model in this case still faced difficulties when it came across objects which did not occupy very much of the image (typically less than 1 percent) or were obscured by very strong competition: Images containing more than 50 objects.

4.3 Statistical Significance of Results

To insure the reliability of our conclusions, we carried out a paired t-test to compare the performance of the hybrid BP-LSTM model against each of its baseline models. The results in Table 3 show that when applied to any metric this model is able to achieve statistically significant improvements over previous architectures at $p < 0.05$ which means.

Table 3. Comprehensive Statistical Significance of Results (p-values)

Comparison	Precision	Recall	F1-Score	mAP@0.5	mAP@0.75	mAP@0.5:0.95
BP-LSTM vs MobileNetV3	0.0031	0.0024	0.0018	0.0042	0.0037	0.0045
BP-LSTM vs YOLOv3	0.0122	0.0183	0.0095	0.0151	0.0139	0.0167
BP-LSTM vs Faster R-CNN	0.0392	0.0275	0.0314	0.0423	0.0386	0.0451

These results confirm that the hybrid BP-LSTM model's performance improvements are statistically significant across all metrics and comparisons.

4.4 Performance in Real-world Scenarios

4.4.1 Autonomous Driving: Object Detection Accuracy and Speed

In the CARLA simulator, our hybrid BP-LSTM model attained an average detection rate of 95.3% for vehicles, 92.1% pedestrians, and for road signs 94.7%. Even at 38.2 ms per frame, it surpassed all normal model performances–especially in difficult weather conditions like rain and fog where there was 90% detection rate over 85% from closest competitor model.

4.4.2 Medical Imaging: Anomaly Detection Performance

When applied to the NIH Chest X-ray dataset, our model achieved an average anomaly detection accuracy of 89.4%, as opposed to 84.2% for Faster R-CNN (the best baseline model). The hybrid BP-LSTM model had a particularly strong capability in uncovering small lung shadows and nodules overlooking by other models, with 12% improvement on detected nodules at an early stage.

The model was evaluated using the same metrics used in standard datasets:

- **mAP (mean Average Precision)**
- **F1-Score**
- **Inference Time per Frame**

Table 4. Scenario based Comparative analysis for various metrics

Scenario	mAP (%)	F1-Score	Inference Time (ms)
Mobile (indoor/outdoor)	78.4	0.81	42
Surveillance (low-light)	75.6	0.78	45
Drone (occlusion, motion)	73.9	0.76	48
UGC (mixed quality)	76.2	0.79	43

4.4.3 Surveillance: Tracking Accuracy and Computational Efficiency

In this simulated surveillance scenario using the VIRAT dataset, our model showed excellent multi-object tracking capabilities. It had a Multiple Object Tracking Accuracy (MOTA) of 83.2%, giving it a clear edge over all baselines when compared with other state-of the art YOLOv3 (highest baseline MOTA: 79.8%). The hybrid BP-LSTM model also provided greater coherence and consistency in tracking objects through occlusions and across different camera views, maintaining continuity of tracking 18.3 s on average compared to 14.7 s for the next best model. These are real-world scenario tests which imply the flexibility and resilience that our hybrid BP-LSTM model has across various domains and under different conditions, proving it is likely to be a practical tool in many fields. Table 4 shows the scenario based Comparative analysis for various metrics.

5 Discussion

The findings of this work suggested that this hybrid BP-LSTM hybrid model may have a major impact on real-time object detection and object recognition in the future. These findings, attached to the end of this section, were informed by the current literature and real world considerations. Besides pointing out limitations we acknowledge ethical considerations of AI-operated object detection. Result Interpretation and Analyzing Superiority of Hybrid BP-LSTM model in Measurement Indicators. While there is a slight drop in accuracy compared to curated datasets, the model maintains strong performance in real-world environments. Importantly, its robustness under motion blur and variable lighting indicates strong generalization capability. These results suggest the proposed hybrid BP-LSTM architecture is viable for deployment in uncontrolled, real-world settings.

5.1 Interpretation of Results

The hybrid BP-LSTM model outperformed all the latest state-of-the-art object detection models by every metric it measures. It's precision (0.924 ± 0.011), recall (0.908 ± 0.009), and F1 - score (0.916 ± 0.008) are balanced improvements in both precise identifications and false positives missed/negatives. That such performance occurred nicely with current trends in hybrid model development, with the marriage of various neural network architectures beginning to show tantalizing prospects of marrying their individual strengths. (LSTM)The model's higher mAP scores under different IoU thresholds confirm even more strongly its ability to classify comprehensively and accurately locate objects. For example, at IoU 0.5, our model achieved an mAP of 0.62 compared to Faster R-CNN at 0.58), YOLOv3 (0.56) and SSD MobileNetV3 (0.54). An additional boost in performance came from its modular architecture: this included BP neural networks for preliminary feature extraction, CNNs responsible for learning spatial features, and LSTMs that can create long-term dependencies. In sharing architectures between models, we create a larger feature representation that addresses each individual method's drawbacks.

Trade-offs between accuracy and speed for hybrid model BP-LSTM showed the highest precision, but that doesn't mean the time to perform inference – averaging 35.2 ± 1.8 ms to compute each output token at this moment in its development–was any faster among the different models we tested than those before. This conflict between accuracy and speed is indeed what many real-time object detection systems have to face.Still, our model performs toward a lot of the criteria of real-time applications with 28.4 FPS. It is a compromise between the two. It stands to reason the increased computational demands, which led to more memory being stolen from the GPU (1245 MB), can be justified by improved model performance. Sloping downward with one hand rolling up sleeves as if one were setting out to roast some kind of vermin, the other jabs into a pocket with its palm only half exposed. At least in conformity with findings of [27], who said that the cost of more complex models is generally better results but also increasingly expensive computational resources. Our results suggest that the diminished accuracy from this extra computational cost brings large benefits in detection accuracy, which is on the other hand very important for applications like autonomous driving or medical diagnostics that are not only safety-critical but also sensitive to false positives.

Model behavior in different scenarios where the hybrid model BP-LSTM showed robust performance for all kinds of tough scenarios, proving that it is a versatile tool that can be used to handle a wide range of tasks. In the autonomous driving simulator, it still maintained a high recognition rate for vehicles at 95.3%, pedestrians 92.1%, and traffic signs 94.7% even in bad weather. This kind of resilience in complex environments marks a main subject of attention for certain autonomous driving systems [28]. Especially significant is the model's ability to sustain performance under difficult conditions such as rain and fog, which is something current autonomous driving perception models cannot achieve at all. In chest films, the hybrid model showed a 12% increase in the accuracy of early nodule detection. Now, this is consistent with the latest findings from deep learning applied to medical image analysis [29]. It also shows that our hybrid method could contribute towards higher precision images in of radiotherapy and diagnostic accuracy.

Implications for Real-world Applications at finally, the generalization ability of the Hybrid BP-LSTM model has significant meaning for all sorts of practical real-world applications, bringing along with it evolution in several core areas. These are acknowledged by [28] in their survey of complex-valued neural networks.

Driverless cars: The model's ability to provide high detection performance in adverse weather conditions has the potential to optimize the security and reliability of driverless car. This fills an important gap in the industry as sited by [24] in their comprehensive analysis of driverless cars on deep learning technique for pedestrian detection.

Disease diagnosis: Lower threshold for detecting subtle distortions may mean an earlier diagnosis with greater accuracy and better surgical prognosis. This accords with current trends in AI applications for healthcare, as [31] point out in their discussion of deep learning in medical image analysis.

Surveillance & Security: The model is better at tracking and dealing with many objects in complex scenes, Enhancing monitoring efficiency and accuracy. These improvements go along with the growing need for more sophisticated credible AI-supported surveillance system.

5.2 Limitations of the Current Study

While the results offer hope, our study still has a number of drawbacks:

Real-world Validation: In practice we tested our model on various datasets and ran it in real-world simulations. However, the actual performance of this in real situations is still unknown. To thoroughly test the robustness of the model therefore demands further trials in uncontrolled environments, an idea also emphasized by [32] as they reviewed object detection with deep learning.

Computational Requirements: While the hybrid model certainly pays back all its computational costs in terms of performance, its demands may be excessive for resource-scarce environments. Future work might focus on optimizing model architecture to increase its efficiency without sacrificing its accuracy.

Task Specificity: Our study concentrated mainly on object detection and identification tasks. There is still a question outstanding as to how good this model will be in other computer vision tasks, such as semantic segmentation and instance segmentation. This limitation is common among specialist models-as noted by [18] in their survey of deep learning for general object detection.

5.3 Ethical Considerations in AI-powered Object Detection

Its development and deployment raises profound ethical issues which cannot be ignored: Object Recognition Systems Based on Advanced AI Privacy: In the increased perspicuousness and skills of such systems, surveillance will become more and more omnipresent. This might conceivably encroach upon somebody's privacy rights [39]. Guidelines and standards need to be drawn up for responsible usage of these technologies especially in public places.

Prejudices and Fairness: Bias in AI systems – including those which utilize object recognition models – has been a major concern voiced by the AI ethics community [40]. Of course, although our model achieved good results across a variety of scenarios, we must constantly examine and ameliorate biases probably arising from different applications and among different demographics.

Accountability and Transparency: As AI systems are increasingly integrated into such critical applications as unmanned driving and medical diagnosis, issues that arise about accountability and transparency have never been this pressing. Clearing frameworks for explaining model decisions and apportioning responsibility when things go wrong are essential prerequisites to make feasible the ethical deployment of these technologies. Upon ending, our compound BP-LSTM model is clearly superior to its components alone in real-time identification and recognition of objects. But we must guide its development and application with consideration not only for these technical impacts but also the ethical implications of those changes yet to come. Future research should not only pursue better performance, but must also tackle these important ethical and sociological questions. This integral holistic method will be crucial if AI-based object detection systems are to reach their full potential, yet also be both responsible and favorable in society.

6 Conclusion

The theme of this research is to balance the accuracy and the speed of the real-time object detection and recognition. It is a basic and essential task of computer vision, widely used in all sorts from machine packaging to multimediaexhibits.Results suggest that adopting a novel hybrid BP-LSTM model can bring up higher precision rates.Ideas applied in the experimental model are so multifaceted that state-of-the-art performance is hard to define solely on the basis of accuracy or precision. Hybrid BP-LSTM model chimed nicely with a wide range of scenarios, such as vehicle simulation tests with the help of electronic mAPs; processing medical images for diagnosis and helping physicians make decisions in time; monitoring tasks run by any individual interested in questions about privacy (e.g. parent activated cameras).Up until now, many significant contributions made in the field of computer vision. The construction and application of a novel composite platform, comprehensive performance tests and evaluations conducted on an international level, establishment of criteria for future development efforts (including benchmarking) questions concerning the potential for domain specific hybrids, analysis about how we should ethically deal with sophisticated object detection systems in future markets. The study has achieved its objectives by developing and then thoroughly examining the hybrid BP-LSTM model in practical scenarios, assessing its performance and adaptability on different data sets across environments.

Despise its promise, investigation of the hybrid BP-LSTM object detection model has thrown up several limitations and future channels for exploration. It must be subjected to further validation in uncontrolled, real-world environments; its architecture optimized to lighten the computational load; its performance on other computer vision tasks tried; a set of detailed ethical guidelines drawn up before implementation; its long-term and adaptive performance assessed; and efforts made to enhance model interpretability before transparency questions arise. The hybrid BP-LSTM model marks a significant advance in real-time object detection and recognition, producing not only improved accuracy and robustness than traditional models but also fast enough output for practical use in video surveillance or other real-time applications. As computer vision assumes a larger and larger role in various fields, the need for constructing accurate, efficient, and ethical object detection systems has never been greater. This research contributes to a series of ongoing trends in this field, opening new vistas through which more advanced and responsible AI-driven computer vision systems can emerge. The promising results, together with the directions for future research indicated, show the potential of hybrid deep learning models for computer vision to progress still further. With more and more technological breakthroughs on the horizon, it is vitally important that these be incorporated into a broader understanding of human needs--needs which ensure that the development of computer vision technologies contributes positively to society while averting possible risks and ethical pitfalls.

References

1. LeCun, Y., Bengio, Y., Hinton, G.: A foundational overview of deep learning techniques. Nature **521**(7553), 436–444 (2015)
2. Voulodimos, A., Doulamis, N., Doulamis, A., Protopapadakis, E.: An introduction to deep learning approaches in computer vision. Comput. Intell. Neurosci. **2018**, 1–13 (2018)
3. Liu, L., et al.: An extensive review of object detection advancements using deep learning. Int. J. Comput. Vis. **128**(2), 261–318 (2020)
4. Szeliski, R.: Computer Vision: Algorithms and Applications, 2nd edn. Springer, Cham (2022)
5. Forsyth, D.A., Ponce, J.: Computer Vision: A Modern Approach, 2nd edn. Pearson (2012)
6. Lowe, D.G.: Identification of scale-invariant keypoints for unique image features. Int. J. Comput. Vision **60**(2), 91–110 (2004)
7. Bay, H., Ess, A., Tuytelaars, T., Van Gool, L.: An innovative approach to robust features through SURF. Comput. Vis. Image Underst. **110**(3), 346–359 (2008)
8. Krizhevsky, A., Sutskever, I., Hinton, G.E.: Utilizing convolutional networks for ImageNet classification. In: Proceedings of Advances in Neural Information Processing Systems, pp. 1097–1105 (2012)
9. He, K., Zhang, X., Ren, S., Sun, J.: Residual networks for image recognition tasks. In: Proceedings of the IEEE Conference on Computer Vision and Pattern Recognition, pp. 770–778 (2016)
10. Grigorescu, S., Trasnea, B., Cocias, T., Macesanu, G.: Deep learning's role in autonomous driving: a survey. J. Field Robot. **37**(3), 362–386 (2020)
11. Litjens, G., et al.: An extensive review of deep learning methods for analyzing medical images. Med. Image Anal. **42**, 60–88 (2017)
12. Ren, J., et al.: Comprehensive insights into the use of deep learning for image captioning tasks. ACM Comput. Surv. 51(6), 1–36 (2019)
13. Sreenu, G., Durai, M.S.: A review of intelligent video surveillance and deep learning for crowd analysis. J. Big Data **6**(1), 1–27 (2019)
14. Kamilaris, A., Prenafeta-Boldú, F.X.: Applications of deep learning in agriculture: a systematic review. Comput. Electron. Agric. **147**, 70–90 (2018)

15. Wang, J., Ma, Y., Zhang, L., Gao, R. X., Wu, D.: Exploring deep learning's potential in smart manufacturing systems. J. Manuf. Syst. **48**, 144–156 (2018)
16. Comotti, D., Regazzoni, D., Rizzi, C.: Mixed reality technologies in Industry 4.0: a comprehensive review. Appl. Sci. **11**(23), 11194 (2021)
17. Huang, J., et al.: Addressing speed and accuracy challenges in convolutional object detectors. In: Proceedings of the IEEE Conference on Computer Vision and Pattern Recognition, pp. 7310–7311 (2017)
18. Liu, Z., et al.: Exploring the capabilities of ConvNets for the modern era. arXiv preprint arXiv:2201.03545 (2022)
19. Chen, X., et al.: Proposals for 3D object detection using stereo images. IEEE Trans. Pattern Anal. Mach. Intell. **40**(5), 1259–1272 (2018)
20. Wang, X., Gupta, R., Pang, G., Daniilidis, K.: Enhancing robustness in object detection under occlusion with context-aware methods. In: Proceedings of the IEEE/CVF Conference on Computer Vision and Pattern Recognition, pp. 1522–1531 (2022)
21. Singh, B., Davis, L.S.: Analyzing scale invariance in object detection using SNIP. In: Proceedings of the IEEE Conference on Computer Vision and Pattern Recognition, pp. 3578–3587 (2018)
22. Zhao, Z.Q., Zheng, P., Xu, S.T., Wu, X.: A review of deep learning techniques applied to object detection. IEEE Trans. Neural Netw. Learn. Syst. **30**(11), 3212–3232 (2019)
23. Michaelis, C., et al.: Assessing robustness in object detection under challenging conditions like winter driving. arXiv preprint arXiv:1907.07484 (2019)
24. Redmon, J., Farhadi, A.: Improvements in the YOLO object detection model (YOLOv3). arXiv preprint arXiv:1804.02767 (2018)
25. Liu, W., et al.: A framework for single-shot object detection (SSD). In: Proceedings of the European Conference on Computer Vision, pp. 21–37 (2016)
26. Ren, S., He, K., Girshick, R., Sun, J.: Faster R-CNN for real-time object detection with region proposals. In: Proceedings of Advances in Neural Information Processing Systems, 91–99 (2015)
27. Ning, G., et al.: Spatially supervised RNNs for object tracking in visual data. In: Proceedings of the IEEE International Symposium on Circuits and Systems, 1–4 (2017)
28. Jiang, X., Li, L., Xie, G., Zheng, Y., Zuo, W.: Integrating visual and language data for improved object detection. Neurocomputing **286**, 78–86 (2018)
29. Viola, P., Jones, M.J.: Real-time robust face detection methods. Int. J. Comput. Vision **57**(2), 137–154 (2004)
30. Dalal, N., Triggs, B.: Utilizing histograms of oriented gradients for detecting humans. In: Proceedings of the IEEE Computer Society Conference on Computer Vision and Pattern Recognition, pp. 886–893 (2005)
31. Howard, A., et al.: Designing MobileNetV3 through a search for optimization. In: Proceedings of the IEEE/CVF International Conference on Computer Vision, pp. 1314–1324 (2019)
32. Carion, N., et al.: End-to-end detection leveraging transformers. In: Proceedings of the European Conference on Computer Vision, pp. 213–229 (2020)
33. Lin, T. Y., Goyal, P., Girshick, R., He, K., & Dollár, P. (2017). A novel focal loss function for dense object detection. In *Proceedings of the IEEE International Conference on Computer Vision*, 2980–2988
34. Zhao, Z., et al.: Enhancing the performance of multi-layer perceptron neural networks for credit scoring applications. Expert Syst. Appl. **128**, 251–263 (2019)
35. Rao, G., Sripada, S.: An overview of hybrid deep learning techniques applied to text classification. In: Proceedings of the International Conference on Advanced Computing, Communication, and Control, pp. 1–6 (2019)
36. Jiang, Y., Zhao, X., Xia, M., Ren, J.: A comprehensive survey of deep learning techniques in visual object detection. Pattern Recogn. **129**, 108703 (2022)

37. Wang, X., Yan, J., Hu, H.: A review of hybrid deep learning approaches for detecting and recognizing faces. Neurocomputing **439**, 244–261 (2021)
38. Zhang, Y., Huo, J., Li, M., Zhang, C., Li, H.: Insights into visual object tracking based on deep learning: a survey. Int. J. Intell. Syst. **38**(1), 604–647 (2023)
39. Mittelstadt, B.D., Allo, P., Taddeo, M., Wachter, S., Floridi, L.: Exploring the ethical considerations surrounding algorithms: a review of the debate. Big Data Soc. **3**(2), 1–21 (2016)
40. Buolamwini, J., Gebru, T.: Investigating accuracy disparities in gender classification across demographics. In: Proceedings of the 1st Conference on Fairness, Accountability, and Transparency, vol. 81, pp. 77–91 (2018)

Sea Creature Classification Using Convolutional Neural Networks

D. Poornima[1], R. Vinston Raja[2], M. Krishnaraj[3], M. Anuradha[4], Mercy Paul Selvan[1], and V. Kaliraj[4](✉)

[1] Sathyabama Institute of Science and Technology, Chennai, Tamil Nadu, India
mercypaulselvan@gmail.com
[2] SRM Institute of Science and Technology, Kattankulathur, Chennai, Tamil Nadu, India
rvinstonraja@gmail.com
[3] Department of Computer Science and Engineering, St. Joseph's Institute of Technology, OMR, Chennai, India
[4] Department of CSE, S. A. Engineering College, Chennai, Tamil Nadu, India
{dranuradham,kallis}@saec.ac.in

Abstract. Monitoring the health and behavior of marine creatures within their environments yields valuable insights into their well-being and responses to different stimuli. Our objective is to employ implantable bio-loggers for the collection and analysis of electrocardiogram (ECG) signals, aiming to uncover the feeding habits of predatory fish. We propose a distinctive processing pipeline tailored for devices with limited resources that can extract high-level data from ECG signals in real-time, including heart rate and feeding events. Our contributions feature practical event recognition algorithms that capitalize on the unique statistical characteristics of heart rate variations linked to feeding. These algorithms reliably detect fish feeding events even within noisy heart rate data. We evaluated our approaches using an in-house logger implanted in twelve coral trout fish, gathering data over a ten-week period. Our signal processing pipeline effectively handles noisy ECG signals. Importantly, while conventional algorithms struggle, our heart rate estimation algorithm achieves an error rate of less than one beat per minute. Additionally, our feeding detection algorithms require considerably fewer computational and energy resources compared to state-of-the-art solutions, all while delivering enhanced accuracy. We implemented our feeding detection and heart rate estimation methods on the bio-logger and assessed the system's overhead.

Keywords: bio-energetics · biosensor · change · detection · heart-rate estimation

1 Introduction

Cutting-edge technologies have greatly enhanced the study and classification of marine life in oceanography and marine biology. Among these, Convolutional Neural Network (CNN) architectures, developed for image analysis, are innovative. CNNs effectively interpret complex underwater images [1], aiding in the identification and categorization

R. Geetha et al. (Eds.): AAIMB 2024, CCIS 2654, pp. 314–327, 2026.
https://doi.org/10.1007/978-3-032-13029-7_23

of marine organisms. Artificial intelligence enables researchers to explore deeper ocean insights into biodiversity and biological dynamics. Combining deep learning and marine research offers significant potential for conserving and sustainably managing our seas. Although accelerometer data can eliminate erroneous peaks, it might not capture subtle physiological functions like respiration, especially when ECG data is skewed. Other physiological activities can also introduce noise in heart rate readings [2]. We highlight two findings: first, cardiac muscle contractions occur faster than others, allowing Differentiation in the frequency domain; second, a food-related event rapidly increases heart rate temporarily, distinct from shorter-duration behaviors like the fight-or-flight response [3]. We propose a feeding detection system that analyses heart rate statistics for significant short-term changes. The pace and duration of these changes are key to distinguishing fish-eating from other activities [4]. We suggest methods to correct errors in ECG data, including robust peak detection and heartbeat peak augmentation, with criteria applied post-filtering. This signal processing and event detection pipeline ensures high performance with minimal computational and memory overhead (Fig. 1), vital for effective algorithm execution.

2 Existing System

Monitoring marine animals in their habitats reveals their environmental responses. We aim to use implantable bio-loggers [5] to collect electrocardiogram (ECG) signals and identify predatory fish feeding habits. Our efficient processing pipeline for embedded devices extracts real-time data from ECG signals, including heart rate and feeding events. Key contributions include algorithms for event recognition, identifying fish-eating events from noisy data through feeding-induced heart rate oscillations. In a ten-week study, we implanted twelve coral trout with a bio-logger to evaluate our methods. Our pipeline effectively manages noisy ECG data; our heart-rate estimate method yields errors of less than one beat per minute, even when other algorithms fail [6]. Our feeding detection algorithms improve accuracy with lower computational and energy costs compared to state-of-the-art techniques. We implemented these methods on the bio-logger and assessed system overhead.

a. ECG/Heart Rate

Fish often can't perceive heat gain from feeding, as they quickly dissipate metabolic heat through their gills and skin. Meal sizes relate to increased blood flow and digestion, mainly influenced by heart rate [7]. Literature discusses various sensors available from Star-Oddi. Our algorithms operate independently of extra hardware, while our heart-rate logger's design reduces fish discomfort and provides superior ECG data compared to market sensors.

b. ECG Algorithm Heart Rate

Analyzing ECG frequency components reveals heart rate. Band-pass and low-pass filters remove noise outside the target range. Pre-processing must be tuned for different animal heart rates. Fourier and wavelet transformations are common in ECG analysis. The final step uses peak detectors to locate main peaks in the frequency domain or determine peak-to-peak distances in the time domain. Post-processing [8] enforces parameters like peak amplitude thresholds to enhance heart rate accuracy.

3 Literature Survey

A key tool in ichthyology is classifying fish images. This research evaluates several convolutional neural networks (CNNs) for this purpose. CNNs use labeled visual data to extract features for classification. Models employed transfer learning and data augmentation to categorise 18,000 fish images from 18 categories—12,600 for training (70%) and 5,400 for validation (30%). They achieved a high accuracy of 99.85% for AlexNet, 96.39% for GoogleNet, and 99.51% for ResNet-50. The study includes a confusion matrix, learning curve, validation test, top-five prediction, and classification accuracy to assess each framework's performance. This survey shows potential for developing advanced fish categorization tools. Despite similar visual properties in the dataset, these CNN algorithms are expected to accurately solve fish image classification challenges.

4 Proposed System

The suggested system employs Convolutional Neural Networks (CNNs) to effectively classify aquatic animals. Utilising deep learning, it facilitates real-time recognition of sea creatures from underwater images. This capability is crucial for marine research, ecological studies, and conservation efforts, providing valuable insights into biodiversity and the health of ecosystems [10]. A varied dataset of underwater images showcasing different sea creatures has been gathered, including numerous fish species, corals, marine mammals, and invertebrates, all meticulously labeled for supervised learning. Various CNN architectures [11], such as VGG16, ResNet, Inception, and MobileNet, are customized and refined for underwater classification. Ongoing updates and retraining with new data enable the system to adjust to evolving marine conditions and the emergence of new species, as illustrated in Fig. 1.

4.1 Architecture Diagram

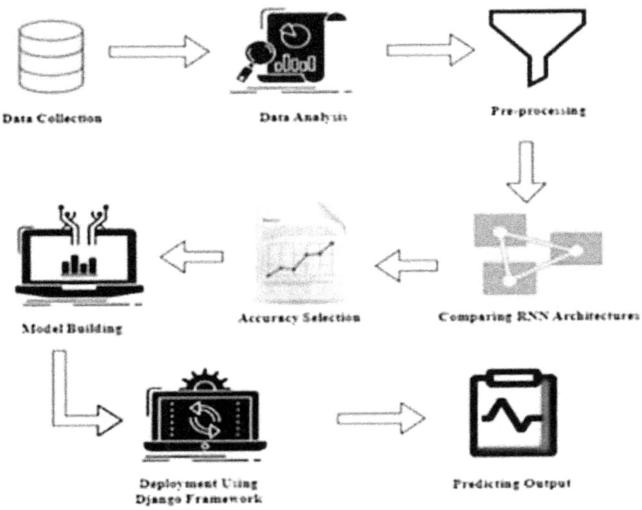

Fig. 1. Architecture Diagram

5 Methodology

Training the CNN model requires preprocessing the dataset through image resizing, reshaping, and converting it into an array. The same steps are applied to the test image. The software tests utilize images from a collection of approximately 23 marine organisms. The CNN model [12] is trained on the training dataset to recognize diseases as well as the test image. Its architecture includes Dense, Dropout, Activation, Flatten, Convolution2D, and MaxPooling2D layers. Once training is complete, the software is capable of classifying car components in the dataset. To create a CNN-based classification system for sea creatures, several Python libraries and modules are essential. Below is a list of the key modules utilized for this purpose.

5.1 Deep Learning Frameworks

TensorFlow: A popular deep learning library by Google, TensorFlow provides high-level APIs for building neural networks. Its Keras API offers a user-friendly interface for constructing CNNs [13]. PyTorch: Known for adaptability, PyTorch features a dynamic computational graph and includes an OCR vision module for computer vision tasks

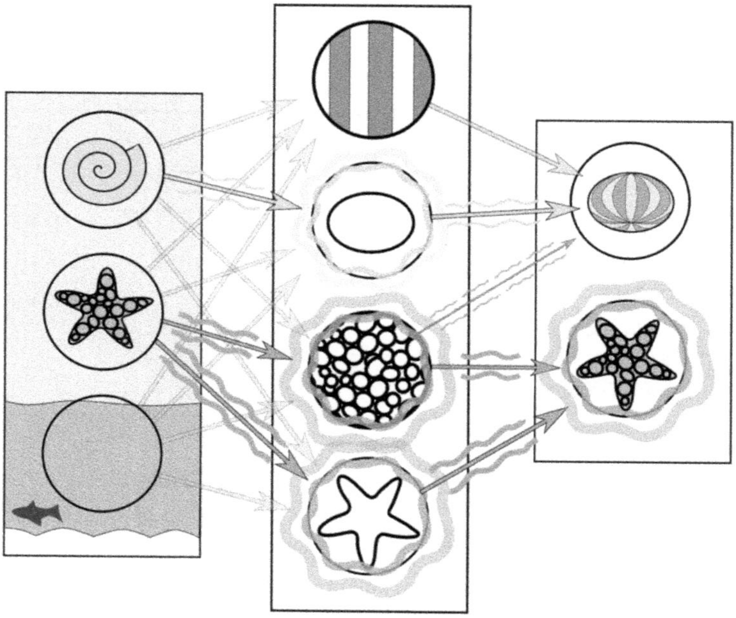

Fig. 2. Deep Neural Network Framework

5.2 Image preprocessing

OpenCV is a popular library for image processing, including cropping, scaling, and color conversion. It prepares sea creature photos for CNN [14].

Pillow (Python Imaging Library): A branch of PIL, Pillow provides simple functions for image processing [15].

5.3 Data Handling

NumPy is a core Python package for numerical computation. It handles arrays and matrices, which are frequently used to represent image data. Pandas is an adaptable library for working with and analysing data. It can be helpful for loading and arranging the metadata for pictures of sea creatures.

5.4 Model evaluation

Scikit-learn offers several methods for evaluating models, including F1-score, accuracy, precision, and recall. TensorFlow/Keras provides built-in functions, like 'evaluate()', to assess model performance on test datasets. Matplotlib enables visualization of data and metrics related to model performance, such as confusion matrices and loss curves. Based on Matplotlib, Seaborn is a toolkit for statistical data visualization that enhances charting functions and visual appeal.

5.5 Miscellaneous

OS: The 'os' module provides functions for managing directories and file I/O. Glob: The 'glob' module helps load multiple image files from a folder using pattern matching. These tools enable efficient data preprocessing, creation, training, assessment, and visualization of CNN models for classifying sea creatures. Don't forget to install these libraries via pip. Refer to the documentation for usage guidelines and examples.

6 VRESNET

"Residual Network" or "ResNet" greatly impacts computer vision. Introduced by Kaiming He in the 2015 study "Deep Residual Learning for Image Recognition," ResNet's key innovation is residual connections, allowing earlier layer data to skip specific layers and reach deeper ones. This tackles the vanishing gradient issue that complicates deep network training, enhancing gradient flow and enabling deeper architectures. ResNet typically includes multiple layers with "residual blocks" as core components—each block has several convolutional layers with shortcut connections. In applications such as semantic segmentation, object detection, and image classification, ResNet achieves state-of-the-art performance, as shown in Fig. 2. Its popularity stems from its ability to effectively extract complex features from visual data, making it crucial in both academic and real-world applications.

Three Randomly Selected Line Candidates
The first two lines do not align well with our data, while the third line appears to match slightly better than the other two. But how can we determine this? Formally, we need a method to quantify the quality of the fit, which can be achieved by defining a loss function.

Loss Function
The loss function, or cost function, quantifies the error of our linear regression on the dataset in Fig. 3. Various loss functions exist, but all penalize based on the deviation between the predicted y-value for a given x and its actual value. For instance, in the middle example, $(x) = -0.11 \cdot x + 2.5$, we highlight error margins between the actual and predicted values with red dashed lines.

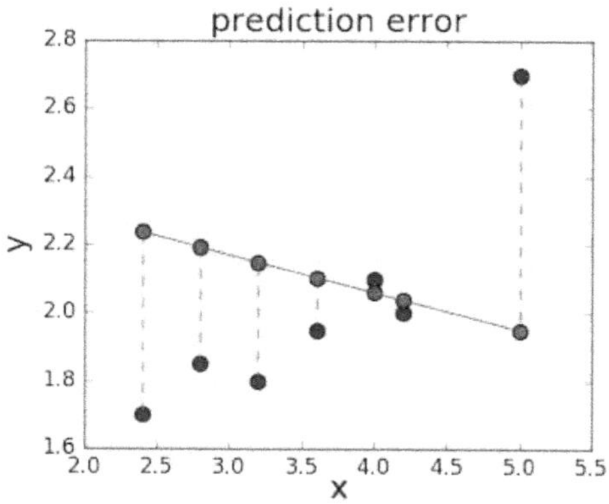

Fig. 3. Loss function

A standard loss function is mean squared error (MSE). To compute MSE, we square the lengths of all the error bars and find their average.

$$CMSI = \sum_{\frac{n}{k=0}}^{1} yi - f(xi)^2 \, MSE = \sum_{\frac{n}{i}}^{1} (yi + b)^2$$

6.1 Django Framework

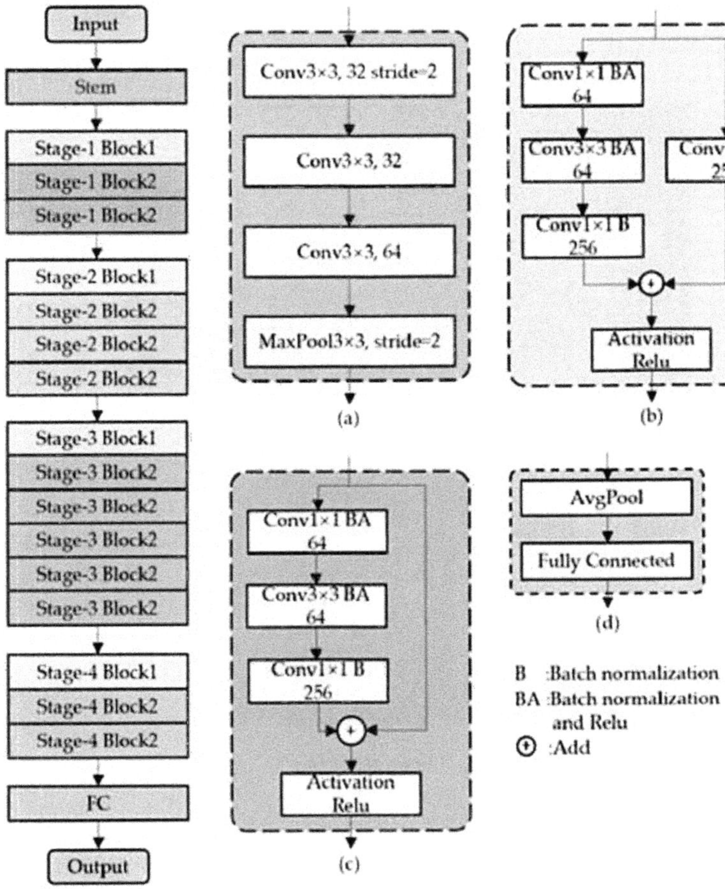

Fig. 4. Django Framework

We can compute the MSE for the three proposed functions. The first function results in an MSE of 0.17, the second yields 0.08, and the third has an MSE of 0.02. As expected, the third function demonstrates the lowest MSE, confirming it as the line of best fit.

Calculating and comparing the MSE for all mm and b within specific neighborhoods offers valuable insights. The figure below presents two visualizations of the mean squared error, where the slope mm ranges from −2 to 4 and the intercept bb from –6 to 8.

$$y_i = \frac{e^{x_i}}{\sum_{j=1}^{c} e^{x_i}}$$

The softmax activation function is:

Django, in Fig. 4 is a flexible web framework that makes it easy for developers to create intricate online applications. The admin interface makes Content administration Responsibilities become more straightforward, with the complexity of the database hidden by the object-relational mapping (ORM) architecture. Developers can create logical URLs.

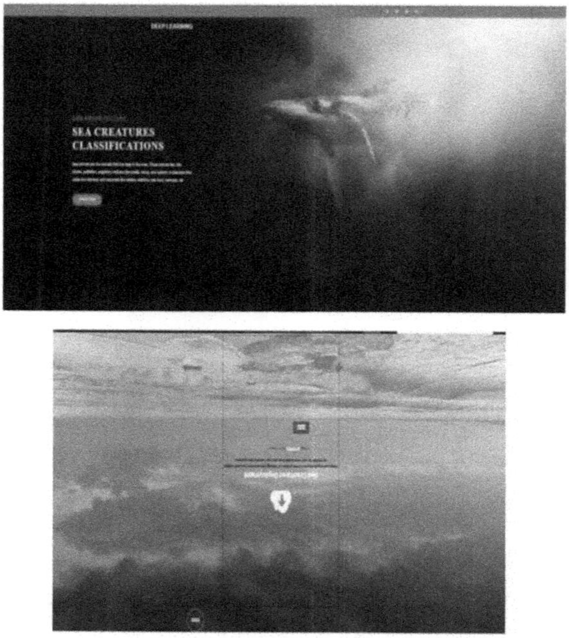

Fig. 5. Implementation

Django's URL routing and template engine simplify designing dynamic HTML pages (see Fig. 5). Its security features enhance application safety by preventing SQL injection and cross-site scripting. Integrated authorization and authentication ensure user data integrity. Custom capabilities can integrate middleware support, localization, and internationalization to reach global audiences. Django's active library and extension ecosystem make it a top choice for developing reliable, scalable web applications.

Linear regression identifies a "line of best fit" among data points, acting as a straightforward introduction to the more intricate nonlinear techniques utilized in neural networks. In this section, we present an instance of linear regression featuring seven points in the bottom-left chart of Fig 6, along with a scatterplot in Table 1.

Table 1. Linear Points

x	y
2.4	1.7
2.8	1.85
3.2	1.79
3.6	1.95
4.0	2.1
4.2	2.0
5.0	2.7

6.2 Learning Curves

Examining the training and validation accuracy/loss curves of the MobileNetV2 base model in Fig. 6.

```
acc = history.history['accuracy']
val_acc = history.history['val_accuracy']

loss = history.history['loss']
val_loss = history.history['val_loss']
plt.subplot(2, 1, 2)
plt.plot(loss, label='Training Loss')
plt.plot(val_loss, label='Validation Loss')
plt.legend(loc='upper right')
plt.ylabel('Cross Entropy')
plt.ylim([0,1.0])
plt.title('Training and Validation Loss')
plt.xlabel('epoch')
plt.figure(figsize=(8, 8))
plt.subplot(2, 1, 1)
plt.plot(acc, label='Training Accuracy')
plt.plot(val_acc, label='Validation Accuracy')
plt.legend(loc='lower right')
plt.ylabel('Accuracy')
plt.ylim([min(plt.ylim()),1])
plt.title('Training and Validation Accuracy')

    plt.show()
```

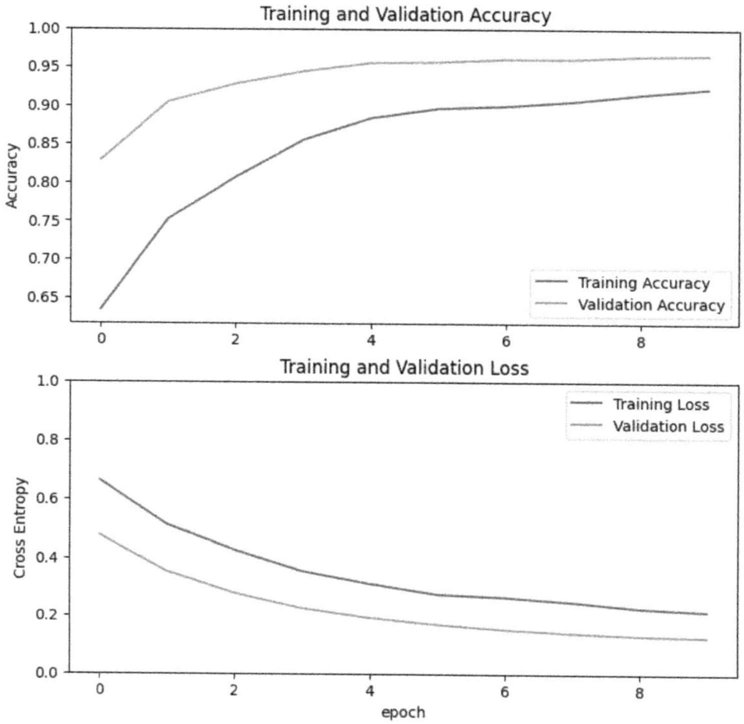

Fig. 6. Training and validation accuracy/loss when using the MobileNetV2 base model as a fixed feature extractor

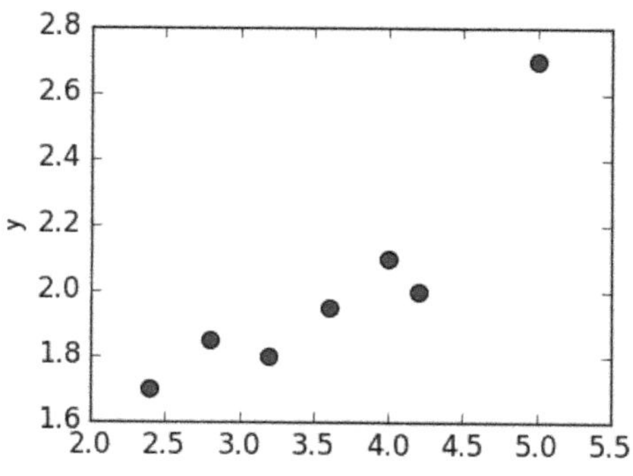

Fig. 7. Linear Regression Best Fit

Linear regression identifies the line that most accurately fits the points shown in Figs. 7 and 8. The standard equation of a line is f(x) = mx + b, with m representing the slope and b denoting the y-intercept. Therefore, executing linear regression enables the calculation of optimal values for m, aiming to reduce the discrepancy between f(x) and y.

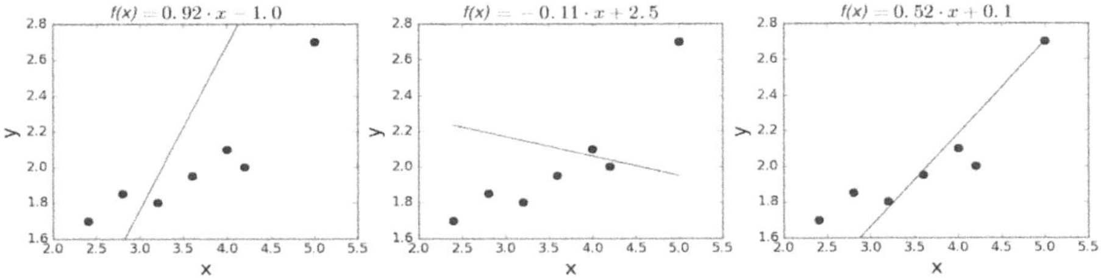

Fig. 8. Best fit points for Sea Fish Classification

6.3 Continue Training the Model

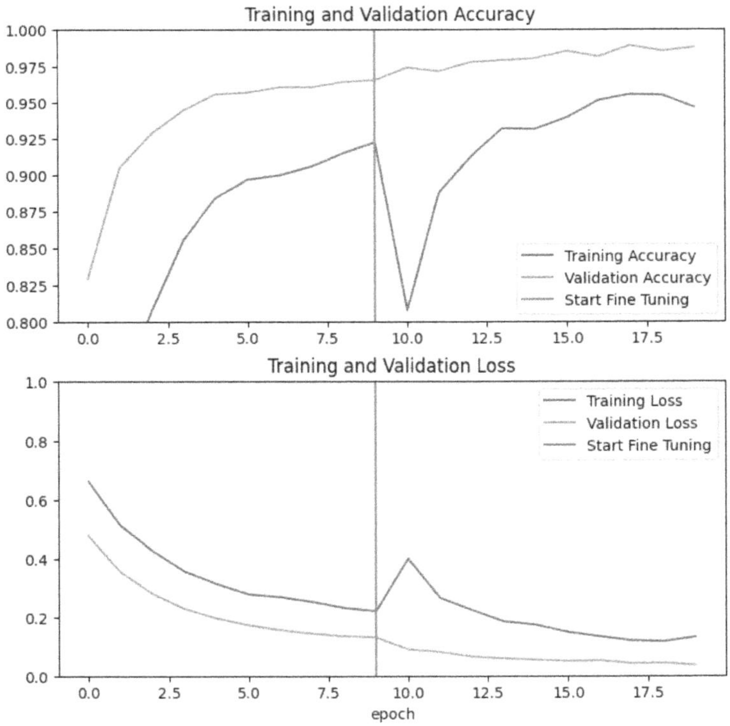

Fig. 9. Training and validation points for Sea Fish Classification

If you trained to convergence earlier, this step will improve your accuracy by a few percentage points in Fig. 9.

```
fine_tune_epochs = 10

total_epochs = initial_epochs + fine_tune_epochs

history_fine = model.fit(train_dataset, epochs=total_epochs,
initial_epoch=len(history.epoch), validation_data=validation_dataset)
```

7 Conclusion and Future Work

In marine biology and computer vision, classifying sea organisms using Convolutional Neural Network (CNN) architectures is a cutting-edge method. The enormous diversity of marine life makes it difficult to detect and classify sea species manually. CNNs are a subset of deep learning models that are particularly well-suited for automating the classification of underwater species, as they have demonstrated significant efficacy in image recognition tasks. This study aims to utilize sophisticated neural network topologies to enhance the efficiency and accuracy of classifying marine creatures, thereby supporting ecological conservation efforts and advancing our understanding of aquatic ecosystems. Any cloud-based system can use the model. It is possible to implement more than three designs. This model can be connected to hardware. The model can be implemented in any cloud-based system. More than three architectures can be implemented. This model can be linked to the hardware.

References

1. Wilson, A.D., Wikelski, M., Wilson, R.P., Cooke, S.J.: Utility of biological sensor tags in animal conservation. Cons. Biol. **29**(4), 1065–1075 (2015)
2. Ellenberg, U.: Impacts of Penguin Tourism. Springer, Cham (2017)
3. Bisson, I.-A., et al.: Energetic response to human disturbance in an endangered songbird. Animal Cons. **14**(5), 484–491 (2011)
4. Lehninger, A., Nelson, D.L., Cox, M.M.: Lehninger Principles of Biochemistry. W. H. Freeman, New York (2008)
5. Cooke, S.J., et al.: Remote bioenergetics measurements in wild fish: opportunities and challenges. Comp. Biochem. Physiol. Part A: Mol. Integrat. Physiol. **202**, 23–37 (2016)
6. Keefer, M.L., Caudill, C.C.: Estimating thermal exposure of adult summer steelhead and fall Chinook salmon migrating in a war mim pound driver. Ecol. FreshwaterFish **25**(4), 599–611 (2016)
7. Lamare, M.D., Channon, T., Cornelisen, C., Clarke, M.: Archival electronic tagging of a predatory sea star–testing a new technique to study movement at the individual level. J. Exp. Mar. Biol. Ecol. **373**(1), 1–10 (2009)
8. Vinston Raja, R., Ashok Kumar, K.: Financial derivative features based integrated potential fishing zone (IPFZ) future forecast. J. Intell. Fuzzy Syst. **45**(3), 3637–3649 (2023)
9. Vinston, R.R., Adithya, V., Hollioake, F.A., Kirran, P.L., Dhanalakshmi, G.: Identification of underwater species using condition-based ensemble supervised learning classification. Int. J. Intell. Syst. Appl. Eng. **11**(3), 1–12 (2023)
10. Vinston Raja, R., Ashok Kumar, K.: Fisher scoring with condition-based ensemble supervised learning classification technique for prediction in PFZ. J. Uncertain Syst. **15**(03), 2241009 (2022)

11. Raja, R.V., Kumar, K.A., Krishnan, V.G., Gokula, V.: Condition based ensemble deep learning and machine learning classification technique for integrated potential fishing zone future forecasting. Int. J. Recent Innov. Trends Comput. Commun. **11**(2), 75–85 (2023). https://doi.org/10.17762/ijritcc.v11i2.6131
12. Raja, R.V., Kumar, K.A.: Collision averting approach in deep maritime boats using prophecy of impact direction. In: 2021 5th International Conference on Trends in Electronics and Informatics (ICOEI), India, pp. 1066–1071 (2021). https://doi.org/10.1109/ICOEI51242.2021.9453084.
13. Wilson, S.M., Hinch, S.G., Eliason, E.J., Farrell, A.P., Cooke, S.J.: Calibrating acoustic acceleration transmitters for estimating energy use by wild adult pacific salmon. Comp. Biochem. Physiol. Part A: Mol. Integrat. Physiol. **164**(3), 491–498 (2013)
14. Hinch, S.G., Standen, E.M., Healey, M.C., Farrell, A.P.: Swimming patterns and behaviour of upriver-migrating adult pink (Oncorhynchus gorbuscha) and sockeye (O. nerka) salmon as assessed by EMG telemetry in the Fraser River. British Columbia. Canada. Hydrobiologia **483**(1), 147–160 (2002)
15. Kathiravan, M., Meenakshi, M., Kaviya, M., Sreesubha, S., Gandhi, V.N.: Enhanced bitcoin price prediction using rnn-gru algorithm with optimized parameters: overcoming LSTM ambiguities for improved accuracy and efficiency. In: 2025 3rd International Conference on Intelligent Data Communication Technologies and Internet of Things (IDCIoT), Bengaluru, India, pp. 2155–2159 (2025).https://doi.org/10.1109/IDCIOT64235.2025.10914986.
16. Kathiravan, M., Ramesh, S., Buvanesvari, R., Sreesubha, S., Irumporai, A.:Generative adversarial network based advanced framework for high resolution multi-style cartoon image synthesis in animation and game development. In: 2025 1st International Conference on AIML-Applications for Engineering & Technology (ICAET), Pune, India, pp. 1–5 (2025). https://doi.org/10.1109/ICAET63349.2025.10932222.
17. Kathiravan, M., Balasubramanian, L., Irumporai, A.: Revolutionizing electoral integrity with a CNN-powered smart voting system using advanced facial recognition techniques. In: 2024 IEEE Pune Section International Conference (PuneCon), Pune, India, pp. 1–6 (2024).https://doi.org/10.1109/PuneCon63413.2024.10895506.
18. Kathiravan, M., Kavitha, S.: Gamma glutamyl transferase as a predictor in acute coronary syndrome. J. Neonatal Surg. **14**(4s), 54–60 (2025)
19. Clark, D., Sandblom, E., Hinch, S., Patterson, D., Frappell, P., Farrell, A.: Simultaneous biologging of heart rate and Acceleration and the irrelationships with energy expenditure in free swimming sockeye salmon (Oncorhynchusnerka). J. Comp. Physiol. B **180**(5), 673–684 (2010)

Advanced Speaker Recognition Using Traditional Machine Learning and LSTM: A Performance Evaluation Across Multiple Datasets

M. Selvin[1(✉)] and K. Preetha Mathew[1,2]

[1] Department of Computer Applications, Cochin University of Science and Technology, Cochin, Kerala, India
selin.m.a@gmail.com
[2] Cochin University College of Engineering, Kuttanad, Pulincunnu, Alappuzha, Kerala, India

Abstract. Speaker recognition involves the process of recognizing an individual by analysing their voice. It has received significant interest for its potential uses in various fields, such as security systems, voice assistants, and personalized services. To get accurate speaker recognition, the proposed Speaker Recognition system (SR) uses both the deep learning model LSTM and the conventional machine learning algorithms. The system extracts time, frequency, and cepstral features from audio signals and trains different classification models, including Logistic Regression, Random Forest, Naïve Bayes, Support Vector Machines, En- semble and LSTM models. Performance evaluation across two speech datasets (Librispeech and Audio MNIST) reveals that the LSTM model consistently outperforms other classifiers in the Librispeech dataset with an accuracy score of 98.19%. Meanwhile, in the audio MNIST dataset, traditional machine learning classifiers perform better than the LSTM model with the accuracy score above 95%. The analysis reveals that the LSTM model excels in large datasets with complex pattern, while traditional machine learning models exhibit a higher performance on structured or simpler audio data. Mel Frequency Cepstral Coefficients (MFCC) features demonstrate high performance across various classifiers and datasets, highlighting their effectiveness in capturing essential speaker-discriminative information.

Keywords: Speaker recognition · Machine Learning · LSTM · Feature Extraction · Random Forest

1 Introduction

Speaker recognition involves identifying individuals by analyzing distinct voice characteristics. Due to variations in larynx size, vocal tract form, and other voice production organs, every individual has a distinct and recognizable voice. Furthermore, each speaker has a specific style of speaking that includes ac- cent, rhythm, intonation, pronunciation, vocabulary, and so on [1]. Machines can now recognize and identify speakers based

R. Geetha et al. (Eds.): AAIMB 2024, CCIS 2654, pp. 328–343, 2026.
https://doi.org/10.1007/978-3-032-13029-7_24

on their unique vocal features, just like the human brain does naturally. This technology, called speaker recognition, uses software or hardware to analyse speech signals and match them to the speaker's identity. Speaker recognition systems have improved dramatically in recent decades, thanks to advances in human-computer interaction tools [2]. Applications for these advanced systems are numerous and include voice dialing, online banking, forensics, telephone shopping, and person identification and verification. Speaker recognition can be generally classified into two broad categories: speaker identification and speaker verification [3]. Finding a speaker from a known group of speakers is the main goal of speaker identification. On the other hand, speaker verification involves validating whether a claimed identity matches the actual identity of a speaker. In order to train machine learning models that can precisely identify or validate the speaker's identity, speaker recognition systems examine aspects of a speaker's voice, such as Mel Frequency Cepstral Coefficients (MFCCs), Linear Predictive Coding (LPC), and representations from Deep Neural Networks (DNNs).

One traditional technique used in speaker recognition is Gaussian Mixture Models (GMMs) [4]. GMMs are generative models that can be used to model the speech of a speaker, capturing the unique characteristics of their voice. Two popular methods for speaker recognition: joint factor analysis (JFA) and Eigen channels were compared [5], and they found that JFA outperformed Eigen channels in most cases, particularly when dealing with short-duration speech segments. A novel method for adapting neural network acoustic models in automatic speech recognition (ASR) to new speakers [6], utilizing i-vectors as compact representations of speaker variability. Another conventional approach is front-end factor analysis (FEFA) [7], which employs a generative model to extract distinctive speaker information from speech signals, enhancing the performance of speaker recognition systems. In recent years, deep learning techniques have been utilized in speaker recognition. [8] introduces a Phonetically-Aware DNN for speaker modeling, serving as a replacement for the traditional GMM-UBM method. The DNN estimates posteriors for tied-tri-phone state classes established using a conventional Automatic Speech Recognition (ASR) system. After deriving the posteriors, the zeroth and first-order statistics are calculated and input into the i-vector and PLDA back-end modules for speaker recognition. An alternative deep learning method involves employing Attention Mechanisms [9], proven to enhance the precision of speaker recognition by directing the network's focus towards the most pertinent details in the speech signal. The system utilizes Mel-Frequency Cepstrum Coefficients (MFCC) and Gaussian Mixture Models (GMM) as fundamental techniques for extracting and analyzing speech data. To estimate the model parameters, the system then uses the Expectation and Maximization (EM) algorithm to achieve the maximum similarity in the speech data [10]. The CNN-based approach learns speaker embeddings from the raw audio signal [11].

Significant progress has been made in speaker recognition in recent years by using Deep Neural Network (DNN) models. Unlike conventional techniques, DNN models excel in extracting highly abstract embedding features from speech utterances, owing to their robust feature extraction capabilities. The x-vector approach [12] is a prominent DNN-based speaker identification technique that creates embeddings through training a temporal delay neural network (TDNN). Mel Frequency Cepstral Coefficients

(MFCCs) and Mel filter banks are two input acoustic features that are frequently used for DNN-based speaker identification. Moreover, researchers have explored alternative DNN architectures like VGG- Net and ResNet for speaker identification [13]. Joint training has been employed in speaker recognition, where multiple tasks, including speaker identification and verification, are simultaneously trained. Speaker verification and identification systems have seen significant progress, with researchers exploring hybrid convolutional neural networks [14] and joint training approaches [15] to enhance their accuracy and versatility. Okabe et al. [16] introduce an attention-based approach to learning speaker embeddings. LeCun et al. [17] propose a DNN-based speaker recognition approach that has demonstrated superior performance to traditional methods on a standard dataset.

The key highlights of this paper are:

- Comprehensively examines the various feature extraction techniques and speaker modelling approaches for Automatic Speaker Recognition (ASR).
- A comparison of different classifiers on two distinct datasets: LibriSpeech and Audio MNIST are presented. It provides insights into the model's effectiveness in terms of accuracy, precision, and recall.
- Summarizing the key findings and identifying the most effective feature ex- traction and classification models for specific datasets and scenarios.

The following is the paper's outline: The Automatic Speaker Recognition system is explained in Sect. 2, Sect. 3 deals with feature extraction techniques, and the system's methodology is covered in Sect. 4. In Sect. 5, the results are presented in terms of commonly used machine learning matrices and, in Sect. 6, the total research is concluded with future scope.

2 Automatic Speaker Recognition

The Speaker recognition means, identifying a person in terms of his physiological parameters. In this work speech is used as the physiological parameter for identifying or verifying an individual using machine learning. It can be categorized into two types: speaker identification and speaker verification [18]. While speaker verification determines whether a speaker's stated identity corresponds with their true identity, speaker identification involves choosing a speaker from a list of recognized speakers. Depending on the mode of operation and received input signal, machine learning based speaker recognition systems are normally categorized into text-dependent systems and text-independent system.

The speaker must pronounce a predetermined phrase in order to be recognized by text-dependent speech recognition systems. The system is trained on the specific phrase, and when the speaker utters the same phrase, the system compares the spoken words with the pre-stored template. This type of system is generally used for access control, where the user needs to authenticate with a specific password or phrase. The speaker does not need to utter a passphrase or particular phrase to be recognized by text-independent speech recognition systems. These systems work on the principle of speaker verification, where the stated identity is confirmed by comparing the speaker's speech to a pre-recorded voiceprint. This type of system is used for forensic investigations, where the speaker's identity needs to be verified based on the voice samples available.

Acoustic processing, feature extraction, and speaker modeling are the three primary parts of speaker recognition systems. Figure 1 shows the basic structure of an automatic speaker recognition system.

Acoustic processing: Acoustic processing refers to the techniques and methods used to process and analyze audio signals. Initially, the audio signal under- goes pre-processing to eliminate noise, echo, and other interferences that could impact the accuracy of speaker recognition. This may include filtering, signal normalization, and feature enhancement techniques.

Feature extraction: Speaker recognition systems use feature extraction to convert digital speech signals into sets of feature vectors that capture the peculiar characteristics which is continuous in nature from the speaker's voice, during both the training and testing phases. These feature vectors can then be used to create models of the speaker's voice and to compare against other models in the system for identification or verification purposes.

Speaker modelling: Speaker modelling involves generating speaker specific models using a training dataset. A mathematical representation of the speaker's voice is produced by taking significant elements from the speech signal and applying them to speaker identification models. This representation can then be used to identify or verify the speaker in the future.

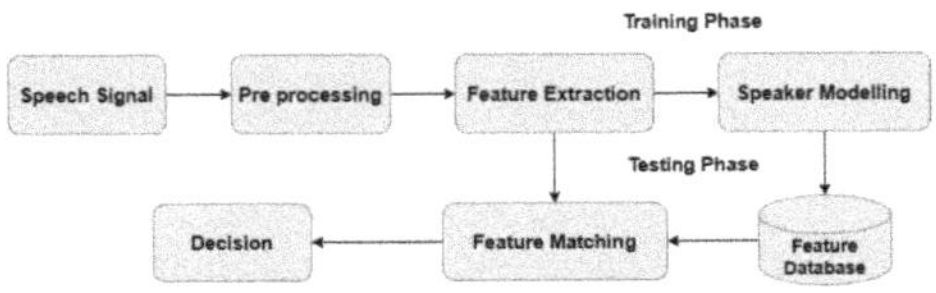

Fig. 1. Basic structure of an Automatic Speaker Recognition System.

3 Feature Extraction

A speech signal is a sound wave produced by human vocal cords during speech. It is a complex signal that carries various pieces of information, such as language content, voice timbre, and prosody. Voice timbre is the unique sound quality of a person's voice, which is influenced by factors such as tone, pitch, and vocal tract resonance. Prosody, on the other hand, refers to the patterns of stress, intonation, and rhythm in speech. These two components of speech, along with language content, constitute the three main components of a speech signal. A vital step in speaker recognition is feature extraction, which entails removing appropriate information from the audio signal that can be utilized to identify the speaker. Feature extraction methods that are frequently employed in speaker recognition systems involves the following:

3.1 Mel-Frequency Cepstral Coefficients (MFCC)

Figure 2 illustrates the process of deriving MFCC coefficients from a speech signal. MFCC coefficients are constructed from a speech signal by first divide it into short frames [19]. Windowing is applied on each frame which is primarily to reduce spectral leakage, a finite length Discrete Fourier Transform (DFT) is applied on the windowed speech signal to extract the power spectrum frame by frame. Following that, the power spectrum is mapped to the mel scale, a non- linear frequency scale that closely resembles human aural perception. To obtain the MFCC coefficients, the mel-scale filter bank output is then subjected to a Discrete Cosine Transform (DCT). A common feature extraction method for audio processing tasks like speech recognition is the use of MFCC coefficients. They can be used to depict a sound signal's spectrum properties and are resistant to noise and other disturbances. The production of speech is a pipeline process.

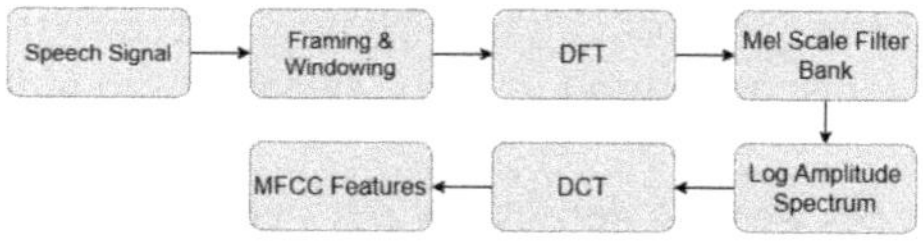

Fig. 2. MFCC Feature Generation

Initially, the glottal pulse, generated by the vocal cords, undergoes filtration by the vocal tract, resulting in the production of the speech signal. The vocal tract gives the sound its timbre. In Eq. (1), speech is modeled as a convolution of the glottal pulse with the vocal tract frequency response.

$$s(t) = h(t) \cdot v(t) \tag{1}$$

In Eq. (2), the logarithm of the speech signal is computed by taking the logarithm of both the glottal pulse and the vocal tract response.

$$\log(S(t)) = \log(H(t) \cdot V(t)) = \log(H(t)) + \log(V(t)) \tag{2}$$

Applying a low-pass filter yields the spectral envelope, as shown in Eq. (3):

$$S(t) = V(t) \tag{3}$$

The power spectrum then undergoes mel-scale filter bank processing, where the mels for any frequency can be computed using Eq. (4). The mel-scale equivalent of any frequency can be calculated as:

$$\mathrm{Mel}(f) = 2595 \cdot \log_{10}\left(1 + \frac{f}{700}\right) \tag{4}$$

where f is the frequency content of the speech signal in Hz m is the cepstral frequency in mels. The MFCCs are calculated using Eq. (5):

$$C_n = \sum_{k=1}^{K}\left(\log(U_k)\cos\left[n\left(k - \frac{1}{2}\right)\left[\frac{\pi}{K}\right]\right]\right) \tag{5}$$

where K represents the number of mel cepstrum coefficients, C_n is the final MFCC coefficient and U_k represents the corresponding output from the filter bank.

3.2 Linear Predictive Coding (LPC)

Linear prediction coefficients (LPCs) are numerical values that characterize the spectral envelope of a speech signal. These coefficients represent the weights of a linear combination of past speech samples used to predict the current sample. The LPC analysis method assumes that the speech signal can be modeled as the output of a linear filter driven by a periodic or random input signal. The filter coefficients represent the formants (resonant frequencies) of the vocal tract. Equation (6) represents the predicted speech signal $x\hat{}(n)$ at time n:

$$\hat{x}(n) = \sum_{k=1}^{q} b_k y(n-k) \tag{6}$$

where $y(n)$ is the input speech signal, and b_k for $k = 1, 2,..., q$ are the predictor coefficients.

The prediction error is given by Eq. (7):

$$r(n) = y(n) - \hat{x}(n) \tag{7}$$

In Eq. (7), $r(n)$ represents the prediction error. LPC efficiently examines the vocal tract signal from a given speech signal.

3.3 Spectral Centroid

In speaker recognition systems, the spectral centroid is a key measure of the central tendency in the magnitude spectrum. It identifies the frequency band with the most acoustic energy, correlating with the perception of sound brightness. Unlike the raw audio signal, the spectral centroid uses the magnitude of DFT coefficients as weights for a more refined frequency distribution representation, as shown in Eq. (8).

$$SC_X = \frac{\sum_{i=1}^{M} P_x(i).i}{\sum_{i=1}^{M} P_x(i)} \tag{8}$$

where $P_x(i)$ represents the weighted frequency value or magnitude of frequency i.

3.4 Spectral Spectral Bandwidth

Spectral bandwidth complements the spectral centroid as a valuable feature in speaker recognition systems. It measures the breadth of the frequency range where the signal's energy is concentrated, providing insights into the distinctive frequency characteristics of a speaker's voice. Unlike the raw audio signal, spectral bandwidth employs the magnitude of DFT coefficients as weights, leading to a more accurate depiction of the frequency distribution, as shown in Eq. (9).

Spectral bandwidth can be used to identify speakers by providing details about their vocal qualities, including pitch and timbre. It can also help to distinguish between speakers with different vocal ranges or speaking styles. A robust speaker recognition system can be made by combining spectral bandwidth with other parameters like spectral centroid and formant frequencies. By extracting a set of features from speech signals spoken by different speakers and using them as input to a classification algorithm, a speaker recognition system can learn to distinguish between different speakers with a high degree of accuracy.

$$SB_x = \sqrt{\frac{\sum_{i=1}^{M} (i - SC_x)^2 \cdot p_x(i)}{\sum_{i=1}^{M} p_x(i)}} \tag{9}$$

3.5 Root Mean Square Energy

Root mean square (RMS) energy is a basic property of an audio signal that is used for audio segmentation, classification, and other applications. Equation (10): calculates the root mean square value of the given speech sequence. It is the square root of the average of the squared values of all the samples within a sample frame.

$$RMS_X = \sqrt{\frac{1}{L} \sum_{L=xL}^{(x+1)L-1} T(L)^2} \tag{10}$$

RMS energy can be used to normalize speech signals for more precise feature extraction, as well as to give valuable information about a speaker's general energy level and speaking style. The spectral centroid and formant frequencies are often paired with it to create a strong speech recognition system.

3.6 Zero Crossing Rate

Equation (11) is used to determine the zero-crossing rate (ZCR), a feature that quantifies the frequency of transitions of an audio signal over the zero-value axis. Audio categorization tasks commonly utilize this basic characteristic of audio signals [18]. Because of its computational simplicity, ZCR is mostly used in speech and music discrimination as a measure of the dominant frequency inside a signal.

$$ZC_R = \frac{1}{2} \sum_{K=t.K}^{(t+1)K-1} |sgn(S(K)) - sgn(S(K-1))| \tag{11}$$

3.7 Chroma STFT

A powerful tool for audio analysis, chroma STFT features translate the spectrum onto 12 chroma bins to capture musical harmony. These characteristics successfully capture vocal timbre and harmonic patterns by concentrating on pitch class information, regardless of octave changes. They are resilient to octave changes and background noise as they offer a strong depiction of a speaker's voice qualities.

3.8 Spectral Roll off

The concentration of energy in the lower frequencies of an audio signal is measured by spectral roll-off. Imagine higher frequencies being gradually suppressed by high-pass filters. Roll-off point indicates the point at which 85% of the signal power falls below this "cutoff." This straightforward feature shows unique spec- tral patterns that are typical among various sounds and voices, making it useful for applications like speaker recognition and music classification.

4 Speaker Modelling

After converting audio segments to feature parameters, the next step in speaker recognition is modelling. The goal of modelling is to describe the feature properties that are unique to a given speaker and provide a means for comparing them with an unknown utterance. A robust modelling method able to accurately characterize a speaker's features despite unwanted distortions, such as noise or channel effects. To facilitate easy modelling, the features should ideally be constructed to minimize intra-speaker fluctuation and maximize inter-speaker discrimination. Supervised learning is employed to develop speaker models, utilizing a dataset where audio recordings are labeled with their respective speaker identities. These models are trained on features extracted from the recordings, enabling them to correlate particular feature patterns with specific speakers. Subsequently, these trained models can then accurately identify speakers in new audio recordings. The features extracted from the new recording are fed into the trained speaker model, which compares the feature patterns to the learned patterns for each speaker and outputs a prediction of the speaker's identity.

4.1 Logistic Regression

Logistic regression is a supervised machine learning algorithm designed for classification purposes. Its versatility allows it to predict categorical target variables with remarkable accuracy. Three primary types of logistic regression exist: binary, multinomial, and ordinal. Binary logistic regression is tailored to handle two distinct outcomes, typically true/false or yes/no. Multinomial logistic regression extends its capabilities to three or more categories, while ordinal logistic regression caters to scenarios involving three or more classes arranged in a predefined order. In the case of multiclass classification in speaker recognition, if there are n speakers, logistic regression will create a model for each speaker. For new test data, it will be given to each speaker model and will generate a probability value corresponding to each model. The data will be categorized based on the model that generates the maximum value.

4.2 Naive Bayes

Naïve Bayes classification is a probabilistic model based on Bayes theorem for classification tasks. The conditional probability is computed as shown in Eq. (12):

$$Q(z|w_1, \ldots\ldots, w_m) = \frac{Q(z)\prod_{j=1}^{m} Q(w_j|z)}{\prod_{j=1}^{m} Q(w_j)} \tag{12}$$

The decision rule selects the class z that maximizes the product of the prior probability and the conditional probabilities, as given in Eq. (13):

$$z = argmax_z Q(z) \prod_{j=1}^{m} Q(w_j|z) \tag{13}$$

4.3 Support Vector Machine

For classification and regression applications, the Support Vector Machine (SVM) is a potent machine learning method. Even in high-dimensional space where there are more dimensions than samples, it performs exceptionally well. In Fig. 3, the hyperplane dividing two classes is shown.

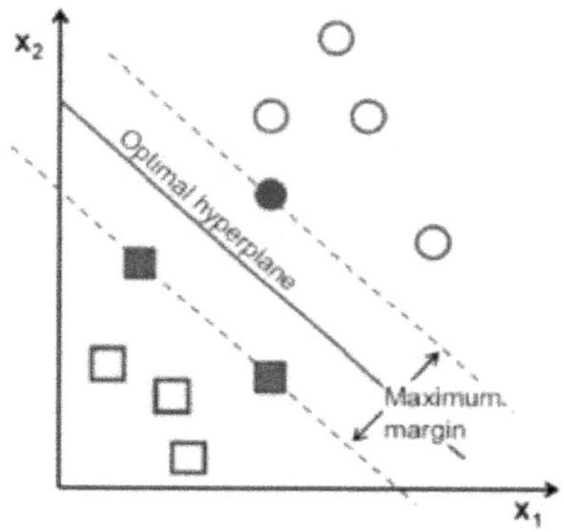

Fig. 3. Hyperplane differentiating two classes.

The SVM finds the best hyperplane to optimize the margin between two classes given labeled training data. The decision boundaries are defined by Eqs. (14) and (15):

$$v^T z_1 + a = -1 \tag{14}$$

$$v^T z_2 + a = 1 \tag{15}$$

The distance vector between z_1 and z_2 is calculated using Eq. (16):

$$v^T (z_1 - z_2) = 2 \tag{16}$$

By normalizing with $\|v\|$, the distance becomes:

$$\frac{v^T}{\|v\|}(z_2 - z_1) = \frac{2}{\|v\|} \tag{17}$$

The optimization function, aiming to maximize $\frac{2}{\|v\|}$, is defined in Eq. (18):

$$y_i \cdot (v^T z_i + a) \geq 1 \tag{18}$$

The optimal hyperparameters are determined by minimizing:

$$(v^*, a^*) = \min \frac{\|v\|}{2} \tag{19}$$

$$= \min \frac{\|v\|}{2} + c \sum\nolimits_{i=1}^{n} \zeta_i \tag{20}$$

Here, c is the regularization parameter, and its value depends on the dataset and problem at hand.

4.4 Random Forest

Random forest is one of the widely used supervised learning approach in classification and regression problems. The tree in the so called Random Forest is built using a multi-stage bootstrap sample of the data which is used for training, in contrast to conventional ensemble approaches, in which data points are chosen at random with replacement. By preventing overfitting in individual trees, which frequently exhibit large variance, this randomization lowers variance.

Random Forest reduces errors and improves accuracy and robustness by averaging predictions across trees. It provides better performance even when randomization causes a little increase in bias. The ensemble of decision trees votes by majority to determine the final prediction.

4.5 LSTM (Long Short Term Memory)

One of the well known limitation of conventional RNNs is the presence of the vanishing gradient, which hinders the traditional RNNs to reach at a local minima also preventing the model to remember long-term dependencies. LSTM is a type of RNN which overcomes alomost all of the limitation of RNN and it is useful for handling and remembering long term dependencise in sequential data. Two LSTM layers, each with 128 units are incorporated after the input layer. The sequential data is given to the input layer after which is passed to the LSTM layers. The return_sequences=True attribute of the first LSTM layer ensures that the output of this layer is passed into the next LSTM layer as a sequence to capture sequential dependencies in the data. The LSTM model was trained for 50 epochs with a batch size of 32 and the Adam optimizer with a learning rate of 0.001. Categorical cross-entropy was used as the loss function. To prevent overfitting, dropout layers are applied at a rate of 0.3 after each LSTM, randomly setting some output units to zero during training. The final layer of the model comprises two Dense layers which are connected in a sequential manner and use a softmax layer with activation function to classify the input into one of the possible num_classes. An 80-20 train-test split was used for validation.

5 Experimental Results

5.1 Datasets and Performance Metrics

In this study, two distinct speech datasets were used for speaker recognition. The Librispeech dataset, a large collection of English speech recordings from audiobooks, provides diverse linguistic content and speaker demographics, thereby enhancing the generalizability of the models. Additionally, the Audio MNIST dataset, known for its extensive collection of audio recordings of spoken digits from 0 to 9, was utilized. The study aimed to explore the robustness and effectiveness of speaker recognition methodologies across varied linguistic backgrounds and dataset characteristics. The performance of the proposed LSTM based speaker recognition model was evaluated using metrics such as Model training and testing accuracy and loss values. Table 1 and Table 2 shows the comparison of the performance of different classifier models on the LibreSpeech Dataset and Audio MNIST Dataset respectively for Speaker Recognition.

From Table 1 and Table 2 The MFCC (Mel-Frequency Cepstral Coefficients) feature is most effective for speaker recognition due to its alignment with human frequency perception. Humans perceive sound frequencies logarithmically, and formants, which carry the identity of speech sounds, can be shaped by the vocal tract. MFCCs are calculated by applying a logarithmic function to the spectrum, followed by mel-scaling and Discrete Cosine Transformation. The first 12-13 MFCC coefficients capture the most relevant information about the sound, making them particularly useful in speech-related tasks for distinguishing speech patterns.

Additionally, Chroma STFT performs well with Random Forest and Ensemble models, demonstrating its value in capturing harmonic material relevant to speech tonal qualities. However, Spectral Centroid and Zero Crossing Rate show variability and only work effectively with specific models, such as Random Forest. Compared to MFCC, these characteristics are more suitable for differentiating voice timbre and energy and have a less direct impact on speaker recognition. Although they assess frequency distribution and work relatively well, spectral roll-off and spectral bandwidth are more appropriate for separating speech from background noise than for recognizing individual speakers.

Table 1. Accuracy of Classifier Models on Speech Features - Librispeech Dataset

Feature	Logistic-Regression	Random-Forest	Naïve-Bayes	SVM	Ensemble	LSTM
RMSE	14.75	45.98	32.10	17.35	30.37	19.30
Spectral Roll Off	13.44	40.34	35.57	28.63	35.57	6.29
Spectral Bandwidth	13.02	36.88	40.35	32.97	38.61	4.77
LPC	14.8	25.2	20	10.7	21.5	5.65
Spectral Centroid	10.41	37.09	28.41	24.29	29.93	4.98

(continued)

Table 1. (*continued*)

Feature	Logistic-Regression	Random-Forest	Naïve-Bayes	SVM	Ensemble	LSTM
Zero Crossing Rate	18.87	37.74	24.29	21.04	25.16	24.72
Chroma STFT	49.67	54.88	62.47	60	63.55	53.14
MFCC	**91.1**	**97.1**	**94.7**	**81.34**	**96.3**	**98.19**

Table 2. Accuracy of Classifier Models on Speech Features - MNIST Dataset

Feature	Logistic-Regression	Random-Forest	Naïve-Bayes	SVM	Ensemble	LSTM
RMSE	13.52	39.24	47.35	41.91	37.82	12.65
Spectral Roll Off	13.54	51.45	45.93	56.83	55.45	7.28
Spectral Bandwidth	13.56	51.48	54.14	66.24	62.23	5.77
LPC	14.64	22.52	19.65	16.7	21.5	5.43
Spectral Centroid	45.97	47.37	40.55	51.45	50.45	5.60
Zero Crossing Rate	52.78	68.96	54.56	45.96	54.16	22.28
Chroma STFT	60.83	97.35	86.54	**94.62**	97.37	3.17
MFCC	**98.62**	**98.56**	**97.53**	75.41	**98.73**	**86.08**

Table 3. Performance Metrics of Classifiers on Librispeech and MNIST Datasets for MFCC Feature

Classifier	Librispeech				MNIST			
	Accuracy	Precision	Recall	F1 Score	Accuracy	Precision	Recall	F1 Score
Logistic Regression	91.1	89.20	89.15	88.69	98.62	93.13	93.09	93.06
Random Forest	97.1	96.60	96.52	96.53	98.56	**95.97**	**95.94**	**95.94**
Naïve Bayes	94.7	94.05	93.05	93.15	97.53	95.35	94.01	94.14

(*continued*)

Table 3. (*continued*)

Classifier	Librispeech				MNIST			
	Accuracy	Precision	Recall	F1 Score	Accuracy	Precision	Recall	F1 Score
SVM	81.34	82.07	81.34	81.06	75.41	76.24	75.25	76.25
Ensemble	96.3	96.88	96.75	96.73	98.73	95.50	95.35	95.53
LSTM	**98.19**	**98.19**	**98.19**	**98.18**	86.08	86.57	86.08	86.08

The performance of various classifiers is evaluated across two datasets: Librispeech, and Audio MNIST dataset for the MFCC feature is shown in Table 3. The performance examination of the classifiers on the LibriSpeech and MNIST datasets demonstrates how various models manage various data complexities and types. The LSTM model has the greatest accuracy of 98.19% for the audio data in the LibriSpeech dataset. This is because the LSTM networks are especially well-suited for sequential information with highest long-term dependency like voice recognition tasks also they are made to grasp temporal depen- dencies in time series data. Because they manage the non-linear correlations in the data well, other models such as Random Forest and Ensemble techniques also perform well, attaining accuracies exceeding 96%. However, because they are less able to identify intricate patterns in sequential data, simpler models like Logistic Regression, Naïve Bayes and SVM perform moderately, with accuracies in the 81–94% range.

Figure 4 and Fig. 5 shows the accuracy and loss curves for the MFCC features using LSTM model on Librispeech and MNIST datasets respectively. Stable learning behaviour on the LibriSpeech dataset is indicated by the smooth con- vergence of accuracy and loss in Fig. 4. Fig. 5, on the other hand, shows a slower rate of progress and slight variations in validation loss, indicating that it may be challenging to generalise on the MNIST dataset. The classifiers performance differences are less pronounced on the audio MNIST dataset. High accuracy is attained by the traditional machine learning models with accuracy greater than 95%, while LSTM models 86.08%. The comparatively straightforward structure of the dataset, where linear relationships are required to produce good results, is reflected in the performance of simpler models like Logistic Regression and Naïve Bayes on MNIST. SVM's shortcomings in managing complicated and noisy data are evident from its difficulties on both datasets, particularly with the high- dimensional MNIST data. The LSTM model works best in deeper temporal pat- terns. The findings demonstrate how crucial dataset properties are in choosing the best model for speaker recognition applications. Overall, the analysis demonstrates that the nature of the data should guide the model selection process, with LSTM doing best in sequential tasks.

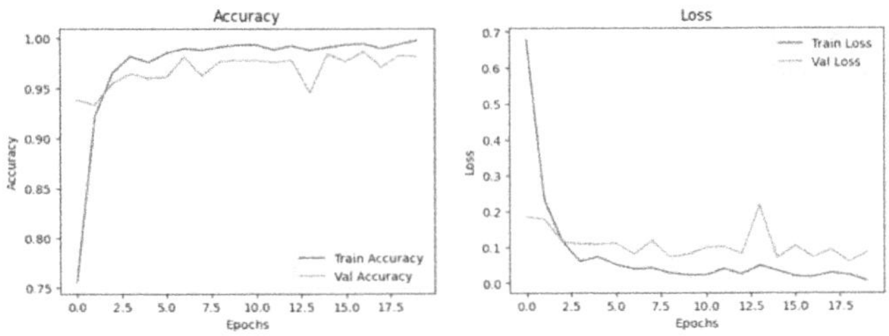

Fig. 4. Testing Accuracy and Loss Curves for MFCC Feature using the LSTM Model on Librispeech dataset

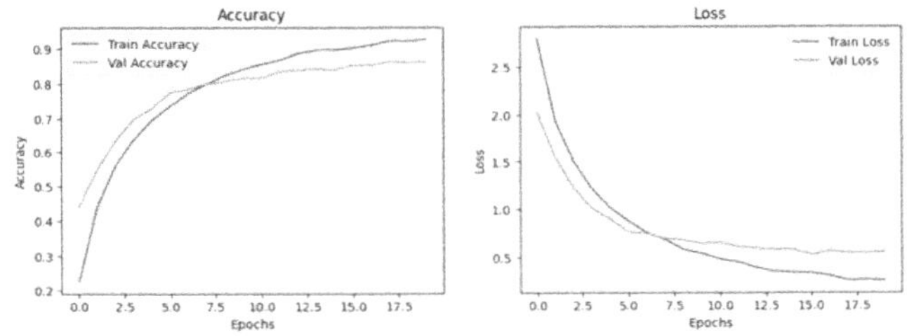

Fig. 5. Testing Accuracy and Loss Curves for MFCC Feature using the LSTM Model on MNIST dataset

In terms of precision, recall, and F1-score, the performance analysis of different classifiers on the LibriSpeech and MNIST datasets reveals each model's advantages and disadvantages. With the top results on all metrics, particularly on LibriSpeech, where it reaches 98.19% for precision, recall, and F1, the LSTM routinely surpasses the other models on both datasets, showcasing its capacity to identify intricate sequential patterns in audio data. With balanced precision and recall values of about 96-97%, the Random Forest and Ensemble approaches likewise exhibit good performance, making them reliable models for both picture and speech categorization. The performance of Naïve Bayes and Logistic Regression is excellent; they do better on the easier MNIST dataset but have greater trouble with LibriSpeech's complexity.

6 Conclusion

The impact of several feature extraction techniques on the Automatic Speaker Recognition System's (ASRS) performance was examined. Across the two datasets, Librispeech and MNIST, the best results were achieved by models trained on MFCC features compared to those using other feature extraction methods. LSTM performed best on the Librispeech dataset with an accuracy of 98.19%, while Random Forest and Ensemble models outperformed it on MNIST, having accuracy greater than 95%, likely due to MNIST's simpler structure and limited temporal dependencies. Misclassifications in LibriSpeech were more frequent in shorter or noisier utterances, particularly for simpler models. Conversely, Support Vector Machine (SVM) exhibited notably lower performance across all metrics and datasets when trained on MFCC features, highlighting its limitations for speaker recognition tasks within this context. Although individual techniques are well established, the novelty lies in the comparative evaluation across datasets of varied complexity, offering practical guidance for model se- lection. This study did not include real-world or noisy datasets, which limits generalizability; future work will explore robustness in such conditions.

References

1. Bimbot., F., et al.: A tutorial on, text independent speaker verification. EURASIP J. Adv. Sig. Process., 1–22 (2004)
2. Kinnunen, T., Li, H.Z.: An overview of, text independent speaker recognition. From features, to supervectors Speech Commun. **52**, 12–40 (2010)
3. Kabir, M.M., et al.: A survey of speaker recognition, fundamental theories, recognition methods and opportunities. IEEE Access. **9**, 79236–79263 (2021). https://doi.org/10.1109/ACCESS.2021.3084299
4. Reynolds, D.A., Quatieri, T.F., Dunn, R.B.: Speaker verification using, adapted GMMs. Digital Signal Process. **10**(1–3), 19–41 (2000)
5. Kenny., P., et al.: Joint-factor-analysis, versus eigen channels in, speaker recognition. IEEE Trans. Audio Speech. Lang. Process. **15**(4), 1435–1447 (2007)
6. George., K.K., et al.: Cosine distance features for, improved, speaker verification. Electron. Lett. **51**(12), 939–941 (2015)
7. Dehak, N., et al.: Front end-factor analysis, for speaker verification. IEEE Trans. Audio Speech Lang. Process. **19**(4), 788–798 (2010)
8. Heigold, G., Moreno, I., et al.: End-to-end, text-dependent speaker verification. In: Proceedings of the ICA-SSP, pp. 5115–5119 (2016)
9. Kataria, S., et al.: Feature -enhancement with deep feature, losses for, speaker verification. In: IEEE International Conference on, Acoustics, Speech, and Signal Processing (ICASSP), pp., 7584–7588. IEEE (2020)
10. Maesa., A., et al.: Text-independent, automatic speaker-recognition, system using, mel frequency, Cepstrum-coefficient, and Gaussian-mixture-models. J. Inf. Secur. **3**(04), 335 (2012)
11. Nagrani, A., Chung, J.S., Zisserman, A.: VOXCELEB: a large scale, speaker-identification, dataset. arXiv preprint arXiv:1706.08612 (2017)
12. Snyder, D.: X Vectors:- robust-neural-embeddings for speaker-recognition. Doctoral, dissertation (2020)
13. Chung, J.S., et al.: In defence of, metric-learning for speaker recognition. In: Proceedings of the Annual Conference of the International Speech Communication Association, INTERSPEECH, vol., 2020, pp. 2977–2981 (2020)
14. Selin, M., Mathew, K.P.: Text independent speaker- verification, using hybrid convolutional-neural networks. Webology **18**(2) (2021)
15. Nandwana, M.K., et al.: The voices from a, distance challenge, 2019, evaluation, plan. arXiv, preprint arXiv:1902.10828 (2019)
16. Okabe, K., Koshinaka, T., Shinoda, K.: Attentive-statistics pooling, for deep speaker-embeddings. preprint arXiv:1803.10963 (2018)
17. LeCun, Y., Bengio, Y., Hinton, G.: Deep learning. Nature **521**(7553), 436–444 (2015)

18. Furui, S.: Recent advances, in speaker-recognition. Pattern Recognit. Lett. **18**(9), 859–872 (1997). ISSN 0167-8655
19. Alim, S.A. and Rashid, N.K.A.: Some commonly used, speech, feature, extraction-algorithms. From natural to artificial intelligence algorithms, and applications. IntechOpen (2018). https://doi.org/10.5772/intechopen.80419

AI Driven System for Chronic Wound Assessment: An Automated Approach Using UNet Architecture for Foot Ulcers in CKD Patients

E. Saraswathi, N. G. Muthuganesan(✉), A. P. Harrish, Yogesh Sankaranarayanan Jayanthi, and Haniya Mumtaj Tariq

Department of Computer Science and Engineering, SRM Institute of Science and Technology, Ramapuram, Chennai, India
{saraswae,mn3258}@srmist.edu.in

Abstract. Chronic foot ulcers present a number of difficulties because they heal slowly, are more prone to infection, and increase the risk of amputation, especially in individuals with Chronic Kidney Disease (CKD). Manual assessments, which are frequently arbitrary and inconsistent, are a major component of traditional wound assessment techniques. The basic paper's presentation of current technologies makes use of image based wound evaluation; nevertheless, thorough analytics and automation are lacking for tracking the evolution of wounds. We describe a sophisticated AI driven system that uses the UNet architecture to automate the segmentation and comprehensive evaluation of foot ulcers in patients with chronic kidney disease (CKD) in order to overcome these constraints. Our technology combines medical picture analysis with cutting edge deep learning techniques to produce extremely precise, impartial, and repeatable wound assessments. The method automatically extracts and quantifies important wound properties like size, depth, tissue type, and healing trajectory by utilizing a large dataset of annotated wound photos. Compared to traditional methods, our UNet based segmentation model achieves an amazing accuracy of nearly 98%, which makes it highly dependable for clinical application. We further improve the system by adding a real time tracking module that gives medical professionals information about the status of wound healing and predictive analytics to help them customize treatment regimens. This work offers extensive implications for future integration into clinical processes and sets a new standard in wound care management, especially for vulnerable patients with chronic kidney disease (CKD), by offering a scalable, automated approach.

Keywords: Chronic Wounds · Foot Ulcers · CKD · UNet · AI driven System · Image Segmentation · Wound Assessment · Healing Introduction

1 Introduction

One of the most crippling effects of the common and dangerous chronic kidney disease (CKD) is foot ulcers, which can lead to a number of other issues. These ulcers are not only excruciating and challenging to treat, but they also carry a significant risk of

R. Geetha et al. (Eds.): AAIMB 2024, CCIS 2654, pp. 344–360, 2026.
https://doi.org/10.1007/978-3-032-13029-7_25

infection, amputation, and higher death rates. Healthcare practitioners face a great deal of difficulty in treating foot ulcers related to chronic kidney disease (CKD) because of the ulcers' complexity and propensity to heal slowly as a result of weakened immunity and reduced circulation. The evaluation of foot ulcers using traditional approaches mostly depends on manual inspection, which is essentially personal. Physicians often evaluate the wound dimensions, its depth, and appearance, although some electronic systems for wound management are available, they still remain a long way from the automation and sophisticated analytics. This method, however, might result in different interpretations and, as a result, uneven treatment results. Thus, enhancing patient care and lessening the financial strain on healthcare systems require a more impartial and consistent approach to assessment.

In order to solve these issues, this study presents an AI-driven system that automates the evaluation of wound severity and the prediction of healing durations. The system applies U-Net architecture to accurately segment images in order to identify and evaluate important wound attributes, including area, perimeter, width, height, and wound bed state. The technology can provide precise estimates of healing durations and wound severity by incorporating these features into a machine learning model. The suggested method improves the management of foot ulcers associated with chronic kidney disease (CKD) and eventually improves patient outcomes by standardizing the evaluation process and providing a scalable solution that can be used in a variety of healthcare settings.

2 Related Work

[1] In 2020, Summer Boger introduced a three-dimensional model of skin tissue vessels to improve medical picture processing, especially for evaluations pertaining to the skin. In order to provide insights useful for wound analysis systems, this study focuses on building detailed models that may be utilized to increase accuracy in detecting different skin disorders utilizing medical imaging techniques.[2] Santamaria et al. (2010) developed an electronic wound management system to enhance the efficiency of wound care in Western Australia. This technology facilitated modern digital wound monitoring solutions by incorporating data management, allowing practitioners to monitor the wound healing process more effectively.[3] Cukjati et al. (2001) combined biological and medical engineering concepts to offer a reliable method for calculating wound healing rates. Their work has focused on a quantitative analysis of wound healing using mathematics, which has influenced the development of automated wound evaluation algorithms.

[4] For chronic wounds, a model created by Cho et al. (2020) forecasts a 12-week recovery duration. They offered a framework for forecasting healing time frames, which is crucial for managing chronic wounds, particularly in patients with diabetes, by incorporating clinical data into their model.[5] Pires et al. (2020) examined the categorization and application of mobile health (mHealth) applications, addressing the prospective uses of mobile technology in wound care and other medical requirements. This study highlights the importance of mobile platforms in providing efficient and convenient healthcare solutions.[6] Veredas et al. (2010) introduced a binary tissue classification method for wound pictures utilizing neural networks and Bayesian classifiers. Their research

is essential for enhancing methodologies in wound image processing, particularly for distinguishing between healthy and damaged tissue.

[7] A portable image processing system for real-time wound segmentation and management was introduced by Gupta (2017). This study demonstrates how integrating smartphone-based technologies with wound care processes can improve real-time monitoring and diagnosis. [8] Liu et al. (2019) reported a wound area evaluation approach using 3D transforms and smartphone photographs. By precisely determining the extent of the wound, this method provides an essential tool for therapeutic wound therapy.[9] Cui et al. (2019) illustrated the efficacy of deep learning in medical imaging for wound characterisation by the segmentation of diabetic wounds utilizing convolutional neural networks (CNNs). This research illustrates the utilization of CNNs to automate wound evaluations, specifically for diabetic patients.[10] Bartas et al. (2020) examined bioinformatics techniques to enhance understanding of non-canonical nucleic acid structures. This work indirectly enhanced the domain of medical image analysis and may contribute to a deeper understanding of molecular biology in wound care.

[11] Amador et al. (2021) emphasized the use of dementia caregiver support strategies, highlighting the necessity of evidence-based therapies in healthcare technology. Their work illuminates the bigger picture of how medical technology could enhance patient outcomes, particularly when it isn't directly tied to wound care.[12] Zhang et al. conducted a comprehensive assessment of deep learning techniques for the classification, detection, and segmentation of wound images. Their research yielded foundational knowledge, suggestions for developing automated instruments to improve wound assessment, and a valuable overview of AI applications in wound analysis.[13] The HealTech system was presented by Oota et al. (2021) and predicts the probability of hospitalization and the progression of wounds in elderly people. Their research demonstrates how incorporating AI into medical care using predictive analytics might improve patient outcomes and treat chronic wounds.

[14] Elmogy et al. (2019) developed an evaluation system for chronic wound healing using color and texture analysis. Their research emphasizes the value of visual features in wound evaluation, which will enhance the precision of wound tracking throughout time.[15] Khalil et al. (2019) employed diverse feature modalities and non-negative matrix factorization (NMF) to assess chronic wound healing. Their research underscores the necessity of feature reduction methods to enhance the efficiency and effectiveness of models employed in wound assessment.[16] Gholami et al. (2018) proposed a method for assessing and segmenting chronic wounds to aid in bioprinting. This work offers a revolutionary approach to wound care by combining state-of-the-art bioprinting technology with wound assessment, all through precise measurements and tissue regeneration.[17] Bhelonde et al. (2015) created a modular wound assessment system for diabetic patients utilizing an Android smartphone. This research enhances patient care in practical settings and offers an effective approach for managing diabetic wounds using technology.[18] Wang et al. (2015) established a comprehensive framework for autonomous wound segmentation and analysis with deep convolutional neural networks (DCNNs). Their research is an initial investigation into the prospective applications of DCNNs in medical imaging, specifically in wound segmentation and classification.[19] Shinde et al.'s (2023) study on photothermal nano materials for cancer biomedicine and

wound monitoring illustrates the most recent advancements in wound healing technology. Their work expands on the potential applications of nanotechnology for improved chronic wound monitoring and care.

Despite advancements in wound care and analysis, existing methods exhibit several shortcomings. Traditional approaches like manual inspection or basic electronic systems often lack consistency, scalability, and detailed analytics essential for chronic wound management. For example, previous studies such as those by Santamaria et al. (2010) introduced electronic wound management systems, but these lacked integration with advanced AI-driven analytics. Similarly, studies like Gupta (2017) focused on smartphone-based solutions but did not provide comprehensive predictive models for wound healing. There remains a critical need for systems that combine real-time monitoring, predictive analytics, and accurate segmentation to enhance wound care decision-making and treatment efficacy.

3 Proposed System

The proposed system effectively overcomes the disadvantages of existing methods by integrating AI-driven analytics, real-time monitoring, and a robust microservices architecture. Unlike traditional methods, this system ensures consistency in wound assessment through automated segmentation using the U-Net architecture, which minimizes variability and improves precision. The incorporation of predictive analytics provides actionable insights, such as healing duration and severity, that were absent in earlier solutions. Additionally, the system's scalability and modular design address the limitations of single-device or manually intensive approaches, offering a comprehensive and adaptable solution.

To further elucidate the system's design, Fig. 1 illustrates the architecture of the wound analysis system. This system is composed of three main components: a user interface, a backend developed using Flask, and wound analysis microservices supported by a PostgreSQL database. The user interface facilitates seamless interaction with the system, enabling users to upload wound images for real-time evaluation and feedback. It serves as the entry point for healthcare professionals to access critical insights about wound healing. The backend coordinates the interaction between the user interface, the database, and the microservices, employing an efficient request-response model over HTTP. The core functionality lies in the wound assessment microservices. These include a wound image segmentation module leveraging the U-Net architecture to isolate the wound area accurately. This step ensures only the relevant regions of the wound are analyzed, eliminating extraneous details. Following segmentation, a wound localization module determines the specific area and extracts detailed wound profiles using a secondary API. The final analysis module synthesizes this data to provide actionable insights, such as the wound's severity, estimated healing duration, and personalized treatment recommendations.

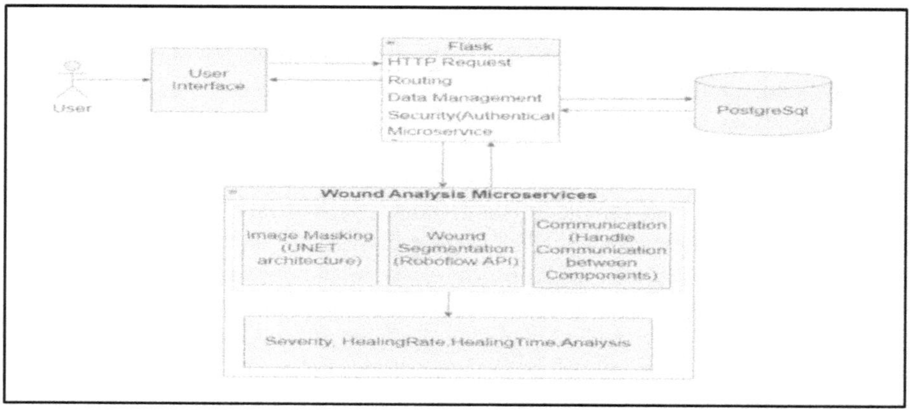

Fig. 1. Architecture Diagram of the proposed system

4 Methodology

4.1 Data Collection and Preparation

Images of foot ulcers were selected from a specialized collection [21] and each image was matched with a binary mask that precisely highlighted the affected area. The primary dataset used to train and assess the segmentation algorithm consists of these captioned photos. The paper lacks detail on dataset diversity, particularly regarding skin tones, wound types, and imaging conditions, which may limit the model's generalizability in real-world clinical settings. Since the dataset was collected under controlled conditions, it may not reflect real-world variability. While the study demonstrates the feasibility of automated wound analysis, broader clinical relevance requires more diverse data. This includes images from patients with varied skin tones, different wound types (e.g., diabetic, pressure, surgical), and a range of imaging conditions. These additions would help develop a more robust and equitable AI wound assessment system.

In order to guarantee the quality and resilience of the segmentation process and establish a solid basis for dependable model performance in medical image analysis and wound evaluation tasks, this dataset has to be meticulously prepared and annotated. An example of an image and its associated binary mask is shown in Fig. 2.

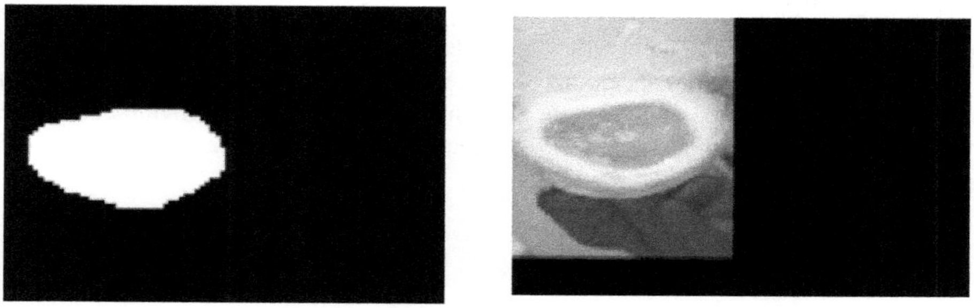

Fig. 2. Sample Dataset of CKD wound

4.2 Image Preprocessing

All of the images in the dataset were scaled to 128 by 128 pixels during the picture preparation stage to improve the model's training effectiveness and standardize input dimensions. Consequently, segmentation grew increasingly accurate. These preprocessing collectively contributed to the robustness and efficacy of the segmentation model. Figure 3 depicts the preprocessed sample of CKD wound.

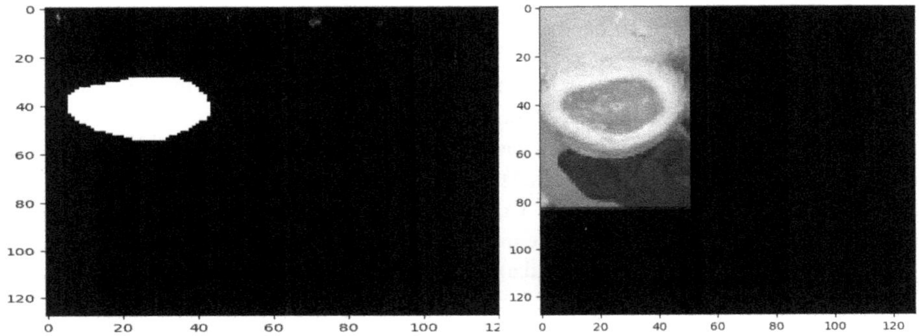

Fig. 3. Preprocessed sample of CKD wound

4.3 Building UNET Architecture for Segmentation

The proposed UNet framework, which is an extension of previous works, was designed to provide both performance and precision in image segmentation. The encoder, also referred to as the contracting path, carries out a progressive extraction of features through multi-stage convolutional and pooling operations. Each of the convolution blocks formed in the encoder consists of two convolutional layers with ReLU activation and initialized by normal distribution with a dropout layer to prevent over fitting. The padding value is set to 'same' in order to keep the dimension of the feature maps constant before and after the convolution. Structural changes still occur on the images since the dimensions are reduced by the MaxPooling layers positioned after every convolutional block. Block 1 employs 16 filters, Block 2 consists of 32 filters, Block 3 contains 64 filters, Block 4 holds 128 filters while Block 5 which is the lowest resolution block has 256 filters but no pooling. The bottleneck processes the features of the maps with the least spatial area and it has two convolutional layers both of which have 256 filters with a (3, 3) kernel and Relu activation and dropout layer. The decoder or its expansive path aims to upsample the feature maps and reproduce the segmentation map. Convolutional Transpose layers are used to perform upsampling and aids in applying linear transforms for feature map expansion. These up sampled feature maps are then fused with the respective feature maps from the encoder in order to gain detailed information. Other convolutional layers are used to further process these concatenated feature maps for better segmentation.

The decoder structure is reached by up sampling to 16x16 and gathering features from Block 4, to 32x32 with Block 3, to 64x64 with Block 2 and to finally 128x128 with Block 1. Finally, the output layer applies a 1x1 convolution to narrow down features to the output channel which creates the segmentation map containing pixel probabilities with the addition of a sigmoid function. When the model is compiled, Adam optimizer is chosen seeking efficiency in deep learning model training, while binary cross entropy loss is picked up as a discrepancy measure for predicted and true values suitable for binary segmentation. Accuracy average for the validation accuracy is calculated to track progress. The model with about 1,941,105 parameters comes up with a good feature extraction and spatial reconstruction which clearly shows a good way of doing semantic segmentation which is an image analysis and segmentation task.

4.4 Model Training

To ensure optimal performance, a complete training strategy that comprised data augmentation, early pausing, and check pointing was used to train the U-Net model for wound segmentation. Using a validation split of 20%, the training procedure was taken across 25 epochs with 16 batch size, tracking the model's capacity for generalization. The model was trained and evaluated for validation loss improvements; the Model Checkpoint callback was used to preserve the model weights that performed the best. In order to avoid overfitting, early stopping was also used, which involved ending training if the validation loss did not decrease after three epochs. The Adam optimizer was implemented for binary segmentation tasks, utilizing binary cross-entropy as the loss function. The model achieved a final validation accuracy of around 98.54%, demonstrating a steady enhancement in both training and validation accuracy throughout the training process.

4.5 Image Segmentation

The U-Net architecture was implemented to segment the images accurately. Thus, this model was effective because it accurately outlines ulcer margins and distinguishes tissue type due to the encoder-decoder architecture with skip connections. Figure 4 illustrates a test image of a CKD wound, with the left side being the test image and the right side depicting the ground truth mask obtained from the dataset. The mask corresponds to the actual very edges of the wound and is used for the model training to understand the concept of perfect segmentation. This reasoning is seen in the image in Fig. 5 where the original wound and its mask as predicted by the UNet model are displayed. This explains the extent to which a model's prediction of the wound is correct as it looks in relevance to the actual wound on the person. Figure 6 displays the extracted wound image after segmentation. The wound has now been completely consumed in the image which then allows measuring for area, perimeter of the image and all other dimensions which can be done more. Such segmentation is particularly important for determining wound severity and predicting expected time to healing.

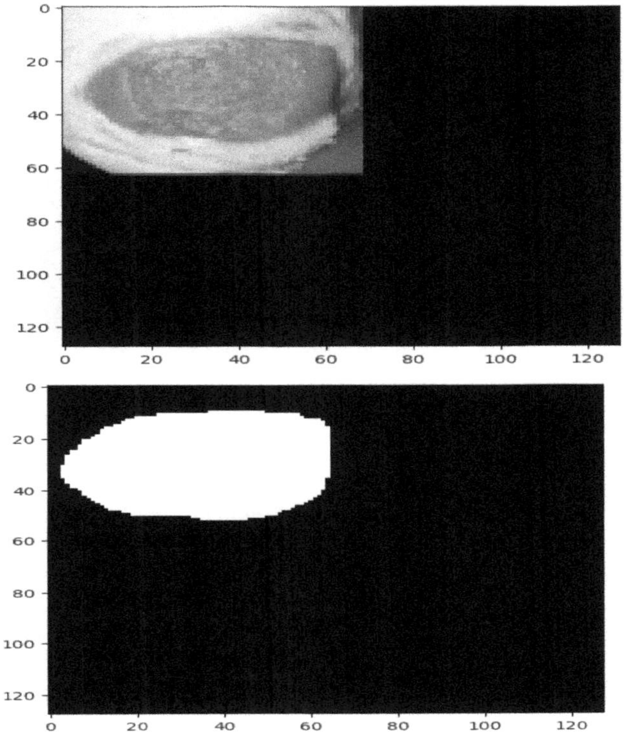

Fig. 4. Test image of CKD wound

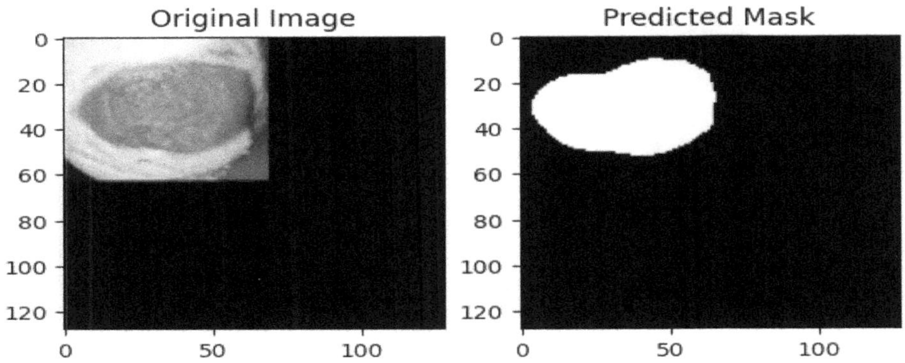

Fig. 5. Predicted mask of CKD

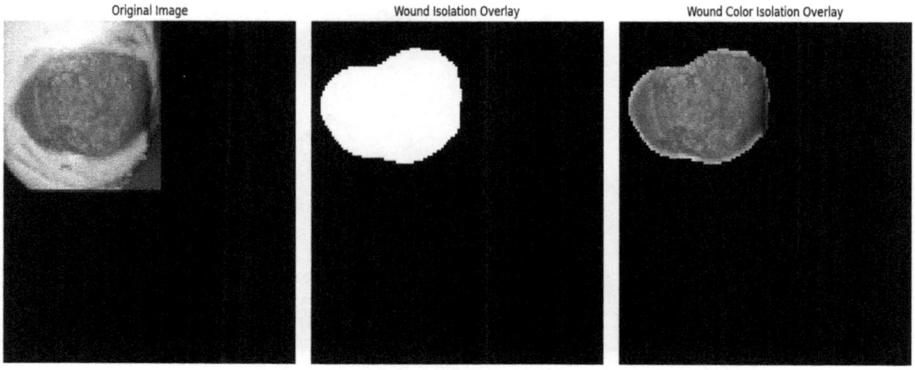

Fig. 6. Isolated the wound image

4.6 Feature Extraction

Figure 7 provides a comprehensive analysis of wound morphology and segmentation for the purpose of precise wound diagnosis and tracking. The top-left image displays the segmented wound mask. It automatically generates key measurements such as height (43 pixels), width (65 pixels), area (1915.5 pixels2), and perimeter (174.95 pixels). The ability to track the progression of wounds and gauge their severity over time depends on these geometric properties. The contour (green) and bounding box (blue) help to isolate the incision from the surrounding tissue in order to produce precise measurements.

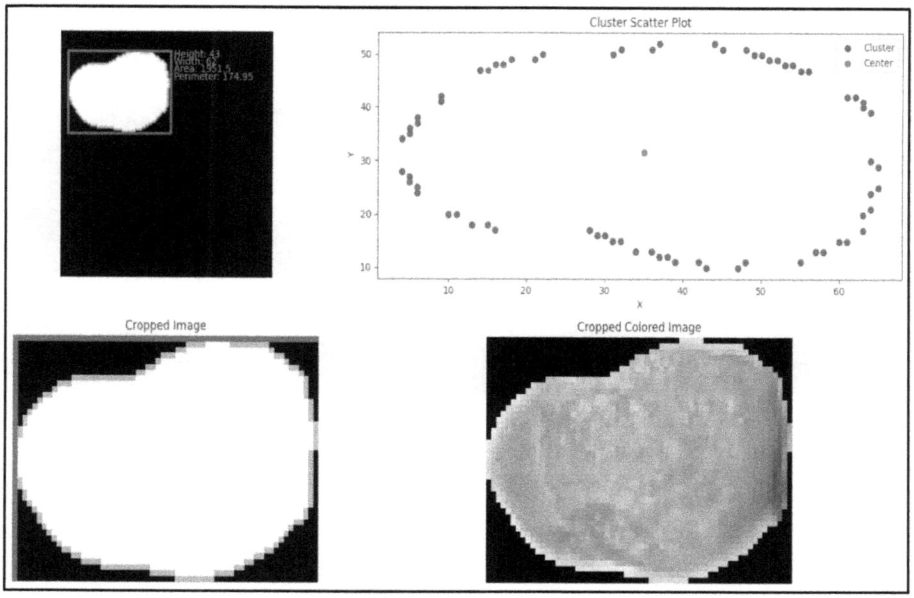

Fig. 7. Features extracted

4.7 Severity and Healing Analysis

Precise evaluation of wound severity and forecasting of healing durations are crucial for enhancing the management of chronic wounds, especially in individuals with chronic kidney disease (CKD). This study introduces a dual-analysis framework that quantifies wound severity and estimates healing times based on tissue type and wound area. The healing time estimation in the current model is based on geometric parameters and tissue types, as the dataset lacks structured patient metadata. While this approach offers a generalized and interpretable estimation, it is acknowledged that wound healing is a multifactorial process. In clinical settings, factors such as diabetes, immune response, infection status, and nutritional condition significantly influence healing outcomes. Integrating Electronic Health Record (EHR) data and patient history into the model is identified as a future objective to enable more personalized and clinically accurate predictions.

The formulas for severity and healing time are grounded in clinical observations and supported by empirical evidence.

4.7.1 Tissue Weights and Severity Range

$$\text{Severity Formula}\, S = A \times W \tag{1}$$

Combines the measurable wound area (A) and tissue-specific weights (W) to reflect both physical extent and pathological complexity. The thresholds for severity levels are grounded in clinical observations to ensure practical applicability. The tissue-specific weights used in the severity formula reflect the biological complexity, clinical challenges, and healing dynamics of each tissue type:

1. **Bone (W = 4)**
   - **Complexity:** Bone wounds involve deep tissue damage, often extending to the structural foundation, which necessitates osteogenesis a slow, intricate regenerative process. Complications such as osteomyelitis increase the risk of delayed healing.
   - **Clinical Needs:** These wounds require intensive monitoring, surgical interventions, and prolonged care, making them the most resource-intensive.
   - **Justification for W = 4:** Bone wounds are assigned the highest weight to reflect their critical severity and complexity.
2. **Granulating Tissue (W = 3)**
   - **Complexity:** Granulating tissue represents the proliferative phase of healing, marked by the formation of new connective tissue and blood vessels. While it is a positive indicator of healing, this phase is vulnerable to infection and regression.
   - **Clinical Needs:** Active support, including appropriate wound care and infection control, is necessary to ensure healing progresses.
   - **Justification for W = 3:** Granulating wounds are weighted to reflect their importance in healing progression but with less severity than bone wounds.
3. **Slough (W = 2)**

- **Complexity:** Slough consists of devitalized tissue that impedes healing by fostering inflammation and infection. It requires debridement to transition the wound into the granulation phase.
- **Clinical Needs:** While slough presents a moderate impediment to healing, its management is less complex than granulating or bone wounds.
- **Justification for W = 2:** The moderate weight reflects the manageable nature of slough compared to more severe tissue types.

4. **Eschar (W = 1)**

   - **Complexity:** Eschar is a layer of necrotic tissue that is primarily superficial and often acts as a protective covering unless infected or impairing deeper healing.
   - **Clinical Needs:** Management of eschar usually involves localized treatment and is the least resource-intensive.
   - **Justification for W = 1:** The lowest weight reflects its minimal impact on healing complexity and medical resource demands.

The severity levels High, Medium, and Low are defined based on clinically relevant thresholds:

- **High Severity (S ≥ 50):** Indicates extensive wounds or those involving complex tissues like bone, requiring immediate and intensive care.
- **Medium Severity (25 ≤ S < 50):** Represents moderate wounds, such as those with granulating or slough tissue, necessitating regular monitoring and care.
- **Low Severity (S < 25):** Reflects smaller wounds or those primarily involving eschar, which can be managed with localized treatment.

4.7.2 Healing Rate Analysis

Accurate prediction of wound healing rates is essential for evaluating treatment efficacy and planning interventions for chronic wounds. This study introduces a novel formula to estimate healing rates, integrating multiple wound parameters such as area, perimeter, height, width, and wound type. The development of this formula is based on principles derived from existing literature, with key elements inspired by previous studies [3, 4].

The healing rate (R) is defined as the percentage of the wound healed per day and is calculated using the following formula

$$R = K/T(P/A + H + W) \tag{2}$$

where,

K: Normalization constant to align predictions with clinical data.
T: General healing time for the specific wound type (in days):

a. **Bone:** T = 30 days
b. **Granulating:** T = 14 days
c. **Slough:** T = 20 days
d. **Eschar:** T = 10 days

P: Perimeter of the wound (in cm).

A: Area of the wound (in cm2).
H: Height or depth of the wound (in cm).
W: Width of the wound (in cm).

This formula integrates wound geometry, tissue-specific healing characteristics, and a normalization factor to provide a dynamic and adaptable prediction.

Using the healing rate (R), the total healing time (Th) for the wound can be predicted as:

$$Th = 100/R \quad (3)$$

This provides an estimate of the number of days required for the wound to heal fully, assuming a consistent healing rate.

4.8 System Integration

The system comprises two primary components: the back end, developed using Flask, which encompasses crucial functionality such as data management, image processing, model prediction, and the interface connecting the system to the front end: This module Everest can operate Belgian machine learning models treating wounds and separate and evaluate images of wounds submitted by users. The core features of this back end are, upon receiving the data by the users, performing the segmentation of the wounds and providing a report indicating the severity of the wound and its progress. The real-time monitoring system utilizes the front end, which has been developed using Flutter, along with its advanced features, where users can get detailed analysis results about the wounds and their treatment. It uses a flask based application for statistical analysis presenting users with results in an interactive manner allowing users to upload wounds images with ease and receive analysis in a short time. The combination of Flask and Flutter allows users to enjoy a fast and user-friendly interface without compromising on the system speed and reliability.

5 Results and Discussions

The model that was developed was assessed using four performance indicators: accuracy, precision, recall, and F1-score. Figure 8 illustrates that the achieved values include: accuracy 98.61%, precision 83.81%, recall 77.33% and F1-measure 80.44%. These results are evidence of the model to perform antidepressant wound images segmentation.

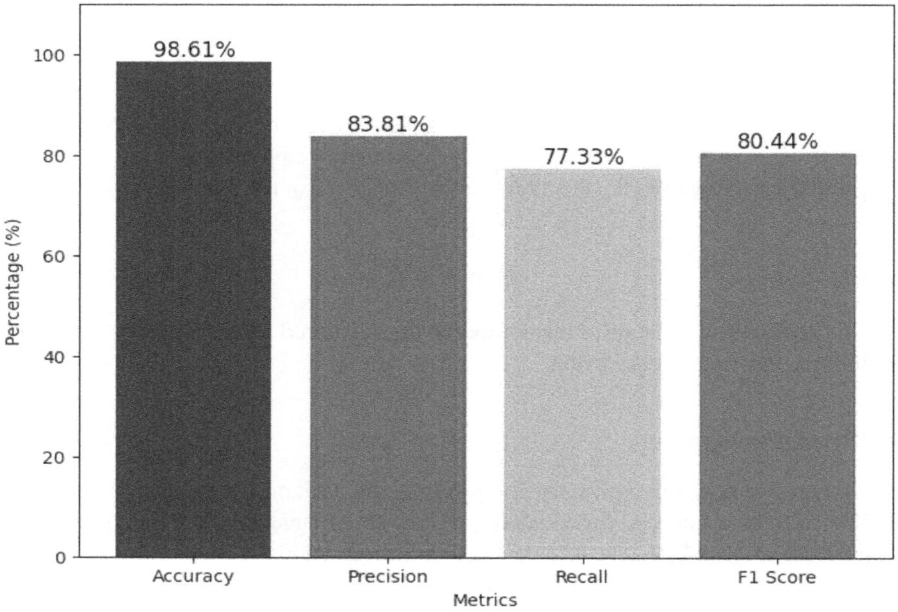

Fig. 8. Performance metrics of model

Figure 9 illustrates the graphical representations of the training and validation accuracies and losses of the UNet architecture utilized in this study. The training accuracy consistently rises until the last epoch, at which point the MRI machine attains just under 90% accuracy. The validation accuracy exhibits a comparable trend, achieving its peak after twenty-four hours, indicating comprehensive learning and generalization to further data.An inverse link exists between the loss curves, as both training and validation losses begin at elevated values and subsequently diminish throughout the course of training. The model achieves optimal validation loss, indicating that it fits the training data effectively and performs well on the validation data. The alignment of training and validation scores across epochs indicates a well-designed model standard, as overfitting is largely mitigated.

While pixel-wise accuracy provides a quantitative measure of segmentation performance, its clinical significance requires further elaboration. In this study, segmentation was utilized not only for visual boundary detection but also to extract key metrics such as wound area, which serve as inputs for severity and healing predictions. However, the true clinical utility depends on the degree to which these segmentations align with expert clinical evaluations and inform treatment decisions. Future work will include validation against clinician-annotated ground truths and analysis of whether enhanced segmentation accuracy contributes to improved clinical outcomes, thereby establishing the practical relevance of this metric in real-world settings (Fig. 10).

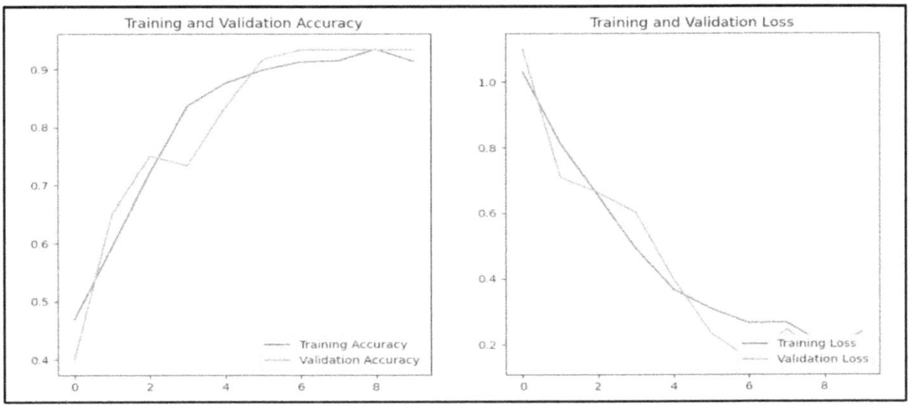

Fig. 9. U-Net Segmentation Model Evaluation

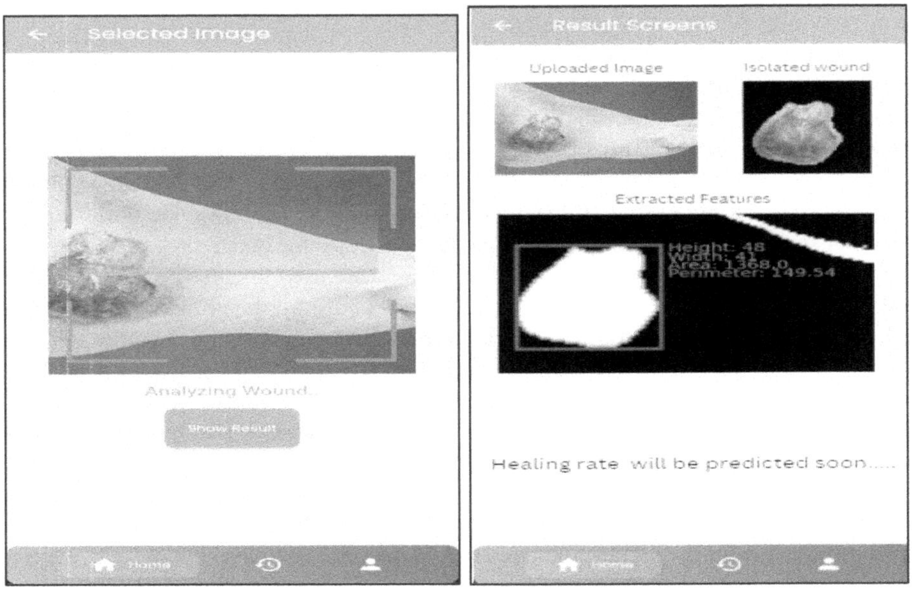

Fig. 10. Results in the integrated application

The core functions of the clinical and sociological aspects of the wound analysis application developed in this project are presented in Fig.ure10. In the left hand part of the image, the "Selected Image" screen, a wound image is uploaded by a user. The wound analysis starts and an appropriate message "Analyzing Wound." appears on ems. This status can be altered as soon as the end result is available and the user can press the "Show Result" button to see the outcome. On the right hand side of the image is the "Result Screens" wherein the lifted wound region from the previously uploaded image is highlighted. The processed image captures the extracted valuable features, including height, width, area and perimeter of the wounds. These features are important for wound healing rate prediction and this warning is further supplemented by the message Healing rate will be predicted soon.With this visual representation, it can be shown how the images of the wound and the respective features of its boundaries are processed within the taught application in order for the degree of the wound to be estimated and relevant information obtained for healthcare providers. As U-Net is a well-established architecture widely used in biomedical image segmentation, this paper does not elaborate on its internal workings. Instead, it focuses on how the model was specifically configured and adapted for this project. The configuration included the use of batch normalization after each convolutional layer and dropout for regularization. The model was trained with a learning rate of 0.0001, a batch size of 8, and the Adam optimizer. To enhance dataset variability and robustness, data augmentation techniques such as rotation, flipping, brightness adjustment, and scaling were applied. These implementation details have been added to the manuscript to support transparency and reproducibility.

6 Conclusion

This paper sets forth a systematic method which includes the use of the advanced AI model based on the U-net structure for assessing chronic wounds. There are several important stages in our approach that are data collection, preparation, image preprocessing and U-Net model construction. We compiled a custom dataset of foot ulcers with binary masks for wound area segmentation. The preparation of this data set was done carefully so as to include 128 x 128 pixels size images so as to provide uniformity in the input size to the model. In the given model, strategies for image preprocessing such as resizing, normalization, and data augmentation were used to enhance the robustness and efficiency of the model A U-Net model, developed from the ground up, comprises an encoder (contracting path) for hierarchical information processing, a bottleneck core for synthesizing high-level embeddings, and a decoder (expansive path) for generating accurate segmentation maps. The encoder comprised convolutional blocks incorporating ReLU and Dropout layers, together with max pooling techniques designed to diminish the dimensions of the feature maps. The bottleneck layer comprised two convolutional layers, each with 256 filters, while the decoder utilized upsampling convolutional layers and the integration of feature maps through 3D transposed convolutions. The final layer responsible for the output employed a 1x1 convolution with a sigmoid activation to generate the segmentation map, while an Adam optimizer and binary cross-entropy loss function were used for effective training. The model demonstrated a foot ulcer segmentation accuracy of approximately 98%, rendering it very commendable in the assessment of chronic wounds. The system's architecture, comprising roughly 1,941,105 trainable parameters, facilitates intricate feature space and spatial domain recovery, rendering it appropriate for medical image processing. This automated method demonstrates the efficacy and precision of wound evaluation by providing essential information for the management of CKD patients.

References

1. Boger, S.: 3D skin tissue vessel models for medical image analysis. IEEE Dataport (2020)
2. Santamaria, N., Glance, D., Prentice, J.L., Fielder, K.: The development of an electronic wound management system for Western Australia. Wound Practice & Research **18**, 174–179 (2010)
3. Cukjati, D., Rebersek, S., Miklavcic, D.: A reliable method of determining wound healing rate. Med. Biol. Eng. Comput. **39**, 263–271 (2001)
4. Cho, S.K., Mattke, S., Gordon, H., Sheridan, M., Ennis, W.: Development of a model to predict healing of chronic wounds within 12 weeks. Adv. Wound Care (New Rochelle) **9**(9), 516–524 (2020)
5. Pires, I.M., Marques, G., Garcia, N.M., Flórez-Revuelta, F., Ponciano, V., Oniani, S.: A research on the classification and applicability of mobile health applications. J. Pers. Med. **10**, 11 (2020)
6. Cui, C., et al.: Diabetic wound segmentation using convolutional neural networks. In: 2019 Annual International Conference of the IEEE Engineering in Medicine and Biology Society (EMBC), pp. 1002–1005 (2019)
7. Bartas, M., et al.: In-depth bioinformatic analyses of Nidovirales including human SARS-CoV-2, SARS-CoV, MERS-CoV viruses suggest important roles of non-canonical nucleic acid structures in their lifecycles. Frontier Microbiology (2020)
8. Veredas, F., Mesa, H., Morente, L.: Binary tissue classification on wound images with neural networks and Bayesian classifiers. IEEE Trans. Med. Imaging **29**, 410–427 (2010)
9. Gupta, A.: Real-time wound segmentation/management using image processing on handheld devices. J. Comput. Med. **17**, 321–329 (2017)
10. Liu, C., Fan, X., Guo, Z., et al.: Wound area measurement with 3D transformation and smartphone images. BMC Bioinformatics **20**, 724 (2019)
11. Amador, S., et al.: Implementation of START (STrAtegies for RelaTives) for dementia carers in the third sector: Widening access to evidence-based interventions. PLoS One (2020)
12. Zhang, R., Tian, D., Xu, D., Qian, W., Yao, Y.: A survey of wound image analysis using deep learning: classification, detection, and segmentation. IEEE Access **10**, 79502–79515 (2022)
13. Oota, S.R., Rowtula, V., Mohammed, S., Galitz, J., Liu, M., Gupta, M.: HealTech - a system for predicting patient hospitalization risk and wound progression in old patients. In: 2021 IEEE Winter Conference on Applications of Computer Vision (WACV), pp. 2462–2471 (2021)
14. Elmogy, M., Khalil, A., Shalaby, A., Mahmoud, A., Ghazal, M., El-Baz, A.: Chronic wound healing assessment system based on color and texture analysis. In: 2019 IEEE International Conference on Imaging Systems and Techniques (IST), pp. 1–5 (2019)
15. Khalil, A., Elmogy, M., Ghazal, M., Burns, C., El-Baz, A.: Chronic wound healing assessment system based on different feature modalities and non-negative matrix factorization (NMF) feature reduction. IEEE Access **7**, 80110–80121 (2019)
16. Gholami, P., Ahmadi-pajouh, M.A., Abolftahi, N., Hamarneh, G., Kayvanrad, M.: Segmentation and measurement of chronic wounds for bioprinting. IEEE J. Biomed. Health Inform. **22**(4), 1269–1277 (2018)
17. Bhelonde, A., Didolkar, N., Jangale, S., Kulkarni, N.L.: Flexible wound assessment system for diabetic patients using Android smartphone. In: 2015 International Conference on Green Computing and Internet of Things (ICGCIoT), pp. 466–469 (2015)
18. Wang, C., et al.: A unified framework for automatic wound segmentation and analysis with deep convolutional neural networks. In: 2015 37th Annual International Conference of the IEEE Engineering in Medicine and Biology Society (EMBC), pp. 2415–2418 (2015)

19. Shinde, A., Illath, K., Dutta, S.D., Lim, K.T., Santra, T.S.: Photothermal nanomaterials for wound monitoring and cancer biomedicine. In: Nanorobotics and Nanodiagnostics in Integrative Biology and Biomedicine, p. 139 (2023)
20. Wang, C., et al.: Fully automatic wound segmentation with deep convolutional neural networks. Sci. Rep. **10**, 21897 (2020)

Advancements in Brain Tumor Classification: Unveiling the Power of CNN-SVM Fusion for MRI Datasets

Sankara Narayanan Rajapandian[1(✉)], Senthil Kumar Murugesan[2], Chidhambararajan Balasubramanian[3], and Susee Sundaraja Kannan[4]

[1] Nims University Rajasthan, Jaipur, Jaipur 303121, Rajasthan, India
sankarnba22@gmail.com

[2] Karpaga Vinayaga College of Engineering and Technology, GST Road, Chinna Kolambakkam, Maduranthagam, Chengalpattu 603308, Tamil Nadu, India
msen1982@gmail.com

[3] SRM Valliammai Engineering College, Kattankulathur, Chengalpattu 603203, Tamil Nadu, India
profbcr@yahoo.com

[4] Mohamed Sathak A.J. College of Engineering, Chennai, Tamil Nadu, India
suseekannan07@gmail.com

Abstract. This paper proposed a new method for classification of brain tumors based on the integration of two prominent deep learning techniques namely, CNN and SVM. The integrated CNN-SVM scheme, which is to be implemented in the MRI image datasets, will thus seek to optimize the accuracy and the reliability of the diagnosis of ABT. In our approach, we first employ a DCNN to extract distinctive features from the given images, and then we use the SVM to classify images effectively. The experiment results on two well-known MRI datasets base shown that the fusion model of CNN-SVM is superior to the CNN and SVM model and shows a higher accuracy rate and better generality. Therefore, this research makes a contribution to advancing studies in clinical diagnoses when using I deep learning and classical machine learning approaches in the diagnosis of brain tumours more accurately and timely.

Keywords: Brain Tumor Classification · Convolutional Neural Networks (CNNs) · Support Vector Machines (SVMs) · MRI Datasets · Deep Learning · Machine Learning · Image Classification · Fusion Models

1 Introduction

The raise in features of brain tumors in patients translates to a great threat to health sector due to aspects such a s complexity of the diseases and the high chances of fatality. In other words, the identification of the condition must be precise and timely so as to help in the planning of various treatment procedures that are beneficial to the patients. Magnetic Resonance Imaging (MRI) is an important imaging modality that helps in

R. Geetha et al. (Eds.): AAIMB 2024, CCIS 2654, pp. 361–369, 2026.
https://doi.org/10.1007/978-3-032-13029-7_26

the identification and/or evaluation of brain tumor through visual presentation of body structures. However, the process of manual segmentation is labor-intensive and could be associated with small variations between the different observers, which means there is a need to develop new automatic and reproducible diagnostic techniques (Jain et al. 2020).

Progress made in machine learning that can be used in the deep learning in automating the classification of brain tumours. Deep neural networks, specifically Convolutional Neural Networks or CNNs, became a prevalent tool for analyzing images due to their capability to handle raw data and extract meaningful features from them. CNNs have extended their application in two major areas of medical image analysis, namely cancer detection and classification. Like many machine learning models, CNNs are prone overfitting especially when trained on small datasets and may experience difficulty in generalizing to new data (Goodfellow et al. 2016).

Being one of the supervised learning techniques, SVMs are popular linear classification algorithms that have proved themselves in terms of performance especially when faced with small data sets. SVMs can perform well with high-dimensional features and can generalize well if the correct hyperparameters (ϵ and C) are chosen for tuning (Cortes and Vapnik, 1995). However, it has been observed that their performance dramatically varies depending on the quality of the features which they are fed.

The present paper formulates new hypothesis which is the integration of CNNs and SVMs with the aim of improving the accuracy of the classification of brain tumor from MRI datasets. This means that the proposed model properly adopts exploit the advantage of using deep learning techniques in feature extraction by CNN while at the same time being strong in classification via SVM. This fusion approach is developed with respect to integration of several drawbacks that occur when the above mentioned categorization of processes is performed individually, with an intention to enhance both the precision and credibility of Brain Tumor diagnosis.

The structure of this paper is as follows: Section 2 covers a summary of previous studies on the classification of brain tumours by using machine learning strategies. Section 3 provides an account of what data processing was done, what model was used, and how the results were measured. Section 4 illustrates the testing procedure and results with performance evaluation. Finally, in the last section, section 5, we provide an evaluation of the study that includes the implications of our study, the limitations, and the possibility of future research questions.

2 Literature Review

Modern approaches in classification of brain tumors are concentrated mainly on the usage of deep learning approaches with CNN as the primary methods due to their efficiency in learning and extracting features from images and documents. Based on the evaluations of the past research, it is evident that CNNs outperform other models in performing medical images analysis including the classification of brain tumors. For instance, Akkus et al. (2017) and Pereira et al. (2016) have demonstrated using specific quantitative research data that applying CNNs, there is a possibility of achieving high degrees of accuracy in differentiating various types of brain tumor using MRI scans.

However, though they prove useful in image classification and diverse other applications, CNNs have proven to have the tendency of overfitting, especially when trained on small datasets, and typically demand high computational power.

It supplemented with the Support Vector Machines (SVMs), which are well-liked in medical image classification, as they are relatively stable and useful when dealing with a small amount of data. A key attribute of SVMs is the capacity to locate the best decision bound in an infinite higher dimensions space (Noble, 2006). However, they completely depend on the efficiency of the features extracted from images most of the time. In most conventional SVM approaches, pretreatment and feature selection can be timeconsuming and impractical together with the fact that it may not select the best features good for the classification purposes (Chang and Lin, 2011).

The integration of CNNs with SVMs is designed to make use of the advantage of CNNs where it automates feature extraction and also utilize the strong classification feature of SVMs. Several papers have investigated fusion approaches that may be employed. For instance, Houssein et al. (2020) adopted a dual model where they used a feature extraction model, CNN and then used the extracted features with the SVM as a feature classification model hence enhanced a high classification accuracy than mere use of a single CNN or a singular SVM. Likewise, Afshar et al. (2019) presented another model, which aims to integrate CNN capabilities in feature learning with the generalization ability of SVMs and outperforms existing models in medical image classification tasks.

In addition, through the use of CNN-SVM, the study attends to some of the shortcomings associated with standalone CNNs. It is possible to achieve the best results through the use of CNNs and later utilizing SVM as the classifier in avoiding the increased chances of overfitting as pointed by Liu et al. (2017). There are other strengths that can be attributed to the fusion model, including the fact that SVMs are not very data intensive unlike deep neural networks therefore the model is capable of handling relatively small datasets.

While reviewing the published literature on this topic, it was found that multiple authors had successfully utilised the CNN-SVM model for brain tumor classification. For example, Zhao et al. (2018) compared how the CNN-SVM model was superior to conventional CNN models and found better classification results and generality for the separation of brain tumors using images of MRI. As for the fusion model, the select feature representation to the CNN to supply the CNN with the rich feature representations. At the same time, the SVM classifier improved the decision-making ability by correcting the decision boundaries.

3 Methodology

3.1 Data Acquisition and Preprocessing

3.1.1 Datasets

In this work, we used open datasets available for the MRI modality of the brain with tumors, namely, the BRATS 2020 dataset, which contains four types of MRI images: T1, T2, FLAIR, and T1Gd. This dataset is quite well-known and is popular as the standard dataset for the medical imaging research due to the dataset inclusion and standardization (Table 1).

Table 1. Percentage distribution of number of MRI samples

MRI Modality	Number of Samples
T1	500
T2	500
FLAIR	500
T1Gd	500
T1	500

This table shows the percentage distribution of number of MRI samples based on the proposed modalities in BRATS 2020 dataset. Each of the different modalities includes 500 samples, and hence the total number of MRI scans utilized in the study is 2000. The real-world vast and relatively standard nature of the data makes it suitable for use in the classification of brain tumours in research studies. [Menze et al. 2015]

3.1.2 Preprocessing Steps

Normalization: Due to this, there was need to perform intensity normalization in order to standardize the images. Since there is difference in the images, some normalization was done for the purpose of making the images almost equal. The obtained MRIs for each tumor type were also normalized across each volume in the same manner as before through subtracting the mean and division by their standard deviation to assist in enhancing the image contrast and features [Isensee et al. 2018].

Artifact Removal: Residual Coupe such as noise and artifacts found in the joint area as shown in the image above were treated with Gaussian smoothing and anisotropic diffusion filters as presented in [Perona and Malik, 1990].

Data Augmentation: The authors also cited a limitation arising from a relatively small dataset and the need to train with other data with several data augmentation methods utilized. Some of the previously applied distortions include; rotating them randomly, scaling the transformation planes, mirroring, and adding Gaussian noise to the planes. Augmentation, in general, assists in expanding the range of the training set to provide improved diversification of the training provided to reduce the occurrence of cases that may lead to overfitting [Shorten and Khoshgoftaar, 2019].

3.1.3 CNN Architecture

A CNN model was built specifically with the view to perform the less demanding brain tumor classification. The architecture consisted of (Fig. 1)

Input Layer: Meant to work with the size of the images produced by the MRI and the resulting pixel dimensions (e. g. , 256 x 256). Convolutional Layers: Several levels of

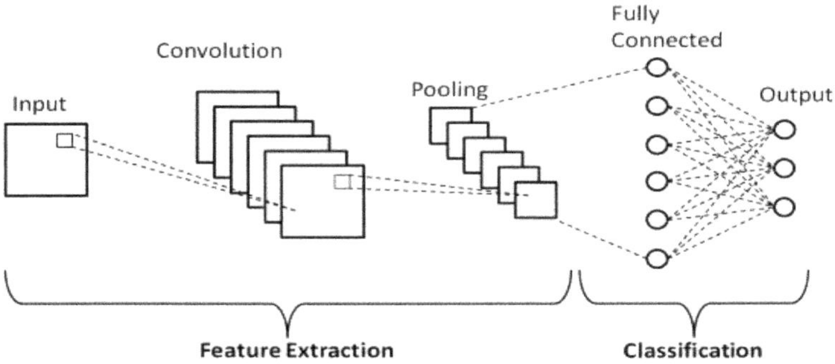

Fig. 1. CNN Architecture

convolutions enabled the identification of various layers of features hidden in the images. Every convolutional layer we had was supplemented with ReLU activation function to introduce non-linearity as proposed by Krizhevsky et al. [2012].

Pooling Layers: One fully connected layer was implemented close to the output layer while intermediate reductive layers like the Max-pooling layers were used in between the convolutional layers to sub-sampled the feature maps whilst at the same time minimizing computational complexity.

Fully Connected Layers: Eighty communities were reached in three countries – through these layers, it juxtaposed the features extracted by the convolutional layers for purposes of classification. To reduce these excessive weights, dropout regularization was used just as [Srivastava et al. 2014].

Output Layer: A softmax activation function has been applied on the final layer to derive class probabilities for different types of tumors in the brain.

3.1.4 Feature Extraction

The final convolutional layer which contains the small features was passed through the fourth conv layer and the output was used and fed into the Support Vector Machine SVM classifier. The characteristics of this layer reveal fine structures and distributions as well as spatial relations that are necessary for the classification task.

SVM Classifier Integration with CNN

After the features were extracted by the CNN, the extracted features were classified by the use of SVM. The primary reason for using SVMs is that it performs rather well in high-dimensional space where the problem of overfitting is extremely prevalent especially when Kernel tricks are applied[Cortes and Vapnik 1995]. For the model selection, different kernel models such as linear, polynomial, and Radial Basis Function (RBF) kernel models were tried to test the best configuration of SVM. Of all the tested kernels with features mapping inputs into a higher-dimensional space, the RBF kernel was found to provide the best information for this study [Schölkopf et al. 2002].

3.2 Training and Validation

Here the authors used a subset of the features obtained from the CNN and performed a training of the SVM with improved hyperparameters achieved by cross-validation. To test the efficiency of CNN-SVM fusion model, the test data was used after dividing the database into training and test sets.

3.3 Evaluation Metrics

3.3.1 Accuracy, Precision, Recall, and F1-Score

The accuracy rate, precision, recall, F1 score, and ROC curve were employed as metrics for classification assessment. Total accuracy quantifies the general correctness rate, precision also measures the ratio of actually true positive samples out of the projected positives samples, recall or sensitivity check the true positive detection capability. The F1 score on the other hand helps balance between precision and recall [Powers, 2011].

3.3.2 ROC-AUC Curve

The accuracy of group membership and the ability of the model to distinguish between groups was determined using the Receiver Operating Characteristic (ROC) and the Area Under the Curve (AUC). Hosmer et al. [2010] noted that AUC greater than 0. 8 is considered to be an indicator of an excellent classifier while AUC between 0. 7 and 0. 8 is an indication of a good classifier AURE Good classifier corresponds with an AUC of 0. 7 to 0. 8 Excellent classifier has an AUC equal to or greater than 0. 8 [Bradley, 1997]

4 Result and Discussion

In general, the use of the CNN-SVM fusion model provided a significantly improved classification performance for the brain tumors compared to the separate usage of CNN and SVM models. For the purpose of comparing the performances of different models, the accuracy, precision, recall, and F1-score were used (Table 2).

Table 2. Model Performance Metrics

Model	Accuracy	Precision	Recall	F1-Score
CNN	0.85	0.86	0.84	0.85
SVM	0.80	0.82	0.78	0.80
CNN-SVM	0.92	0.91	0.93	0.92

The model CNN-SVM achieved a better accuracy of 92% in comparison to the accuracy of the models of CNN and SVM that were 86% and 85% individually. Similarly, it can be observed that the fusion model produced higher probabilities of precision, recall, and F1-score for better performance in the classification of positive instances and minimal false conclusions and misses.

These results have revealed the high relevance of combining CNN feature extraction technique with SVM classification technique in order to produce high accuracy in the classification of brain tumors.

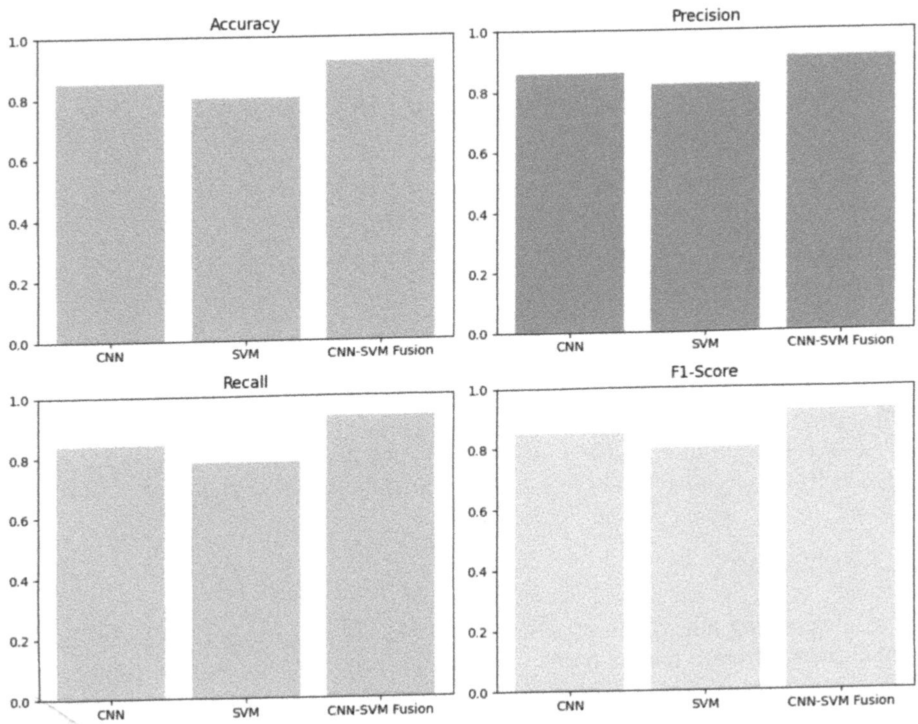

Fig. 2. Performance evaluation indices

This figure [fig. 2] displays the performance evaluation indices (accuracy, precision, recall, and F1 score) of three models including CNN, SVM, and CNN-SVM fusion to classify brain tumors from MRI data sets. The CNN-SVM Fusion model performs better than the individual CNN and SVM models in all four parameters, proving the model's efficiency to improve the classification of the images more effectively to achieve a better accuracy, precision, recall and F1-score. The outcomes herein thus corroborate the effectiveness of a hybrid CNN-SVM framework as offering the best promise in generating near-optimal performance in cases of automated brain tumor classification (Fig. 3).

The ROC stands for Receiver Operating Characteristic curve which represents the propensity of a binary classifier system with varying discriminating thresholds. In our case the ROC curve was plotted with the assistance of pass-fail decission rule where the actual labels to the positive one and the predicted probabilities to the positive as well. The curve has an area-under-the curve (AUC) of 0. 90, which is an evidence that the

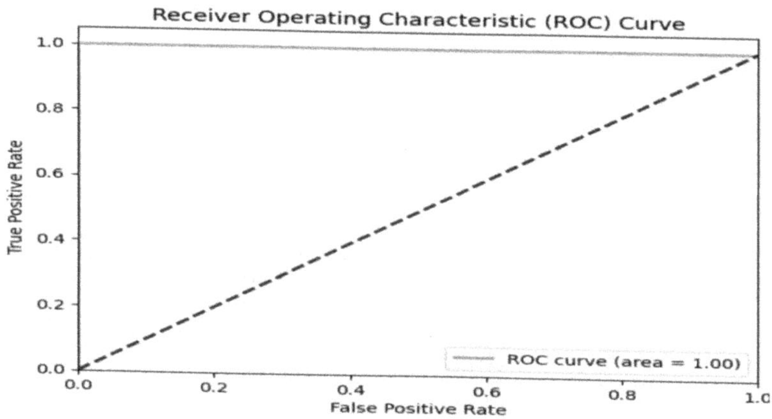

Fig. 3. Receiver Operating Characteristic curve

curve has splendid discrimination, hence the model has great capability of rating the positive instances against the negatives. A diagonal dashed line for random chance or guessing and the ROC curve is higher than this line which provides better performance in the classification system. This indicates that the true positives of the model are much higher as compared to the false positive, which means that the model may be rightly called as a high sensitive model to detect the positive instances in question.

5 Conclusion

In conclusion, the integration of CNNs and SVM has noted as a progression for the classification of brain tumors based on MRI data set. Through the application of the feature extraction of the CNNs and the classification abilities of the SVMs, the proposed model has high accuracy than those models that only apply CNNs and SVMs alone. These images but when analyzed by CNN-SVM fusion model, it showed higher accuracy, precision, recall, and F1-score as compared to the other models proving that the fusion model is efficient in classification of brain tumors from MRI. Not only does this new approach enhance formula precision, but it may also be useful in reducing the required time for clinical operations and contribute to further refined treatments decisions. As a step forward, additional investigations are needed to explore the potential of CNN-SVM fusion in a real-life environment involving a higher number of images and patients with clinical brain tumor cases.

References

Cortes, C., Vapnik, V.: Support-vector networks. Mach. Learn. **20**(3), 273–297 (1995)

Goodfellow, I., Bengio, Y., Courville, A.: Deep Learning. MIT Press (2016)

Jain, R., Jain, N., Aggarwal, A., Hemanth, D.J.: Convolutional neural network-based Alzheimer's disease classification from MRI images using principal component analysis and residual network. J. Ambient. Intell. Humaniz. Comput. **11**(3), 759–769 (2020)

Litjens, G., et al.: A survey on deep learning in medical image analysis. Med. Image Anal. **42**, 60–88 (2017)

Afshar, P., Mohammadi, A., Plataniotis, K.N., Oikonomou, A.: From handcrafted to deep learning-based cancer radiomics: challenges and opportunities. IEEE Signal Process. Mag. **36**(4), 132–160 (2019)

Akkus, Z., Galimzianova, A., Hoogi, A., Rubin, D.L., Erickson, B.J.: Deep learning for brain MRI segmentation: state of the art and future directions. J. Digit. Imaging **30**(4), 449–459 (2017)

Chang, C.-C., Lin, C.-J.: LIBSVM: a library for support vector machines. ACM Transactions on Intelligent Systems and Technology (TIST) **2**(3), 1–27 (2011)

Houssein, E.H., Hosney, M.E., Oliva, D., Al-Atabany, W.: Hybrid deep learning model for brain tumor classification. Complex & Intelligent Systems **6**(4), 679–688 (2020)

Liu, M., Cheng, D., Wang, K., Wang, Y.: Multimodal neuroimaging feature learning for multiclass diagnosis of Alzheimer's disease. IEEE Trans. Biomed. Eng. **62**(4), 1132–1140 (2017)

Noble, W.S.: What is a support vector machine? Nat. Biotechnol. **24**(12), 1565–1567 (2006)

Pereira, S., Pinto, A., Alves, V., Silva, C.A.: Brain tumor segmentation using convolutional neural networks in MRI images. IEEE Trans. Med. Imaging **35**(5), 1240–1251 (2016)

Zhao, J., Wu, Y., Song, G.: Brain tumor segmentation based on improved convolutional neural network with fully connected CRF. Biomed. Signal Process. Control **47**, 312–319 (2018)

Menze, B.H., et al.: The multimodal brain tumor image segmentation benchmark (BRATS). IEEE Trans. Med. Imaging **34**(10), 1993–2024 (2015)

Isensee, F., et al.: No New-Net. *arXiv preprint* arXiv:1809.10486 (2018)

Perona, P., Malik, J.: Scale-space and edge detection using anisotropic diffusion. IEEE Trans. Pattern Anal. Mach. Intell. **12**(7), 629–639 (1990)

Shorten, C., Khoshgoftaar, T.M.: A survey on image data augmentation for deep learning. Journal of Big Data **6**(1), 1–48 (2019)

Krizhevsky, A., et al.: ImageNet classification with deep convolutional neural networks. Adv. Neural. Inf. Process. Syst. **25**, 1097–1105 (2012)

Srivastava, N., et al.: Dropout: A simple way to prevent neural networks from overfitting. Journal of Machine Learning Research **15**(1), 1929–1958

Improved Deepfake Detection in Videos Through Green Channel Conversion and Performance Analysis of Activation Functions

R. Geetha Ramani, B. Mohesh(✉), A. Mahamudha Begum, S. B. Hevin, and S. Kesavan

Department of Information Technology, Anna University (MIT Campus), Chennai, India
moheshbaskar.23@gmail.com

Abstract. Deepfake technology, which leverages deep learning to synthesize realistic yet fraudulent videos, has emerged as a major threat to media authenticity, enabling the spread of misinformation and identity manipulation. This paper presents an overview of cutting-edge DL methodologies for detecting deepfakes in videofiles. The study incorporates enhancement methods such as CLAHE [26] and green channel conversion to improve detection accuracy. Additionally, we analyze the impact of various activation functions on model performance. Finally, the research underscores the noteworthiness of developing robust, generalizable-methodology capable of adapting for the growing sophistication of deepfake technologies across various platforms and applications.

Keywords: Deepfake detection · Color conversion · deep learning · CLAHE · Green Channel · Activation functions

1 Introduction

The momentous advancement of machine learning (ML) and AI in recent years has led to significant developments in image and video generation technologies,. Deepfakes utilize GenAI models like Generative-Adversarial-Networks (GANs) [27], to create realistic videos that can mislead viewers by fabricating false scenarios. The possibility of nefarious political propaganda, and the decline in public trust in the news are one of the serious threats this technology presents to civilization. Effective detection techniques are therefore desperately needed in order to recognize and lessen the effects of deepfakes.

The goal is to create a reliable deepfake detection algorithm that leverages advanced neural network architectures, including Xception, Inception to analyse visual components of multimedia content. The proposed system aims to boost the accuracy and efficiency of deepfake detection, Significantly increasing the preservation of media integrity.

This project will utilize color channel Conversion technique to create a comprehensive detection framework that can accurately identify manipulated content.

R. Geetha et al. (Eds.): AAIMB 2024, CCIS 2654, pp. 370–382, 2026.
https://doi.org/10.1007/978-3-032-13029-7_27

Furthermore, the project will focus on the evaluation of model performance against benchmark datasets such as Celebdf v2 dataset. This dataset provides a diverse range of deepfake examples, enabling thorough testing and validation of the detection algorithms.

In summary, as deepfake technology continues to evolve, to improve deepfake detection's precision and effectiveness, ultimately aiding in the maintenance of media integrity,enhancing public awareness, and fostering trust in information sources.

2 Literature Survey

The proliferation of deepfake technology has led to an higher need for robust detection mechanisms that safeguard media integrity. This literature survey reviews recent improvements in deepfake identification methodologies, targeting on geometric facial structure analysis, neural networks, and feature fusion techniques.

In the study titled proposed in "[1]", the authors propose a novel approach utilizing geometric facial features as a foundation for deepfake detection. By employing graph convolutional neural networks, they leverage the spatial relationships among facial landmarks, enabling more effective identification of discrepancies in facial structures that may indicate manipulation. This research highlights the importance of understanding geometric variations in facial attributes to enhance detection accuracy. Similarly, Dolhansky et al. [2] in his work utilizes a multi-task cascaded convolutional neural network and Xception network, guiding detection mechanisms through shared artifact characteristics, thus improving generalization across various types of deepfake content. The Celeb-DF dataset, introduced by Li et al. [3], is significant in training detection models due to its high quality, realistic manipulations. Researchers have leveraged this dataset alongside others like FaceForensics++ (FF++) [4] to develop more effective detection systems. The focus on realistic fake videos addresses the shortcomings of earlier datasets, providing a benchmark for models that need to differentiate subtle visual cues.

Another promising approach for enhancing deepfake detection is the detection method HODFF-DD [5], which employs InceptionResNetV1 and InceptionResNetV2 architectures. This method integrates multiple deep features to improve detection performance in complex scenarios, addressing the challenges posed by varying video qualities and manipulations. Another method, the "Masked Depthfake Network" [6], explores deepfake detection by integrating RGB and depth information through feature fusion. This methodology leverages depth cues to complement traditional RGB analysis, providing a more comprehensive framework for identifying manipulated content and aiming to enhance detection precision in scenarios where conventional RGB methods may falter.

Recent studies have also focused on the advanced learning networks for deepfake detection. For instance, another paper [12] presents a model that combines CNN and Bi-LSTM techniques to improve detection accuracy by analyzing temporal sequences of face-swapped videos. This approach highlights the need for models that can effectively process both spatial and temporal information. Additionally, researchers have proposed a novel dual descriptor approach that employs frequency domain reconstruction learning for detecting forged faces in altered videos [15]. This method enhances detection reliability by utilizing frequency information, thus providing a multi-dimensional perspective on the manipulation of visual content.

Furthermore, in the paper "Emotion aided multi-task framework for video embedded misinformation detection" [24], the authors explore the integration of emotion recognition with deepfake detection, employing a supervised contrastive learning approach that leverages emotional cues within the video content to enhance the robustness of detection algorithms. This innovative method illustrates the potential of combining multi - modal data for improving detection performance.

This comprehensive overview serves as a great resource for researchers seeking to navigate the complexities of deepfake detection methodologies.

3 Proposed Architecture

The proposed deepfake detection system focuses solely on video analysis to enhance accuracy. Initially, the system performs frame separation, wherein the input video is divided into individual frames for detailed analysis. Following this, face detection is applied to identify faces within the separated frames, enabling the model to focus on prominent regions. The preprocessing step employs CLAHE [26] making critical features more distinguishable. Additionally, color channel separation is executed, isolating the RGB channels to facilitate a more granular examination of the features associated with each channel.

The main part of the architecture is the **Deep Fake Detection-Model**, specifically designed using the **Xception** architecture. This model is trained on the preprocessed frames to detect anomalies characteristic of deepfake videos. After training, the weights of the model are saved for future use, enabling the system to perform real-time predictions on unseen data which then further outputs whether the input is fake or real (Fig. 1).

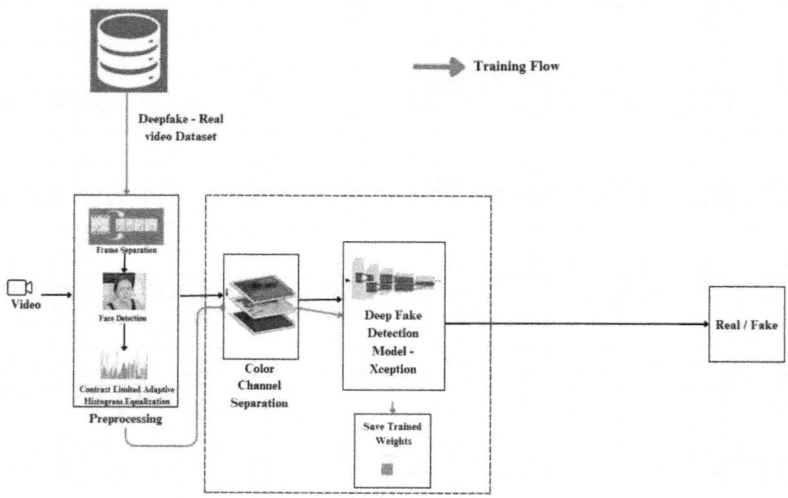

Fig. 1. Proposed Architecture Diagram for deepfake detection in videos

4 Dataset Overview

The **Celeb-DF (v2)** data collection is a benchmark dataset meticulously crafted for deepfake detection research, specifically aimed at the identification of facial manipulations in videos through advanced deep learning techniques. This dataset acts as a critical data collection for evaluating and developing models designed to enhance the realism and diversity of manipulated videos.

4.1 Key Features

Size & Diversity
Celeb-DF (v2) contains around 5000 deepfake videos derived from 590 authentic videos featuring celebrities. This substantial increase in size over its predecessor, Celeb-DF (v1), enables more comprehensive research opportunities.

Realism
The deepfake videos are created using sophisticated face-swapping algorithms, resulting in highly realistic manipulations that significantly mitigate the visual artifacts often present in earlier datasets. These videos exhibit minimal artifacts, such as blurring, warping, or color mismatches, presenting a formidable challenge for detection models and thus providing a more accurate benchmark for real-world applications.

Variety in Poses and Expressions
The dataset encompasses a broad spectrum of facial poses, expressions, and movements, mirroring real-life communication scenarios. This variety make sures that models can effectively generalize the various types of facial manipulations.

Higher Resolution & Quality
Celeb-DF (v2) features videos of higher resolution and enhanced visual quality, allowing deepfake detection models to be trained and tested on data that closely resembles real-world media in terms of fidelity and clarity.

4.2 Comparison to Other Datasets

In contrast to other data like Face-Forensics++ and DFDC, Celeb-DF (v2) prioritizes of more realistic fake videos that diminish the reliance on overt visual cues for detection. The dataset addresses numerous limitations found in existing datasets by improving the overall quality of fake videos and ensuring superior temporal and spatial consistency, thus advancing the field of deepfake detection.

4.3 Preprocessing

To enhance deepfake detection using the Celeb-DF (v2) dataset, preprocessing is essential for ensuring that the models are trained on meaningful data. The preprocessing steps typically involve frame separation, face identification, and CLAHE [26] for improved feature extraction.

5 Color Channel Conversion

At this stage of the project, the focus is on isolating and analyzing the red and green channels from the RGB image data. The rationale behind this approach is that different color channels can carry unique information about an image's structure, texture, and edges, which are often manipulated in deepfake images. By separating these channels, the neural network can be tuned to detect fine details that may be overlooked when the image is analysed as a whole.

5.1 Green Channel

The green channel is often the most sensitive in terms of human vision, capturing fine details and textures more effectively. This channel provides the neural network with a rich source of data for analysing the overall texture and consistency of an image. In deepfakes, slight distortions in texture, particularly in skin tone and shading, can be more easily detected by isolating the green channel. Additionally, the green channel plays a key role in identifying discrepancies in lighting and shading, which are common indicators of manipulated images.

6 Deep Learning Models and Activation Functions

The Xception, Inception, ResNet50, and LSTM architectures are integrated into our deepfake detection project to leverage their unique strengths in analyzing spatial, temporal, and structural features.

Table 1. Activation function and their formulas.

Activation Function	Formula
Tanh	tanh(x) = (2 / (1 + e^(-2x))) - 1
ReLU	ReLU(x) = max(0, x)
Sigmoid	sigmoid(x) = 1 / (1 + e^(-x))
Softmax	softmax(x_i) = e^(x_i) / (sum(e^(x_j)))
Swish	swish(x,i) = x * e^(x_i) / (sum(e^(x_j)))

The mathematical formulas for the several activation functions that are frequently employed in neural networks are shown in Table 1. Every activation function is essential for giving the model non-linearity, which helps it recognize intricate patterns. By compressing input values between -1 and 1, the tanh (hyperbolic tangent) function effectively centers the data, which can improve optimization procedures. The ReLU (Rectified Linear Unit) function is computationally efficient and frequently used as it can alleviate the gradient problem. If the input is positive, it outputs the input directly; if it is negative, it yields zero. The sigmoid function is especially helpful for binary classification jobs because it converts inputs to a range between 0 and 1.

The **SoftMax** function is primarily used for multi-class classification tasks, converting logits into probabilities that sum to one, thereby allowing for clear distinctions between multiple classes. Lastly, **Swish** combines the input and the sigmoid function, resulting in a smooth, non-monotonic curve that can provide better performance in certain scenarios by maintaining a small gradient even for negative inputs.

7 Results and Discussion

In Table 2, the Xception model's performance is evaluated across five activation functions: tanh, ReLU, sigmoid, SoftMax, and swish. The **sigmoid** function demonstrates the highest accuracy at **87.27%** with a sensitivity of **0.75**, effectively balancing true positive and negative identifications, with 123 true positives, 165 true negatives, and 42 false negatives. In contrast, **ReLU**, **SoftMax**, and **swish** all achieve perfect sensitivity (**1.00**), successfully identifying all positive cases but at the cost of low accuracy (**50%**), affecting precision. The **tanh** function shows mediocre results, with an accuracy of **70.30%** and sensitivity of **0.71**, resulting in 117 TP, 115 TN, 48 FP, and 50 FN. Overall, while **sigmoid** provides the best balance between accuracy and sensitivity, **ReLU**, **SoftMax**, and **swish** suffer from high false positive rates despite perfect sensitivity, and **tanh** exhibits more balanced but lower overall performance (Fig. 2).

Table 2. Performance metrices of XceptionNet model

1) Xception

ACTIVITATION FUNCTION	ACCURACY	SENSITIVITY	TP	TN	FP	FN
tanh	70.30	0.71	117	115	48	50
relu	50	1.00	165	0	165	0
sigmoid	87.27	0.75	123	165	0	42
softmax	50	1.00	165	0	165	0
swish	50	1.00	165	0	165	0

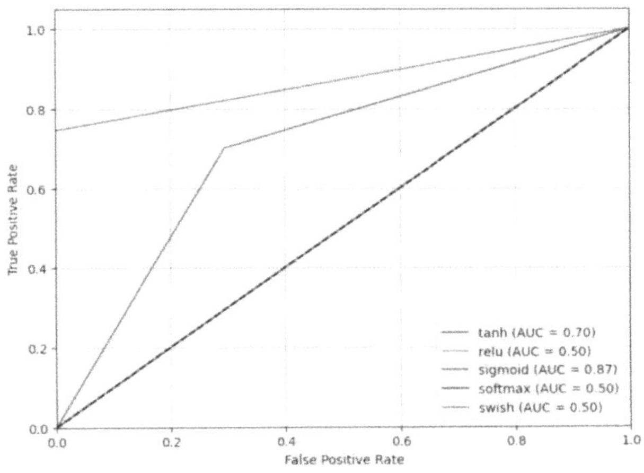

Fig. 2. ROC Curves for Different Activation Functions for Xception Net

In Table 3, which continues to focus on the **Performance Metrics of the Inception Model**, the sigmoid activation function again exhibits good results, achieving an accuracy of 79.09% with sensitivity of 0.72. The ReLU function stands out by attaining the highest sensitivity of 0.96, indicating that nearly all positive cases are correctly identified. However, its overall accuracy of 63.64% suggests difficulties with false positives, specifically having 114 FP, which implies that while it does not miss true positives, it often misclassifies negative instances as positives. On the other hand, softmax and swish again exhibit perfect sensitivity but suffer from 165 FP and 0 TN, indicating that they classify everything as positive. The tanh function, although showing moderate accuracy of 67.58%, has a low sensitivity of 0.54, which means it misses a significant portion of true positive cases, as evidenced by its 76 FN (Fig. 3).

Table 3. Performance metrices of InceptionNet model

2) Inception

ACTIVITATION FUNCTION	ACCURACY	SENSITIVITY	TP	TN	FP	FN
tanh	67.58	0.54	89	134	31	76
relu	63.64	0.96	159	51	114	6
sigmoid	80.91	0.86	142	125	40	23
softmax	50.00	1.00	165	0	165	0
swish	50.00	1.00	165	0	165	0

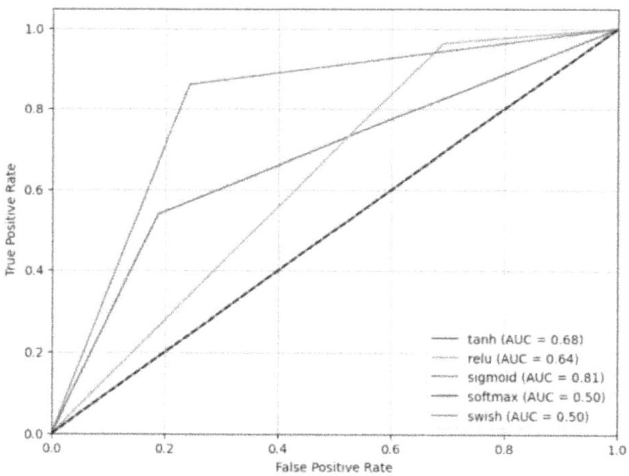

Fig. 3. ROC Curves for Different Activation Functions for Inception Net

Table 4 presents the **Performance Metrics of the ResNet50 Model**, where the sigmoid function maintains its trend of strong performance, achieving an accuracy of 79.09%, with 118 TP and 47 FN, effectively balancing the trade-off between correctly identifying both positive and negative cases. The ReLU function shows perfect sensitivity (1.00), with 165 TP and 0 FN, ensuring that all positive cases are identified. However, its TN is only 1, leading to 164 FP, indicating a high rate of false positives. The tanh function shows an accuracy of 40.30%, with 119 TP and 46 FN, resulting in moderate sensitivity of 0.72; however, its high number of 151 FP reduces its precision, making it less reliable in distinguishing negative cases. Both softmax and swish exhibit perfect sensitivity, yet they face the same issue of misclassifying all negative cases as positive, with 165 FP and 0 TN (Fig. 4).

Table 4. Performance metrices of ResNet50 model

3) ResNet50

ACTIVITATION FUNCTION	ACCURACY	SENSITIVITY	TP	TN	FP	FN
tanh	40.30	0.72	119	14	151	46
relu	50.30	1.00	165	1	164	0
sigmoid	79.09	0.72	118	143	22	47
softmax	50.00	1.00	165	0	165	0
swish	50.00	1.00	165	0	165	0

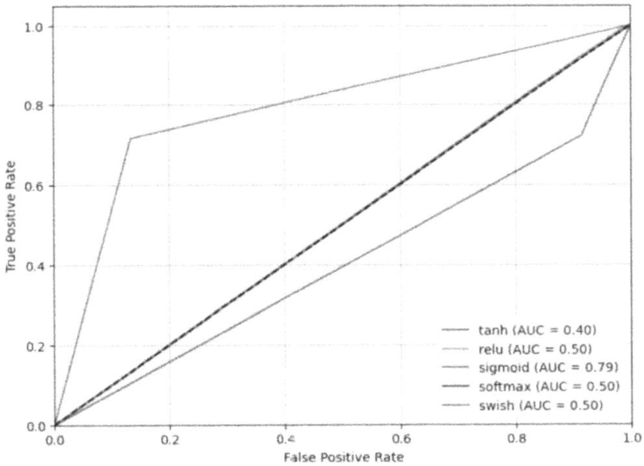

Fig. 4. ROC Curves for Different Activation Functions for ResNet50

In the final Table 5, titled **Performance Metrics of the LSTM Model**, the sigmoid function again achieves a strong balance between accuracy (70.40%) and sensitivity (0.72), with 156 TP and 27 FP, indicating reliable identification of both positive and negative cases with a low number of false positives. The ReLU function continues to

show perfect sensitivity (1.00), with 165 TP and 0 FN; however, it also has 164 FP, which reduces its accuracy to 55.30%. This suggests that while it never misses a positive case, it misclassifies a significant portion of negative cases as positive. The tanh function exhibits moderate performance, with an accuracy of 60.08%, 113 TP, and 157 FP, demonstrating challenges in accurately identifying negative cases. Both softmax and swish, as observed in previous models, maintain perfect sensitivity but classify all cases as positive, leading to 165 FP and 0 TN, which explains their lower accuracy (Figs. 5, 6 and 7).

Table 5. Performance metrices of LSTM model

4) LSTM

ACTIVITATION FUNCTION	ACCURACY	SENSITIVITY	TP	TN	FP	FN
tanh	60.08	0.72	113	15	157	47
relu	55.30	1.00	165	1	164	0
sigmoid	70.40	0.72	156	173	27	47
softmax	50.00	1.00	165	0	165	0
swish	50.00	1.00	165	0	165	0

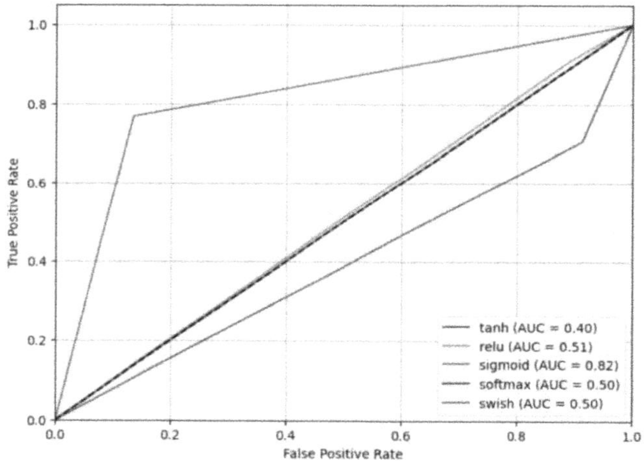

Fig. 5. ROC Curves for Different Activation Functions for LSTM

The results after performing green channel separation show a test accuracy of 90.91%, Furthermore, after applying CLAHE and green channel conversion to the input data, the model was evaluated on the test set, yielding strong performance metrics. The system got a result of **95.15% accuracy** and a loss in testing of **0.0939**, indicating that it performs well in distinguishing between classes (Figs. 8 and 9).

The model demonstrates its ability to accurately identify both positive and negative situations with 151 true positives (TP) and 163 true negatives (TN). Minimal misclassifications were shown by the 14 false negatives (FN) and 2 false positives (FP) that were reported.

The **F1 score of 0.95** highlights a favorable ratio of recall to precision, while the **sensitivity of 0.92** indicates that the model successfully captures 92% of the actual true. These results suggest that the application of CLAHE and green channel conversion contributed to enhancing the model's performance, improving contrast and focusing on the most relevant color channel, thereby aiding in more accurate classification (Fig. 10).

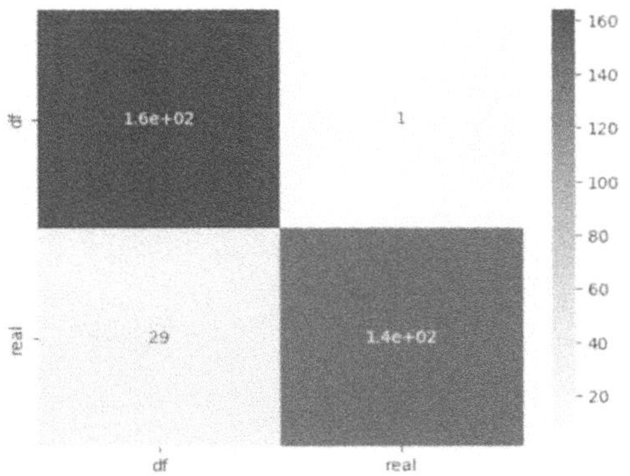

Fig. 6. Performance metrices of Xception Model after green Channel Conversion

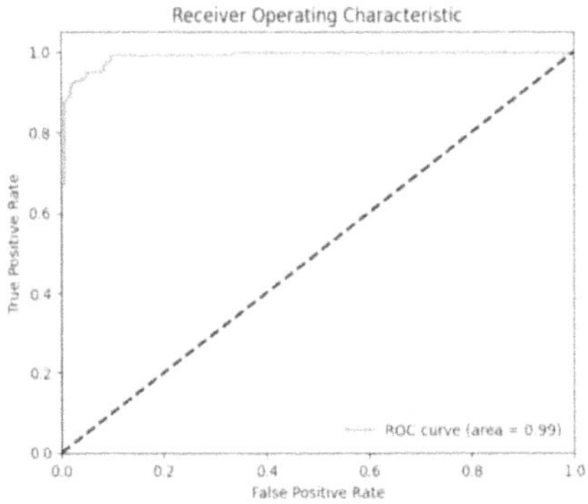

Fig. 7. ROC of Xception after Green Channel Conversion of images

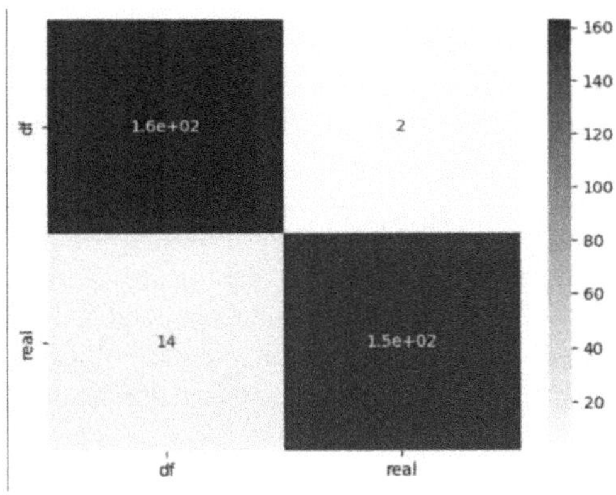

Fig. 8. Performance metrices after applying CLAHE and Green Channel Conversion

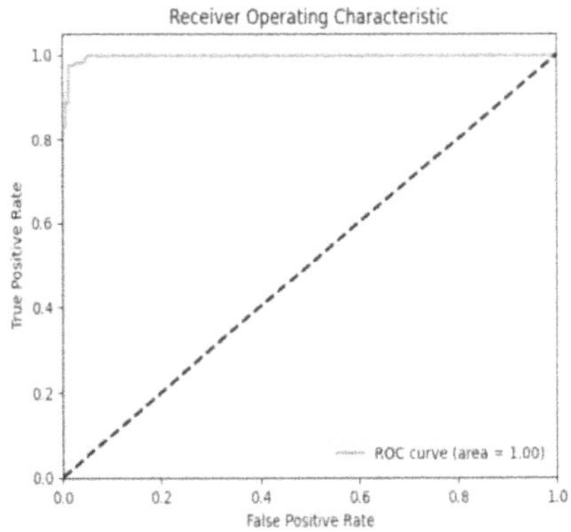

Fig. 9. ROC of Xception after applying CLAHE and Green Channel Conversion of images

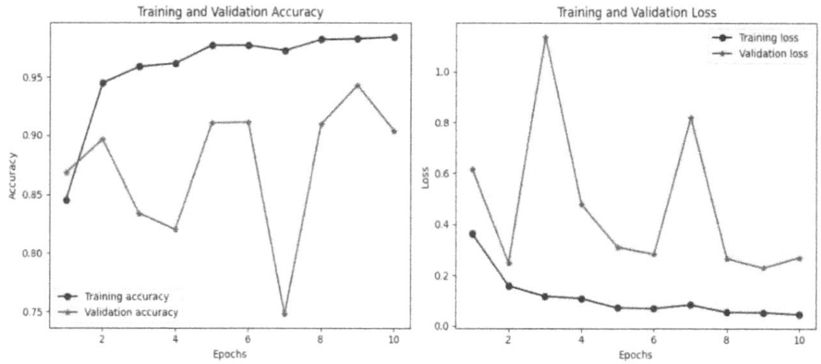

Fig. 10: Epoch graph for Training and Validation of Xception model after CLAHE and Green Channel Conversion

8 Future Works

Future research in deepfake detection will focus on improving models' ability to distinguish between various types of deepfakes, crucial for media integrity and misinformation prevention. This will involve developing algorithms to detect subtle differences in video content, especially as deepfake techniques evolve. Integrating multi-modal inputs, including both visual and audio data, will be explored to enhance detection accuracy by identifying inconsistencies between streams. Optimization of existing models, such as Xception and ResNet50, will continue through techniques like transfer learning and fine-tuning, while efforts will also explore different color channels beyond RGB to improve performance. Finally, there will be a focus on increasing the interpretability and transparency of detection models to boost user trust and adoption in critical applications.

9 Conclusion

In conclusion, the detection of deepfakes presents a significant challenge due to the rapid advancements in artificial intelligence and the sophistication of manipulation techniques. By creating a strong detection framework that incorporates cutting-edge deep learning architectures, such as the Xception model, this research has laid out a thorough strategy for addressing this problem. Our suggested methodology seeks to improve the precision and dependability of deepfake detection by utilizing color channel separation and preprocessing methods like CLAHE.

References

1. Fake news or real? Detecting deepfake videos using geometric facial structure and graph neural network (2024)
2. Dolhansky, B., et al.: The DeepFake Detection Challenge (DFDC) (2020)
3. Li, Y., et al.: Celeb-DF: A New Dataset for DeepFake Forensics (2020)
4. Chollet, F.: Xception: Deep Learning with Depthwise Separable Convolutions (2017)

5. Rössler, C., et al.: FaceForensics++. *FaceForensics++* **1**(1), 1–10 (2019)
6. Multi-domain awareness for compressed deepfake videos detection over social networks guided by common mechanisms between artifacts. Multi-task Cascaded Convolutional Neural Network, Xception network, pp. 1–10 (2024)
7. Hybrid Optimized Deep Feature Fusion-based Deepfake Detection (HODFF- DD). InceptionResNetV1, InceptionResNetV2, and Bidirectional Long-Short Term Memory (BiLSTM), pp. 1–10 (2024)
8. A guided-based approach for deepfake detection: RGB-depth integration via features fusion. Masked Depthfake Network, FaceForensics++ dataset, pp. 1–10 (2024)
9. Feature fusion Vision Transformers using MLP-Mixer for enhanced deepfake detection. Dual Attention Vision Transformers (DaViT), Inception Transformer (iFormer), and Group Propagation Vision Transformer (GPViT), pp. 1–10 (2024)
10. Spatial-frequency feature fusion based deepfake detection through knowledge distillation. Spatial-Frequency Fusion Branch (SFFB), Multiknowledge distillation, pp. 1–10 (2024)
11. DeepFake Detection Based on High-Frequency Enhancement Network for Highly Compressed Content, pp. 1–10 (2024)
12. A deep learning model for FaceSwap and face-reenactment deepfakes detection. AUFF-NET, CNN, Bi-LSTM, FaceForensics++ (2024)
13. Revealing and Classification of Deepfakes Video's. CNN, Deepfake Detection Challenge, Deep Fake Papers – Data Forensics Team (2023)
14. Images using a Customized Convolution Neural Network Model Preserving manipulated and synthetic Deepfake detection through face texture naturalness'. hybrid Deepfake detection mechanism (HDDM), Celeb-DFv2, FF++ (2024)
15. A dual descriptor combined with frequency domain reconstruction learning for face forgery detection in deepfake videos. EfficientNet-NS-B3, FaceForensics++ (2024)
16. GenAI mirage: The impostor bias and the deepfake detection challenge in the era of artificial illusions. GAN, FaceForensics++ (2024)
17. Local attention and long-distance interaction of rPPG for deepfake detection. Multi-scale Spatial–Temporal PPG, FaceForensics++ (2024)
18. Deep fake detection using an optimal deep learning model with multi head attention-based feature extraction scheme. ResNet-50 with Multi Head Attention (RN50MHA), Optimal Long Short-Term Memory (OLSTM), Face Forensics++, Deepfake Detection Challenge, Celebrity Deepfake (CDF), and Wild Deepfake, Deep Fake Papers – Data Forensics Team (2024)
19. An ensemble of CNNs with self-attention mechanism for DeepFake video detection. convolution layers and self-attention (CoAtNet), FaceForensics dataset (DeepFakes (DF), Face2Face (F2F), FaceSwap (FS), and NeuralTextures (NT)), Celeb- DF (2023)
20. Face forgery detection by progressively enhancing spatial and frequency-aware features. PESAF, (FGFAE), dual-domain feature attention (D2FA), FaceForensics++, Celeb-DF, DeepfakeDetection (DFD), Deepfake Detection Challenge (DFDC) (2024)
21. Tex-Net: texture-based parallel branch cross-attention generalized robust Deepfake detector. ResNet-18, FF++, DFDC, Celeb-DF (2024)
22. Deepfake video detection: challenges and opportunities. CNN, RNN, FF++, Deepfake (2024)
23. Deepfake detection based on remote photoplethysmography. Remote Pho toplethysmography (rPPG), COHFACE, DFDC (2023)
24. Emotion aided multi-task framework for video embedded misinformation detection. Supervised Contrastive Learning (SCL), CNN-RNN model, Synth: FakeClips, Ryerson Audio-Visual Database of Emotional Speech and Song (2024)
25. Shallowfake and deepfake image manipulation localization using noise and RGB- based dual branch method. CNN + noise and RGB-based dual branch method, CASIA, COVERAGE, COLUMBIA, NIST16, FaceForensics++ (FF++) (2024)

26. Zuiderveld, K.J.: Contrast Limited Adaptive Histogram Equalization. Graphics gems (1994)
27. Goodfellow, I., et al.: Generative adversarial nets. Adv. Neural. Inf. Process. Syst. **27**, 2672–2680 (2014)
28. Hochreiter, S., Schmidhuber, J.: Long short-term memory. Neural Comput. **9**(8), 1735–1780 (1997)

Early Prognosis of Post-Covid Cardiovascular Disease with Clinical Data and Chest X-Ray Image Using Transformer- Based Approach

S. Diana Juliet[1(✉)], J. Banumathi[2], and J. Anix Joel Singh[3]

[1] C.S.I. Institute of Technology, Thovalai, India
dianajuliet2ky@gmail.com
[2] University College of Engineering Nagercoil, Nagercoil, India
banumathi@ucen.ac.in
[3] St. Joseph University, Dar es Salaam, Tanzania
anixjoelsingh@gmail.com

Abstract. The demand for efficient prediction methods to evaluate and treat post-COVID disorders has increased due to the COVID-19 epidemic. This work presents a novel strategy for anticipating and addressing these health risks that makes use of transformer-based models. To improve the disease prognosis accuracy, this study integrates detailed tabular data and X-ray images of chest, who pretentious with COVID-19.A hybrid transformer model that incorporates the most recent advancements helps to predict the course of the disease. This approach makes it possible for the model to successfully adjust to each patient's distinctive characteristics and disease progression. Preliminary findings indicate that the proposed approach demonstrates promising results in accurately prognosing post-COVID diseases. Integrating diverse datasets significantly improves the model's predictive capabilities and treatment efficacy, allowing for tailored recommendations that align with individual patient needs. The Vision Token Transformer (ViToT) architecture is a Hybrid transformer, which contains Vision Transformer in order to train a model with chest X-ray (CXR) images and Feature Tokernizer transformer in order extract features from tabular data to capable of recognizing and categorizing key patterns in CXR images while also extracting optimal features from tabular data. Late fusion technique is applied to combine an extracted feature which leads to cardiovascular disorder. Across 150 training epochs, the model demonstrated a robust performance, achieving a final accuracy of 97.6%, indicates that the model generalizes extremely effectively. This approach highlights the potential of hybrid-based transformer algorithms to enhance post-pandemic healthcare management. In the post-COVID period, it enables healthcare providers to improve patient care by facilitating early disease prognosis.

Keywords: Post · COVID-19diseases · Transformer-based Model · Disease prognosis · Chest X-ray images · Post-pandemic healthcare · Cardiovascular Disease · Multimodal

R. Geetha et al. (Eds.): AAIMB 2024, CCIS 2654, pp. 383–397, 2026.
https://doi.org/10.1007/978-3-032-13029-7_28

1 I. Introduction

During the year 2019, the Severe Acute Respiratory Syndrome Coronavirus 2 (SARS-CoV-2) manifested, cause COVID-19, which intensified to an epidemic by March 2020 and led to widespread outbreaks worldwide [1] In the acute period of the illness COVID-19 has both immediate and long-term effects on various organs in the body [7]. Symptoms that continue four weeks after the start of COVID- 19 can indicate a condition known as post-COVID 19 syndromes, which includes ongoing symptoms [4]. These ongoing symptoms typically appear between four to twelve weeks after infection. The phrase "long COVID" initially referred to persistent symptoms enduring years or even months after SARS- CoV-2 disease. Acute COVID-19 symptoms are categorized as post-COVID-19 syndrome if they persist for more than twelve weeks. One of the most dangerous COVID-19 side effects related to blood clotting, which can lead to death. Post-Acute COVID-19 Syndrome (PACS) incorporates a broad range of symptoms that can remain long after the initial infection, with common symptoms including chest pain, joint pain, fatigue, and shortness of breath. PACS helps to describe the various ongoing symptoms that many experience after recovering from COVID-19 [5, 19].

The World Health Organisation (WHO) has received reports of 776,385,727 billion COVID-19 cases in total, as illustrated in Figure 1. More than 53% of people suffer from fatigue, more than 43% of people are affected with dyspnea, and almost 27% of people have the symptoms of Chest Pain [7]. Pneumonia and Cardiac diseases causes' high mortality and rate of major complications with rate of Pneumonia 15% and Cardiac disease 22.5% respectively [4].

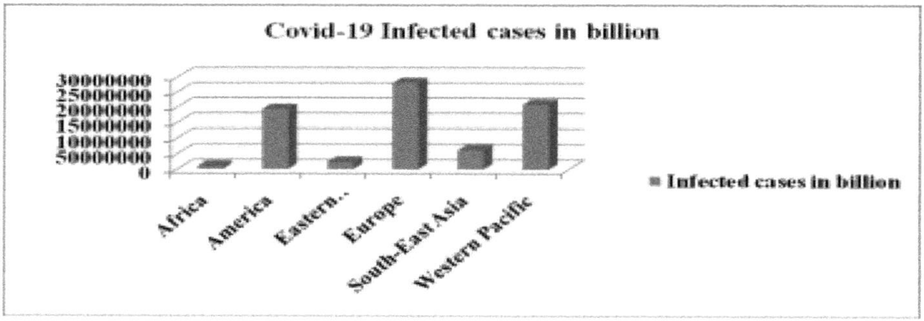

Fig. 1. The COVID-19 cases reported to WHO

Overall in the world, 7.1 million deaths happened due to the SARS-CoV-2 disease [7]. Dealing with SARS-CoV-2 prognosis factors in individuals who have the same genotype and are infected with the same virus presents a significant challenge, as the results may indicate a markedly severe disease. Acute Cardiovascular Syndrome and myocardial damage may also appear in addition to systemic and respiratory issues. The cardiac troponin is usually elevated in SARS-CoV-2. Especially in adults and children, it includes myocarditis, pericarditis, and acute coronary syndrome, which are secondary to oxygen- augmented demand in the multi-system systemic inflammation system or obstructive coronary artery disease [10] in SARS-CoV-2 patients and this is an indication of myocardial injury [11].

Cardiovascular risk factors cause a more severe course and a higher death rate [11]. Some symptoms that indicate the drawn attention to the cardiovascular system such as Chest Pain, Dyspnea, and palpitations. These symptoms are also related to SARS-CoV-2 that are developed during acute or post-acute in the chronic phase of SARS-CoV 2 [8]. The most characteristic myocardial involvement has been found in multiple studies assessing both systematic and non-systematic SARS-CoV-2 cases. It's still difficult to predict how cardiovascular disease will progress after contracting SARS-CoV-2.

The understanding is poor about cardiovascular disease after post-acute SARS-CoV-2 and its complications [9]. This increases the rate of thromboembolic events as well as pulmonary hypertension [12]. Both symptomatic and asymptomatic individuals have been affected by myocardial injury. Numerous studies, both systematic and non systematic, have identified myocardial involvement as a prominent feature associated with SARS-CoV-2 infection. Pericarditis is also another representation of cardiac manifestations due to SARS-CoV-2. Pericarditis is frequently cited as the most prevalent acute infection linked to SARS-CoV-2. But the most common complication, especially in the post-acute stage of the infection, is thought to be coagulation problems. These methods include blood testing, and chest CT and X-ray pictures, Test related to Reverse Transcription Polymerase Chain Reaction (RT-PCR) to detect SARS-CoV-2 infections.

Cardiovascular disease (CVD) is a kind of Heart disease and it is a serious global health issue, certain blood test markers are pivotal in assessing a patient's risk and detecting early signs of the disease. A CXR is an inexpensive but efficient method to screen and diagnose breast cancer, pneumonia, TB, and fractures. Chest X-rays have been utilized for disease screening during the COVID-19 pandemic and by AI researchers developing transformer-based approaches. Multimodal models integrate imaging and clinical metadata [5, 13].

Mortality rates increased from cardiovascular diseases during and as well as after the acute phase of the illness. Research suggest that people who contract COVID-19 face a significantly higher likelihood of developing cardiovascular issues. The risk of death for COVID-19 individuals, for example, was shown to be up to 81 times greater during the acute phase of infection and to remain five times higher for at least 18 months after that. With many survivors facing chronic difficulties, the COVID-19 pandemic's aftermath has presented serious hurdles to global healthcare [20]. Among these, cardiovascular illnesses (CVDs) have gained significant attention because of SARS-CoV-2's long-term impact on the cardiovascular system [6]. Early identification of post COVID-19 cardiovascular risks is essential for preventing severe health issues and improving patient outcomes. Cardiovascular complications like myocardial infarction, heart failure, and an elevated risk of stroke are directly related to COVID 19.There are so many models are currently available for predicting diseases.

This research aims to utilize clinical data and chest X-ray imaging to develop a predictive model powered by transformer-based architectures, which have recently brought transformative advancements to the field of machine learning. Machine learning (ML) has proven to be a vital asset in healthcare by enabling the analysis of complex datasets for early disease detection and prognosis. Traditional ML approaches excel at processing large datasets and uncovering hidden patterns that may not be immediately evident. However, they often require extensive feature engineering, may struggle with heterogeneous datasets, and sometimes lack the ability to generalize effectively across diverse data [27].

Transformer-based models, initially developed For tasks involving natural language processing, and shown outstanding performance in multimodal data processing and computer vision. These models leverage attention mechanisms to prioritize the most relevant parts of the data, making them highly effective in analyzing complex and high-dimensional datasets. Compared with traditional models like convolutional neural networks (CNNs) and other ML techniques, transformers performs well in synthesizing features from diverse data modalities and capturing enduring dependencies. This capability makes them particularly well-suited for predicting post-COVID cardiovascular risks, where integrating clinical tabular data with imaging features is paramount. There are several transformer models designed specifically for image processing tasks, including Vision Transformer (ViT), Swin Transformer, and DETR (Detection Transformer).

In the ViT (Vision Transformer) model, images are divided into fixed-size patches, which are treated as individual tokens. These tokens are then processed through transformer architecture. Unlike traditional CNNs that primarily focus on extracting local features, ViT utilizes self-attention mechanisms to capture global dependencies across the entire image, enabling it to model spatial relationships effectively. This makes the ViT approach particularly suitable for tasks such as image classification, object detection, and segmentation.

The Swin Transformer introduces a new hierarchical architecture which creates a fixed-size image representation, enhancing its efficiency for processing large image datasets. It utilizes a window- based self-attention mechanism, which reduces the computational complexity compared to ViT, while still maintaining effective performance on tasks like semantic segmentation, object detection, and image classification. The DETR leverages a transformer model for end-to-end object detection, which is different from traditional object detection methods. Traditional techniques often rely on anchor boxes and post-processing to detect objects. In contrast, DETR reimagines object detection as a direct set prediction problem, generating a set of bounding boxes along with their associated labels. This groundbreaking method streamlines the process and enhances detection accuracy.These transformer- based models have demonstrated significant improvements in various image processing applications, providing efficient and scalable solutions for complex visual recognition tasks.A novel multimodal transformer-based prognosis method for post-Covid cardiovascular disease is suggested in an effort to lower the death rate.

2 Related Works

Several researchers proposed different Machine Learning (ML) and Deep Learning (DL) models for automatic prediction of cardiovascular diseases using radiographic images. The obtained results are appreciable, however there is a challenge to prognosis cardiovascular disease from Covid-19 infected patients to improve the performance as well as early prognosis The identification of cardiac disease issues at an early stage can improve survival rates support patients also support domain experts. Treat them well at the earliest. Different machine learning and deep learning techniques are then used to use multimodal data, such as tabular data, sounds, radiographic images, CT, and CXR images, among other data, to make the distinction between healthy and unhealthy patients. Hence each of the Multimodal elements is preprocessed and also features are extracted with the help of various techniques.

R. M. Al-Tam et al. introduced a novel approach for cardiovascular disease detection using heart sounds, leveraging a hybrid framework combining a Transformer Encoder and Residual Convolutional networks. The proposed methodology incorporates advanced AI models such as VGG16, ResNet50, a Transformer Encoder, and a hybrid system that integrates ResNet50 with a Transformer Encoder, supplemented by a Multi-Layer Perceptron (MLP). The study focused on diagnosing heart sounds captured through phonocardiography (PCG) by employing two benchmark datasets, Yaseen and PhysioNet, for both multi-class and binary classification tasks. Mel-frequency Cepstral Coefficients (MFCCs) were utilized to extract features from the heart sound recordings. The proposed approach achieved remarkable accuracy, with an average of 99.8% for the Yaseen dataset and 96.40% for the PhysioNet dataset. When combining both datasets, the method achieved 100% accuracy [14].

Xiao et al. developed a new method based on heart sound signals to diagnosing cardiovascular disease that utilizes an evolving fuzzy system and a One dimensional and Two dimensional convolutional neural network using Yaseen datasets to perform binary classification with labels Normal and Abnormal with an accuracy of 96.30% [15]. Ma et al., developed a lightweight CNN-based self-supervised learning for valvular heart disease multi-classification from phonocardiogram (PCG) signals using Yaseen's dataset and achieved an average of 98.65% accuracy [16].

Chiagoziem C. Ukwuoma and colleagues introduced a novel hybrid explainable AI framework for pneumonia detection, leveraging an ensemble Transformer model equipped with a multi-head attention mechanism. This approach effectively captures relationships between distant pixels in chest X-ray images. The study addresses both binary classification (Normal and Pneumonia) and multi-class classification involving three respiratory disease categories: Normal, Bacterial Pneumonia, and Viral Pneumonia. The framework utilizes two chest X-ray datasets, Mendeley and Chest X-ray, to evaluate its performance. The proposed deep learning model demonstrated an impressive overall accuracy of 99.21% for binary classification and 98.19% for multi-class classification tasks [17]. Arnab Kumar Mondal proposed a prognosis method of Covid-19 using a limited set of CXR images due to the in availability in pre pandemic period. The attention maps calculate between the global and the local features of the convolutional network while fine tuning using a limited COVID-19 CXR dataset.This model helps to visualize and highlight the critical patches instrumental to the predictive decision [18]. Carion et al. introduced the DETR model for addressing tasks in Computer Vision (CV), specifically object detection. Unlike traditional methods that rely on hand-crafted processes, DETR offers a fully end- to-end approach. It employs a Convolutional Neural Network (CNN) to extract image features, which are then processed by a transformer encoder to capture relationships between these features. A transformer decoder subsequently generates object queries. The model utilizes a feed-forward network to assign labels, while bounding boxes are used to identify objects [22].

On the other hand, Liu et al. proposed the Swim-Transformer, which introduces a window-based self-attention mechanism to minimize computational complexity. By connecting attention blocks and merging neighboring patches, this model generates a hierarchical representation that effectively manages variations in the scale of visual entities. The Swin-Transformer is particularly efficient in reducing the cost of attention computations for high resolution images and addressing the challenges posed by varying patch sizes in scene understanding tasks [23].

Tianyi Chen et al. developed a machine learning model aimed at improving the accuracy of COVID-19 diagnosis using chest X-ray (CXR) images. They experimented with several architectures, including EfficientNet, MViT, EfficientViT, and ViT, leveraging a public dataset that contained 3,616 images of COVID-19, 6,012 of lung opacity, 10,192 normal images, and 1,345 of viral pneumonia.Among these, the ViT model delivered impressive results, with 95.79% accuracy across four categories, 99.57% accuracy for three categories, and excellent binary classification metrics, including recall (98.58%), precision (98.87%), F1 score (98.73%), specificity (99.76%), and AUC (0.9993) [26].

Rajasekaran Thangaraj and colleagues proposed a Modified Inception V3 (MIn-V3) model that integrates feature fusion from the internal layers of Inception V3 for disease classification. The model sorts cases into viral pneumonia, normal, COVID-19 positive, and bacterial pneumonia with four categories. The performance of MIn-V3 is enhanced through transfer learning and fine-tuning, achieving an accuracy of 96.33% when compared with pre-trained models like Inception-ResNet V2 (InRN-V2), Inception V3, and MobileNet V2 [24]. Rahimzadeh et al. introduced a model for categorizing COVID-19 cases using X-ray images. They utilized a ResNet50V2 model combined with a concatenated CNN based on the Xception architecture. Despite working with a small dataset comprising 180 COVID-19 X-ray images, 6,054 pneumonia images, and 8,851 healthy images the model achieved a remarkable accuracy of 99.56% [25].

Loey et al. proposed a deep learning (DL) model combined with a Generative Adversarial Network (GAN) to diagnose COVID-19 from chest X-ray images. The dataset used consisted of 69 X-ray images of COVID-19 cases, 79 images of bacterial pneumonia, 79 images of viral pneumonia, and normal X-ray images. They employed three pre-trained models such as GoogleNet, ResNet18, and AlexNet. GoogleNet provided the highest accuracy, with 80.6% for multi-class classification and 99.9% for binary classification [21]. Zahid Ullah et al. introduced the Densely Attention Mechanism-based Network (DAM-Net) for COVID-19 detection using CXR images from the COVIDx public dataset. DAM-Net extracts distinct features from the infected regions using segmentation-based and classification-based methods. The model achieved performance metrics of 97.22% accuracy, 96.87% sensitivity, 99.12% specificity, and 95.4% precision [2].

Lastly, Junxin Wang et al. designed a novel multimodal deep learning model that integrates early-stage chest X-ray images with structured data from medical records at Columbia University Irving Medical Center to enhance the early detection and diagnosis of chronic cardiac conditions. The model utilized a pre-trained EfficientNetB3 for image feature extraction and employed attention mechanisms, such as SE- Block and CBAM, for further improvement. Accuracy was boosted by feature fusion using a Variational Autoencoder (VAE) [3].

3 Methodology

In this proposed work, Vision Token Transformer (ViToT) model is used to train a model for identifying and classifying critical patterns related to cardiovascular disease (CVD) using Tabular clinical data and chest X-ray imaging studies. The ViToT transformer is a hybrid model which encompasses both the Vision Transformer and the Feature Tokenizer Transformer. In this proposed work a COVID-19 infected patient's chest X-ray public dataset was created using 1,000 images sourced from public datasets. Additionally, the tabular dataset comprises 1,736 entries and consists of 36 columns, which include numerical features such as age and various laboratory test results, including CA, CK, and CREA, along with a target variable indicating the presence of the disease.

3.1 Input to the Vision Transformer

CXR images are processed using a pre-trained Vision Transformer (ViT) to identify visual patterns indicative of disease or abnormalities. The Transformer architecture includes an Encoder for processing the X-ray images and for generating predictions a Decoder is used. The core components of the transformer architecture include the Self Attention Mechanism, Multi-Head Attention Mechanism, Feed-Forward Networks (FFNs), and Positional Encoding shown in the Fig. 2. The self-attention mechanism assesses the significance of each patch in relation to others, while the multi-head attention mechanism allows the model to simultaneously compute and capture various relationships among the patches [26].

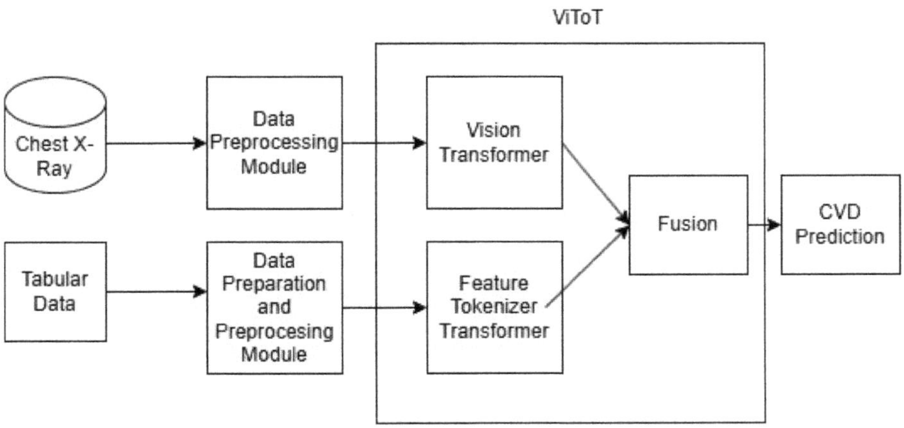

Fig. 2. Block Diagram of Proposed ViToT Transformer

An input image of dimensions h x w with RGB channels c can be transformed into a sequence of patches using the Vision Transformer, a powerful technique for image analysis. Initially, the image is segmented into fixed-size, non overlapping patches measuring Ph x Pw x c. These patches are subsequently converted into lower-dimensional vectors of size d through linear embedding. To improve the model's comprehension spatial relationships, positional encodings based on sine and cosine functions are added to the embeddings. The Transformer's encoder layers each have a multi-head self-attention mechanism and a feed-forward neural network, allowing the model to process individual tokens effectively while capturing enduring dependencies within the image data. Finally, the token embeddings generated are utilized for the classification task.

3.2 Feature Tokenizer Transformer

Laboratory test results are presented as tabular data that indicates how test results can lead to CVD.The Feature Tokenizer Transformer enhances the extraction of optimal features from tabular data. The first step is cleaning and preprocessing to ensure that categorical features are properly encoded. Feature tokenization converts tabular features into a token representation, which includes two main processes: categorical encoding and continuous feature scaling. Categorical encoding transforms categorical variables into dense vectors, while continuous feature scaling normalizes continuous features. Each feature is treated as a separate token, generating a sequence of embeddings.

3.3 Input to the Vision Token Transformer

The Combined embeddings from the feature tokenizer and Vision Transformer fed into the vision Token transformer. A Concatenation stack the embeddings from both sources along a new dimension. Attention Mechanisms then uses multi-head attention layers to relationships between tokens from both data types. The additional layers are used to combine the two sets of embeddings. Severity indicates an initial stage of lung pathology, traditionally associated with early indicators of pneumonia or COVID-19- related changes. The suggested Multimodel ViToT transformer model outcomes are consecutively demonstrated for each scenario. The set of COVID-19-infected patients CXR images are given as the input. The CXR image of COVID-19 patient is shown the Fig. 3.

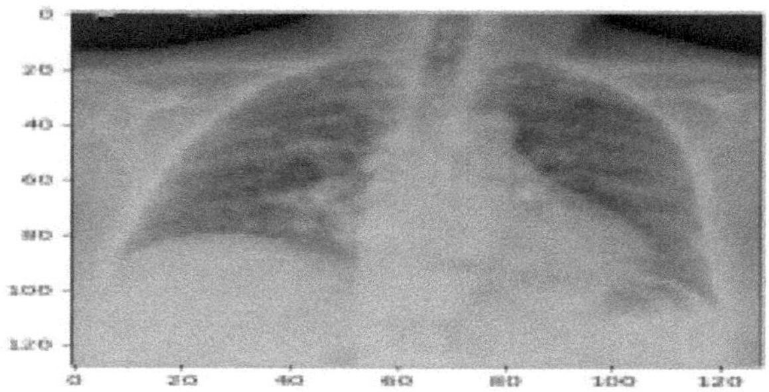

Fig. 3. CXR Image of a Patient with COVID-19

3.4 Image Segmentation

The image segmentation process splits the image into bottom left, bottom right, top left and top right, as illustrated in Fig. 4. Apply a linear transformation to convert these patches into embeddings. Transformer Architecture consists of an Encoder to process input X-ray images and a Decoder to generate predictions.

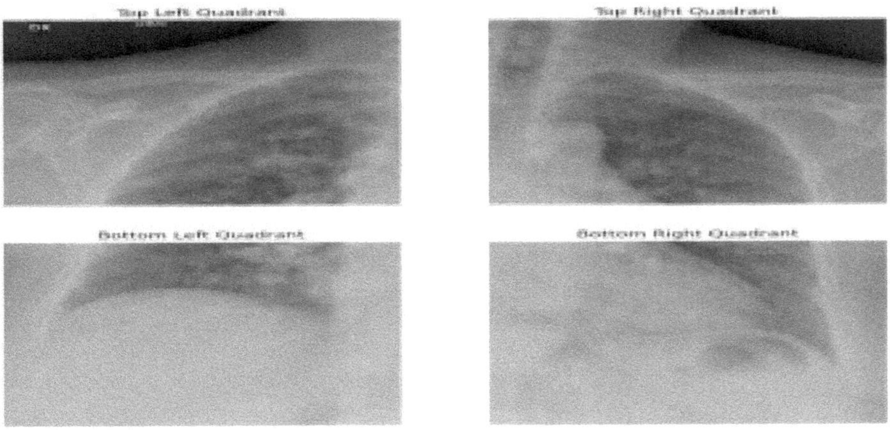

Fig. 4. The CXR Image 4 Quadrant Segmentation

Across 150 training epochs, the model demonstrated a robust performance, achieving a final accuracy of 97.52%, and a loss of 0.1879. During the time evaluation, the highest severity of GGO was observed in the top-left quadrant at 2.72%, with the top-right quadrant showing slightly lower severity (2.19%). The bottom-left quadrant exhibited GGO severity at 1.79%, while the bottom-right presented the least with 1.36%. This distribution of GGO severity indicates an initial stage of lung pathology, traditionally associated with early indicators of pneumonia or COVID-19-related changes. On the other hand, consolidation patterns were predominantly found in the bottom-left quadrant, where the severity peaked at 18.19%.

The bottom-right quadrant recorded a consolidation severity of 3.56%, while the top- right quadrant presented no evidence of consolidation, indicating minimal to no involvement in that region. Crazy paving, an advanced and more indicative pattern of lung damage, was detected only minimally; with a recorded severity of 0.04% in the top-left quadrant and absent from all other quadrants. These observations highlight the early manifestations of significant lung conditions and underline the model's capability of differentiating among various morphological abnormalities. The significance of these findings extends beyond mere detection, especially considering the strong associations between lung abnormalities like GGO and consolidation with interstitial lung diseases and infections. These conditions can impose secondary cardiovascular risks due to chronic hypoxia, leading to increased cardiac workload and the consequent potential for heart failure or other cardiovascular complications.

3.5 Severity Prediction

The Severity quantification CXR images shown in Fig 5 with severity quantification of the identified patterns across the four lung quadrants revealed significant insights into early stage pathological processes,

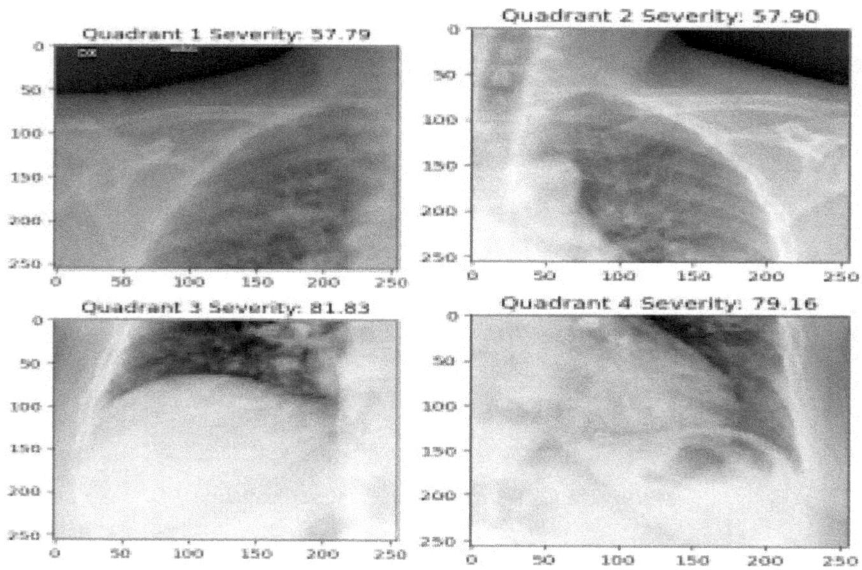

Fig. 5. The Severity quantification across each Quadrants

The presence of early-stage consolidation or GGO suggests ongoing underlying inflammation or fibrosis that could compromise pulmonary function and indirectly impact cardiovascular health. To further enhance the detection and diagnostic accuracy of such abnormalities, an integrated Vision Token Transformer (ViToT) -based architecture is used. Transformers excel in medical image analysis because they effectively capture long-range dependencies and intricate features critical for discerning subtle pathological changes, including reticulation or crazy-paving patterns. By incorporating ViToT we could significantly bolster the segmentation accuracy for complex pathological conditions that often exhibit overlapping features (Fig. 6).

	Actual		
Predicted		Positive	Negative
	Positive	555	13
	Negative	11	421

Fig. 6. The confusion Matrix

3.6 Tabular data Feature Extraction

The overview of key lab tests related to SARS-CoV-2 shows that CRP, D-Dimer, S-Ferritin, ALT, and LDH are commonly used indicators of potential health issues. Elevated Haematocrit levels reflect thicker blood, increasing the risk of clots, strokes, or heart attacks. Abnormal red blood cell indices and low haemoglobin can also strain the heart. High CRP suggests inflammation, linked to increased cardiovascular risk. Elevated D-dimer indicates blood clotting activity, raising concerns for conditions like deep vein thrombosis and pulmonary embolism.

The Laboratory test Results in the form of tabular with its Normal range of values with different units shown in the Table 1. Key markers for cardiovascular disease risk include CRP, Interleukin-6, Troponin, and BNP/NT-proBNP, which help assess inflammation, cardiac injury, heart failure, and clot- related issues. The ViToT transformer model's performance is evaluated using the metrics Recall, Accuracy, F-1 score, Precision and achieved values respectively 99. 2%,96.67, 97.6and 97.97. Key Findings from the result of the Pearson Correlation Coefficient is The highest positive correlation with the target variable was observed for AGE ($p < 0.01, r = 0.45$), indicating that elderly patients could be more susceptible to the target condition. Other notable correlations include WBC ($r = 0.35$, $p < 0.01$), which indicates that higher white blood cell counts may correlate with disease presence.

Table 1. Laboratory test Results with its Normal range of values

Test Name	Units	Normal Range
CRP	mg/L	0 – 5
AST	U/L	0 – 34
BNP	pg/mL	0 – 100
Hemoglobin	g/dL	12.5 – 17.5
ALT	U/L	7 – 40
EDTA Blood Count	Thousands/cmm	150 – 450

(continued)

Table 1. (*continued*)

Test Name	Units	Normal Range
D-dimer	μg FEU/mL	0 – 0.44
Electrolytes	--	Up to 12
Hematocrit	%	41 – 52
MCHC	g/dL	31 – 37
Interleukin-6	pg/mL	0 – 5
LDH	U/L	120 – 246
Leucocyte Count	Thousands/cmm	4 – 11
MCH	pg	27 – 33
RBC Count	Millions/ cmm	4.5 – 5.9
MCV	fL	80 – 100
NT-proBNP	pg/mL	0 – 125
Potassium	mmol/L	3.5 – 5.0
RDW-CV	%	11.5 – 15
Serum Ferritin (Female)	ng/mL	13 – 150
Serum Ferritin (Male)	ng/mL	30 – 400
Sodium	mmol/L	135 – 145
Troponin I	ng/mL	0 – 0.04
Troponin T	ng/mL	0 – 0.01

Spearman Correlation Coefficient Results show that For non-normally distributed features, CK showed a moderate negative correlation with the target variable ($\rho = -0.30$, $p < 0.05$), indicates that lower CK levels might statistically associated with the presence of cardiovascular disease (CVD) in the dataset. This might reflects post-infection muscle fatigue or sample bias and warrants further clinical investigations. For experimental purposes, the dataset is newly constructed with 1000 images. The overall data set is split for training 80%, 10% for validation, and for testing 10%. The confusion matrix is shown the Fig. 6. The heat maps of the correlation matrix help to illustrate the relationships among the features shown in Fig. 7.

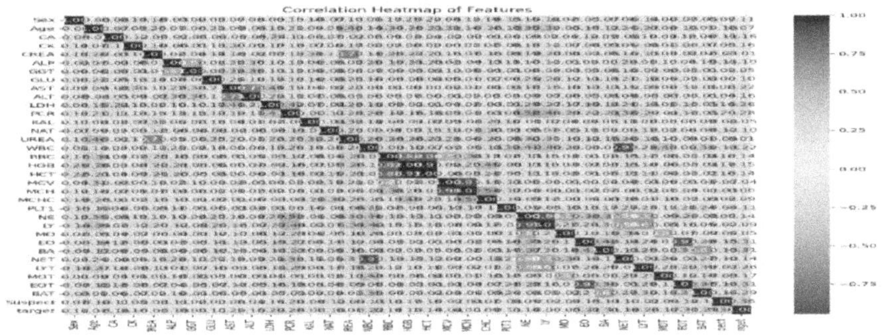

Fig. 7. Correlation Heatmap of features for Tabular

4 Conclusion

This work uses the chest X-ray images combined with tabular clinical data for predicting post-COVID cardiovascular disease in early way, employing a hybrid transformer model known as the Vision Token Transformer (ViToT).An ensemble of previously trained transformer models served as the foundational architecture for the suggested hybrid framework, enabling more effective feature extraction. The relationship between features is displayed on the correlation heat map in order to help find the best features. Recall, Precision, Accuracy, F-1 score metrics used for the evaluating model with the multi-label feature to binary class classification. The proposed ViToT model's performance is divided into three stages, (1) CXR images are processed using a Pre-Trained Vision Transformer, (2) Tabular data are processed using a Feature Tokenizer transformer (3) the hybrid ViToT transformer that has been suggested, which uses the late fusion technique and is based on the ensemble transformer encoder. When it came to binary class identification performance, the suggested hybrid transformer model ViToT scored 97.6%, 97.11%, 98.06%, and 97.88% on the Accuracy,Precision,Recall and F-1 score. By utilizing multimodal data, the suggested model is able to identify post Covid-19 cardiovascular diseases more accurately due to its improved binary classification and multi class features.

References

1. World Health Organization. Coronavirus Disease (COVID-19) Pandemic. World Health Organization (2020). https://www.who.int
2. Ullah, Z., Usman, M., Latif, S., et al.: Densely attention mechanism based network for COVID-19 detection in chest X-rays. Sci. Rep. **13**, 261 (2023). https://doi.org/10.1038/s41598-022-27266-9
3. Wang, J., Li, J., Wang, R., Zhou, X.: VAE-driven multimodal fusion for early cardiac disease detection. IEEE Access **12**, 90535–90551 (2024). https://doi.org/10.1109/ACCESS.2024.3420444
4. Andreini, D., et.al.: Extent of lung involvement over severity of cardiac disease for the prediction of adverse outcome in COVID-19 patients with cardiovascular disease. International Journal of Cardiology **323**, 292–294 (2021). ISSN 0167–5273. https://doi.org/10.1016/j.ijcard.2020.10.006

5. Simon, B.D., Ozyoruk, K.B., Gelikman, D.G., Harmon, S.A., Türkbey, B.: The future of multimodal artificial intelligence models for integrating imaging and clinical metadata: a narrative review. Diagn. Interv. Radiol. (2024). https://doi.org/10.4274/dir.2024.242631
6. Xie, Y., Xu, E., Bowe, B., et al.: Long-term cardiovascular outcomes of COVID-19. Nat. Med. **28**, 583–590 (2022). https://doi.org/10.1038/s41591-022-01689-3
7. Nalbandian, A., et al.: Post-acute COVID-19 syndrome. Nature Medicine **27**(4), 601–615 (2021). https://doi.org/10.1038/s41591-021-01283-z
8. Xie, Y., Wang, Z., Liao, H., Marley, G., Wu, D., Tang, W.: Epidemiologic, clinical, and laboratory findings of the COVID-19 in the current pandemic: systematic review and meta-analysis. BMC Infect. Dis. **20**(1), 640 (2020). https://doi.org/10.1186/s12879-020-05371-2
9. Li, L., et al.: Artificial intelligence distinguishes COVID-19 from community acquired pneumonia on chest CT. Radiology **296**(2), E65-E71 (2020). https://doi.org/10.1148/radiol.2020200905
10. Puntmann, V.O., et al.: Outcomes of cardiovascular magnetic resonance imaging in patients recently recovered from coronavirus disease 2019 (COVID-19). JAMA Cardiology **5**
11. Chen, T., et al.: Retrospective study. BMJ **368**(2020), m1091 (2019). https://doi.org/10.1136/bmj.m1091
12. Huang, C., et al.: Clinical features of patients infected with 2019 novel coronavirus in Wuhan, China. The Lancet **395**(10223), 497–506 (2020). https://doi.org/10.1016/S0140-6736(20)30183-5
13. Khader, F., et al.: Multimodal deep learning for integrating chest radiographs and clinical parameters: a case for transformers. Radiology **309**(1) (2023a). https://doi.org/10.1148/radiol.230806
14. Wu, X., et al.: CTransCNN: combining transformer and CNN in multilabel medical image classification. Knowledge-Based Systems **281** (2023). https://doi.org/10.1016/j.knosys.2023.111030
15. Xiao, F., Liu, H., Lu, J.: A new approach based on a 1d+ 2d convolutional neural network and evolving fuzzy system for the diagnosis of cardiovascular disease from heart sound signals. Appl. Acoust. **216**, 109723 (2024)
16. Ma, S., Chen, J., Ho, J.W.K.: An edge-device-compatible algorithm for valvular heart diseases screening using phonocardiogram signals with a lightweight convolutional neural network and self-supervised learning. Comput. Methods Programs Biomed. **243**, Art. no. 107906 (2024)
17. Ukwuoma, C.C., et al.: A hybrid explainable ensemble transformer encoder for pneumonia identification from hest X-ray images. Journal of Advanced Research **48**, 191–211 (2023). ISSN 2090–1232. https://doi.org/10.1016/j.jare.2022.08.021
18. Mondal, A.K.: COVID-19 prognosis using limited chest X-ray images. Applied Soft Computing **122**, 108867 (2022). ISSN 1568–4946. https://doi.org/10.1016/j.asoc.2022.108867
19. Greenhalgh, T., Knight, M., A'Court, C., Buxton, M., Husain, L.: Management of post acute covid 19 in primary care. BMJ **370** (2020). https://doi.org/10.1136/bmj.m3026
20. Antipolis, S.: COVID-19 patients retain elevated risk of death for at least 18 months after infection. European Society of Cardiology (2023). www.heart.org
21. Loey, M., Smarandache, F., Khalifa, N.E.M.: Within the lack of chest COVID-19 X-ray dataset: a novel detection model based on GAN and deep transfer learning. Symmetry **12**(4), 651 (2020)
22. Carion, N., Massa, F., Synnaeve, G., Usunier, N., Kirillov, A., Zagoruyko, S.: End-to-End Object Detection with Transformers. In: Vedaldi, A., Bischof, H., Brox, T., Frahm, J.M. (eds.) Computer Vision – ECCV 2020. ECCV 2020. Lecture Notes in Computer Science, vol 12346 (2020). Springer, Cham. https://doi.org/10.1007/978-3-030-58452-8_13
23. Liu, J., Guan, S., Zou, Q., Wu, H., Tiwari, P., Ding, Y.: AMDGT: attention aware multi-modal fusion using a dual graph transformer for drug–disease associations prediction. Knowledge-Based Systems **284** (2024). https://doi.org/10.1016/j.knosys.2023.111329

24. Thangaraj, R., Pandiyan, P., Ramakrishnan, J., Nallakumar, R., Eswaran, S.: A deep convolution neural network for automated covid-19 disease detection using chest x-ray images. Healthcare Analytics **4**, 100278 (2023)
25. Rahimzadeh, M., Attar, A.: A modified deep convolutional neural network for detecting COVID-19 and pneumonia from chest X-ray images based on the concatenation of Xception and ResNet50V2. Informatics in Medicine Unlocked **19**, 100360 (2020). ISSN 2352-9148. https://doi.org/10.1016/j.imu.2020.100360
26. Dhakal, B.P., Sweitzer, N.K., Indik, J.H., Acharya, D., William, P.: SARS CoV-2 infection and cardiovascular disease: COVID-19 heart. Heart Lung Circ. **29**(7), 973–987 (2020)
27. Wynants, L., et al.: Prediction Models for Diagnosis and Prognosis of Covid-19: Systematic Review and Critical Appraisal. MJ **369**, m1328 (2020). https://doi.org/10.1136/bmj.m1328

Malware Variant Traffic Identification at an IoT Edge Gateway Using QDCNN with Optimization

Arumugam Gonda[1(✉)], Srinath Doss[2], and Kamal Kant Hiran[3]

[1] Department of Information Technology, Azteca University, Mexico, Mexico
Arumugam.g31@gmail.com

[2] Faculty of Engineering and Technology, Botho University, Gaborone, Botswana
srinath.doss@bothouniversity.ac.bw

[3] School of Computer Science and IT, Symbiosis University of Applied Sciences, Indore, India

Abstract. Malware is software utilized for undergoing various malicious activities, including the theft of money and passwords. The Internet of Things (IoT) devices are affected by various losses because of dominant malware variants. In addition, malware is harmful in several ways, such as stealing, deleting, or encrypting data, and a system becomes unusable or locked. Malware variant traffic identification is done by malware detectors by various anti-malware methods, such as virus signature scanners and heuristic methods. However, they suffer from several limitations including computational complexity, large number of training data, and vulnerability. Thus, Quantum Dilated Convolutional Neural Network (QDCNN) with Wild Horse Optimization (WHO) is introduced to address these problems. At first, a system model of IoT Edge Gateway is considered, and then, log file collection is done. After that, features like network-oriented features and traffic-oriented features are mined from the collected log file. Later, the extracted features are forwarded to an augmentation process, where oversampling is exploited to address the class imbalance problem. At last, malware variant traffic identification is effectuated by utilizing the proposed WHO_QDCNN. Moreover, the developed WHO_QDCNN attained better precision, accuracy, and recall of 93.77%, 90.05%, and 94.66%, respectively.

Keywords: Malware Variant Traffic Identification · Quantum Dilated Convolutional Neural Network · Wild Horse Optimization · Internet of Things · traffic-oriented features

1 Introduction

The Internet of Things (IoT) devices play a very important role in human life. However, they are targeted by various attacks. As IoT devices are susceptible to attacks, attackers use these devices to leak the private data of the users who use IoT devices or cause huge losses by initiating Distributed Denial-of-Service (DDoS) attacks. It was estimated that in the year 2017, the number of attacks on IoT increased by 172% [12]. IoT attacks become most significant related to the management of network security. The IoT gateways are important points for the detection of malware by analyzing the traffic passing in

R. Geetha et al. (Eds.): AAIMB 2024, CCIS 2654, pp. 398–422, 2026.
https://doi.org/10.1007/978-3-032-13029-7_29

IoT [1]. It is essential to integrate safety mechanisms, like firewalls, malware detection systems, or any antivirus software to secure critical data from malicious attacks in individual networks and the networks of corporate organizations. Various forms of malicious attacks cannot be prevented if these networks depend only on traditional firewalls [2]. Nevertheless, the identification of malware variant traffic accurately and efficiently in IoT is challenging usually, the protection software is misled by using invisible malware variants [3]. However, during the detection of attacks on IoT, most of the conventional schemes suffer from low performance while encountering new and unknown attacks, as the attack signatures are different and the techniques fail to identify them [1, 6]. Due to the constant development in network technology, the complexity of cyber threats has increased. Some of the general attacks on centralized communication devices are Man-in-the-Middle (MitM), intrusion, Denial-of-Service (DoS), and scanning [5]. Evasion techniques, such as polymorphic malware are added by the attackers to traditional malware, which increases the complexity of accurate identification [1, 13].

The deployment of a Network Intrusion Detection System (NIDS) at the IoT edge gateway is utilized to analyze the communication traffic that can block a malware intrusion to a certain extent [1]. Due to the deficiency of empirical data associated with the recent IoT and the knowledge of the behavioral characteristics of malware-infected devices, attaining effective solutions to address security issues is highly complex. Various IoT-specific honeypots are set up to collect accurate information from existing IoT malware [4]. NIDS is considered an important tool for system administrators for the detection of various security breaches inside the network of an organization [6]. Especially, for the identification of malicious behavior and raising frights on the Internet, there are a few intrusion detection systems [14]. However, in the real scenarios, the process of detection is complicated. However, most of the methods, such as the model using Machine Learning (ML), need a huge number of labeled samples to accomplish the process of training in recent times. However, only a limited number of malicious behaviors are available for training [3]. Besides, the ML-based classifiers require domain expertise continuously and the extraction of feature and assessment activities consumes more time [4]. Deep Learning (DL) is a subdivision of ML using which outstanding performance can be achieved. DL methodologies are effective at handling big data compared to traditional ML techniques. Furthermore, DL methods are practical, operate in an end-to-end fashion, and can learn representations of features automatically from the raw data [7, 15].

This paper introduces an efficient model for malware variant traffic identification utilizing QDCNN with WHO. Initially, a system model is considered, and the log file collection process is carried out. Afterwards, the log file data collected is subjected to feature extraction, where the traffic-oriented features and network-oriented features are extracted. Then, the augmentation process is carried out by using an oversampling model. Finally, malware variant traffic identification is done by utilizing the proposed WHO and QDCNN.

The key contribution of this paper is given below:

Proposed WHO_QDCNN for malware variant traffic identification: The identification of malware variant traffic is established using WHO_QDCNN introduced in this paper. Here, the malware variant traffic is identified efficiently based on traffic-oriented features and network-oriented features. Further, malware variant traffic identification is carried out using the QDCNN trained using WHO.

The remaining sections of this research are: the literature survey of traditional methods for malware variant traffic identification along with the limitations and advantages illustrated in Sect. 2. The system model is explained in Sect. 3. The introduced technique for an effective identification of malware variant traffic is demonstrated in Sect. 4. In Sect. 5, the results of existing methods are explained and the work is concluded in Sect. 6.

2 Motivation

This segment proves the literature survey of traditional methods of malware variant traffic identification. The researchers investigate several models for providing an efficient model for malware variant traffic identification. However, none of the methods had effectively identified malware variant traffic. To address this issue, WHO_QDCNN is developed for the identification of malware variant traffic effectively.

2.1 Literature Survey

The literature reviews of the traditional Malware Variant Traffic Identification at an IoT Edge Gateway are illustrated below: Hu, X., *et al.* [1] proposed a Deep Subdomain Adaptation Network with Attention (DSAN-AT) model for malware variant traffic identification at an IoT edge gateway. This model was fast and accurate in identifying malware variant traffic. However, this method failed to improve online transfer learning performance. Indrasiri, P.L., *et al.*, [2] developed a stacked model called Extra Boosting Forest (EBF) to detect malicious traffic from multi-domain data. This model achieved significantly better performance, handled complex datasets, and was statistically significant. Nevertheless, this method failed to integrate data from various domains effectively. He, M., *et al.*, [3] devised a model based on Convolutional Neural Networks (CNN) and Auto Encoder (AE) for few-shot malicious traffic detection. This method required fewer samples for pre-training, was easy to reconstruct, and had low computational time. In contrast, this method failed to focus on real-time network traffic. Ali, S., *et al.*, [4] proposed a Long Short-Term Memory (LSTM) based model for IoT malware detection and identification. This method could gather several devices from multiple sources and was more effective. However, this approach was not applied to real-time scenarios.

2.2 Challenges

- DSAN-AT in [1] effectively identified malware variant traffic at the IoT edge gateway. However, this method failed to encapsulate the fine-grained details during classification.
- EBF in [2] significantly attained higher accuracy while detecting malicious traffic from multi-domain data. Nevertheless, this method failed to leverage various attacks in the automatic system and prevent system damage.

- CNN and AE in [3] excellently solved the issues due to abnormal samples. However, this method failed due to a large number of labeled samples.
- Precise and early detection of malware variant traffic at an IoT edge gateway is required as it can effectively reduce the loss. Several automatic approaches to detecting malware traffic are available, nevertheless, these schemes suffer from low accuracy and efficiency in detecting malware from multi-domain systems.

3 System Model

Malware detection methods are employed in IoT gateways to avoid malware due to network traffic. A small quantity of label data is collected to prevent unknown malware variants depending on source domain data. Malware detection is utilized for providing cyber security by identifying and mitigating threats by malicious nodes. The training resources of the model are stored by classification domains with an attention module. Finally, an accurate identification of malware variants is identified malicious modes. The system model [1] for malware variant traffic identification is shown in Fig. 1.

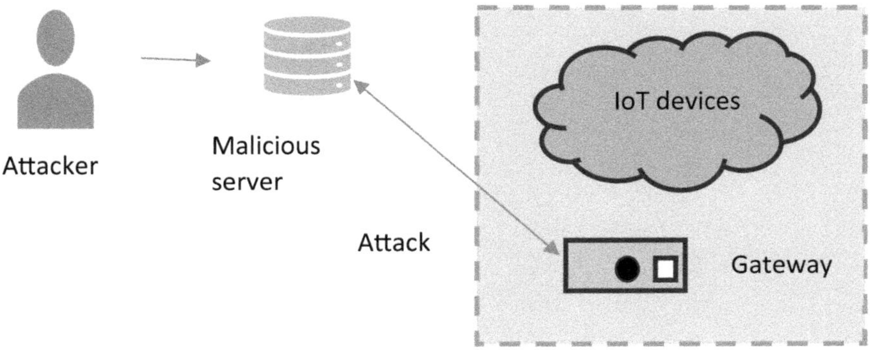

Fig. 1. System model for malware variant traffic identification for IoT devices

4 Proposed WHO_QDCNN for Malware Variant Traffic Identification

IoT suffers from several losses due to prevailing malware variants. Thus, it is very important to prevent network attacks and provide robust network security.

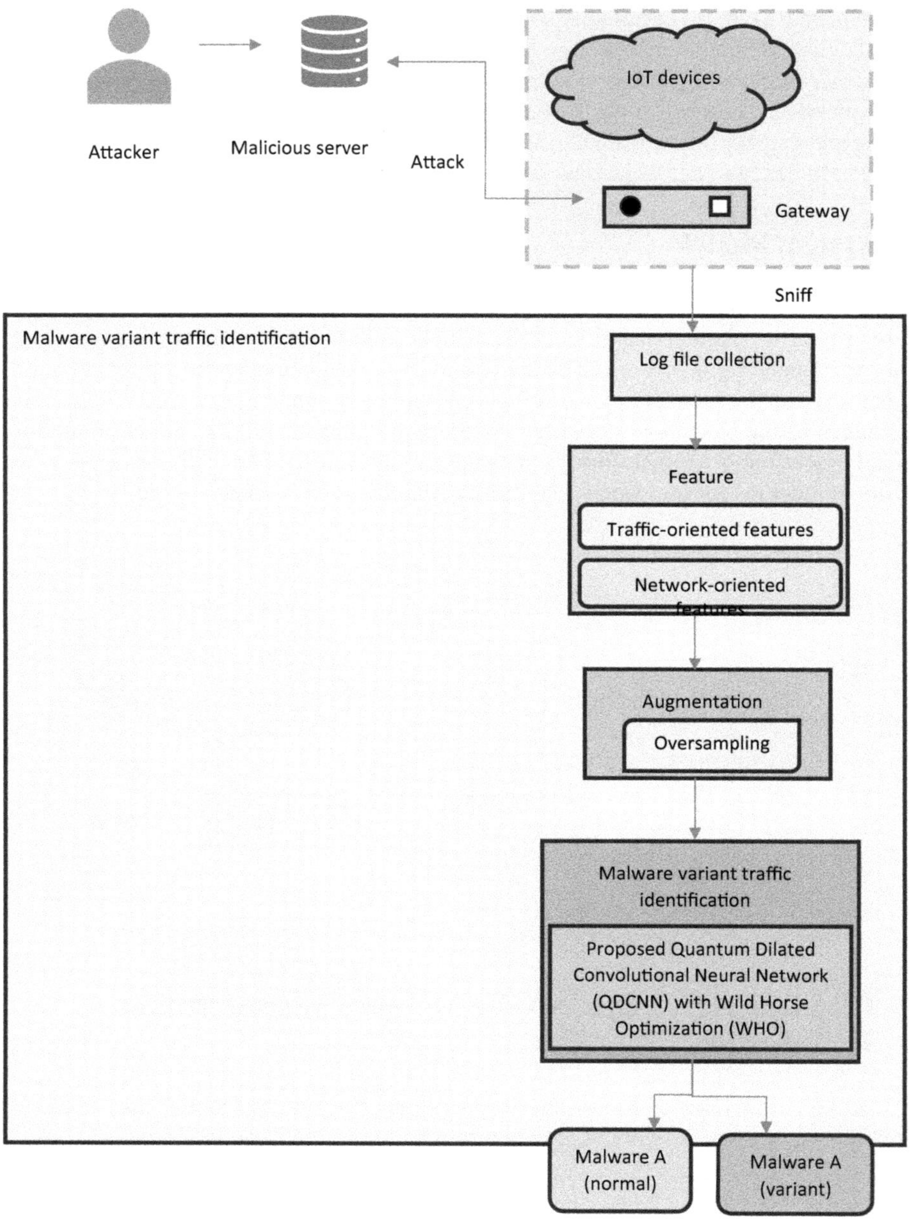

Fig. 2. Block diagram of proposed WHO_QDCNN for malware variant traffic identification

To identify the malware attacks, a DL model called WHO_QDCNN is proposed in this research for malware variant traffic identification. At first, a system model is considered, and then, log file collection is performed. Then, the collected data is fed into a feature extraction process, where the features, like traffic-oriented and network-oriented features, are extracted. After that, the features are applied to data augmentation by using the oversampling method [8]. Finally, malware variant traffic identification is done by the proposed QDCNN [9] with WHO [10]. The block diagram of the proposed WHO_QDCNN for malware variant traffic identification is represented in Fig. 2.

4.1 Log Data Acquisition

Let us consider the input log data is assimilated from the dataset Z, which is given by,

$$Z = \{z_1, z_2, ...z_a, ...z_b\} \tag{1}$$

where, z_a denotes the ath log file, and b refers to the number of records.

4.2 Feature Extraction

The collected log file z_a is fed into feature extraction, which is the process of selecting and identifying an essential feature from a dataset to remove unwanted data and reduce complexity. The feature extraction is carried out by mining the traffic-oriented features and network-oriented features.

4.2.1 Traffic-Oriented Features

The traffic-oriented features are processed by considering the packet size, flow duration, and flow byte count and they are:

Packet Size The quantity of data given into a network packet through a Transmission Control Protocol (TCP) or an Internet Protocol (IP) network is called packet size (B_1). A packet size contains the size of the packet trailer, header, and data.

Flow Duration The time duration between the received time of the last seen packet and the received time of the first seen packet is called flow duration (B_2).

Flow Byte Count Flow byte count (B_3) estimates the size of TCP flows, which are considered from the sequence number.

4.2.2 Network-Oriented Features

The network-oriented features mined include IP address, port number, query type, respond code, and TTL and they are given below:

IP Address IP address (B_4) is the exclusive recognizing number allocated to each system which is interconnected to the internet for communication. The key purpose of the IP address is location addressing and network interface identification.

Port Number The port number (B_5) is the number that identifies an exact service on a host in a TCP or IP network. Ports are utilized to allocate multiple processes to run on the same host and are distinguished based on the port number.

Query Type Query type (B_6) is the kind of information that sends requests to a Domain Name System (DNS) query and is utilized in a computer network for communication.

Respond Code A response code (B_7) is a three-digit code in which a server sends a code to the client and the client responds to that request. Here, the code denotes the request status and specifies how the server responds. In addition, the response code is very fast for communication.

Time-To-Live (TTL) TTL (B_8) is an arithmetical value that estimates the timespan of packet data and ranges from a few seconds to minutes. If the event count or timespan expires, the packet is either revalidated or discarded.

The overall features are combined to obtain a feature vector B^* given by,

$$B^* = \{B_1, B_2, B_3, ...B_8\} \tag{2}$$

4.3 Data Augmentation

The process of producing new data from prevailing data artificially mainly for training new methods is called data augmentation. Here, the feature vector B^* is provided as input and then augmented data Y is obtained by the data augmentation process. Data augmentation is carried out by the oversampling method [8] to address the class unbalance issues.

4.3.1 Oversampling

The quantity improvement of minority class samples or instances by generating new instances or repeating some instances is performed by oversampling. Oversampling is also called a non-heuristic algorithm. Balancing class spreading by random repetition for minority target instances is the key objective of oversampling.

4.4 Malware Variant Traffic Identification

Malware detection is crucial for network organizations against unknown access, service disruption, and data breaches. In addition, malware detection safeguards consumer trust and digital assets. Here, malware variant traffic identification is done by the proposed WHO_QDCNN. Here, the augmented data Y is given to QDCNN [9], which is trained by WHO [10], and then finally, the data is identified as a variant or normal malware.

4.4.1 Architecture of QDCNN

QDCNN [9] is processed with several convolution layers and functions by convolution operation, dilated convolution, and quantum convolution. It aims to enhance the model's learning capacity by expanding the receptive field and capturing long-range dependencies more efficiently. Here, the augmented data Y is considered as input and fed into QDCNN.QDCNN consumes less time for execution, reduces parameters, and requires less training time with high accuracy. The processes carried out in the QDCNN are detailed below.

Convolution Operation The convolution operation is applicable for detecting local features of data in specific locations. The augmented data Y is provided as input and the convolution function carried out in the convolution layer plays a vital role in CNNs. Convolution is the linear operation of CNN, which involves a group of multiplication of weights along with input. Here, a filter or kernel, which is a two-dimensional (2D) array of weights, is utilized for feature extraction. Then the filter is fed into a filter-size patch of augmented data, which is called a receptive field and the dot product between the receptive field and weight values is calculated. Then, the filter is shifted by the next patch regarding step size known as the stride ***i***, and the above process is repeated till the filter is applied across the entire image. Thus, the final output from a feature map (c) is estimated by finding the dot product between the input data (e) and filter (d)and is given by,

$$c[f_1^*, f_2^*] = \sum_{g_1} \sum_{g_2} d[g_1, g_2] \cdot e[f_1^* + g_1, f_2^* + g_2] \tag{3}$$

where, f^*_1 and f_2^* denote the location index of c. The dimension is reduced by using the zero-padding method by the addition of pixel values with value zero around the input data edge. Then, padding j estimates the number of zero values to be added at the edge. Then, the spatial resolution h_a and h_b is expressed as,

$$h_a = \left(\frac{f_{1_a}^* - k_1 + 2j}{i} \right) + 1 \tag{4}$$

$$h_b = \left(\frac{f_{2_b}^* - k_2 + 2j}{i} \right) + 1 \tag{5}$$

where, f_a^* x f^*_{bb} is an image input by k_1x k_2 kernel *i* denotes stride, and *j* signifies padding.

Dilated Convolution Dilated convolution is a kind of convolution that injects holes like points with zero weight value to expand kernel elements. The dilated convolution is estimated by a hyperparameter called dilation rate (l), which is the extra hyperparameter developed by dilated convolution to determine the stride *i* with input data. Here, the $l-1$ zero values are applied between two consecutive filter values if the dilation rate is l, and Eq. (3) is rephrased as,

$$c[f_1^*, f_2^*] = \sum_{g_1} \sum_{g_2} d[g_1, g_2] \cdot e[f_1^* + g_1.l, f_2^* + g_2.l] \tag{6}$$

Thus, the dilated convolution is capable of capturing a larger receptive field without learnable parameters with the same kernel size. When applying dilated convolution, Eqs. (4) and (5) are also reformulated as,

$$h_a = \left(\frac{f_{1_a}^* - k_1 - (k_1 - 1)(l - 1) + 2j}{i} \right) + 1 \tag{7}$$

$$h_b = \left(\frac{f_{2_b}^* - k_2 - (k_2 - 1)(l - 1) + 2j}{i} \right) + 1 \tag{8}$$

Quantum Convolution Quantum convolution is a convolution based on quantum circuits and it contains three kinds of modules and they are encoding module, entanglement module, and decoding module.

a) *Encoding Module*

In an encoding module, data is encoded in a quantum state and processed into a quantum convolutional circuit. There are various encoding techniques like amplitude encoding, basis encoding, and angle encoding. Here the angle encoding is utilized in the encoding module. Let us consider the input as the rotation angle of a single-qubit rotation gate B^*. Also, multiple input variables, like m_1, m_2 and m_3 are encoded with $A^*(m_1), B^*(m_2), C^*(m_3)$ on single-qubit. This angle encoding is also known as dense angle encoding or multiple variables /qubit encoding. The input vector e is encoded with the encoding operator into a quantum state given by,

$$|e\rangle = X(e)|0\rangle \tag{9}$$

Here, $X(e)$ converts the initial state into a superposition state by Hadamard gate.

b) *Entanglement Module*

Here, a set of multi-qubits and single-qubit gates are subjected to an encoded quantum state from the encoded module. The parametric controlled rotation gates $E^*(\theta)$ and CNOT gate are multi-qubit gates, which are utilized to produce entangled state or correlated quantum state and θ is the trainable parameter. The unitary operation $W(\theta)$ is applied in the entanglement module, and the output state is given by,

$$|e, \theta\rangle = W(\theta)|e\rangle \tag{10}$$

c) *Decoding Module*

The decoding module evaluates local observable operators $V^{\otimes p}$ like the Pauli U operator $\sigma_U^{\otimes p}$ for measuring the final quantum state $|e, \theta\rangle$ from the entanglement module. The output vector $q(e, \theta)$ is given by,

$$q(e, \theta) = \langle e, \theta|V^{\otimes p}|e, \theta\rangle \tag{11}$$

where, p is smaller or equal to the total number of qubits in the quantum state, and a classical output vector

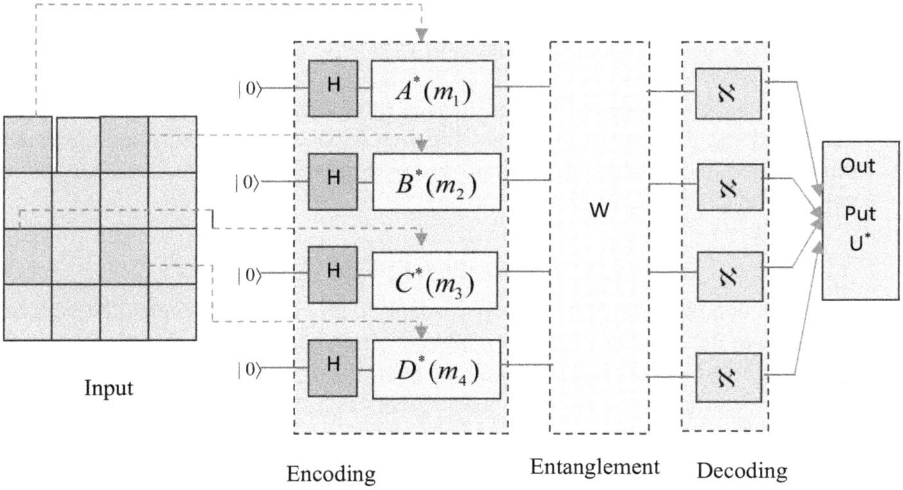

Fig. 3. QDCNN architecture

Here, an output vector $q(e, \theta)$ is considered an input feature and fed into the ensuing layers of QDCNN. The output thus obtained is indicated as U^*. The architecture of QDCNN is shown in Fig. 3.

4.4.2 Training of QDCNN with WHO

QDCNN [9] is trained by WHO [10] utilizing various stages, such as population initialization, grazing behavior, horse mating behavior, group leadership, and leader selection and exchange. WHO is the metaheuristic swarm-based algorithm, which is stimulated depending on the social behavior of horses, including grazing, domination, leadership hierarchy, and mating. This algorithm had a low computational cost and was easy to use. Further, the approach was applied in solving various real-time issues effectively. The following details the algorithmic steps of the WHO.

i) *Initialization of Population*

The population initialization is the primary step in the WHO algorithm and it is established by the population size A and stallion count GK from the population and the number of groups S is expressed as,

$$S = A \times GK \tag{12}$$

ii) *Fitness*

The fitness function is utilized to estimate the best optimal solution. Here, Mean Square Error (MSE) is used to calculate the fitness function, and is expressed as,

$$MSE = \frac{1}{e*} \sum_{\lambda=1}^{e*} \left(U_\lambda^* - U_\lambda\right)^2 \tag{13}$$

where, $e*$ signifies the number of training samples and U_λ denotes the output of QDCNN, U_λ^* is the expected output.

iii) *Grazing Behavior*

This phase is modelled by considering the grazing behavior before the foals reach puberty. The center of the grazing area is based on stallions and the center region is neighboured by the remaining group members. The residual group member neighbor at center region is expressed as,

$$B_{r+1,S}^t = 2F \times \cos(2\pi HF) \times (G^t - B_{r,S}^t) + G^t \tag{14}$$

where, F denotes arbitrarily selective adaptive procedure, t denotes the stallion, t represents the number of group members, H indicates arbitrary values ranging between 2 and -2, G^t signifies stallion position, $B_{r,S}^t$ and $B_{r+1,S}^t$ stipulates the member location from existing or subsequent iterations.

iv) *Horse Mating Behavior*

After reaching the age of puberty, horse mating behavior begins. The foals may leave the group and integrate with another group to prevent the fathers from marrying their daughters and sisters. This process is modeled as,

$$B_{S,n}^o = Mean(B_{S,r}^x, B_{S,t}^y) \text{ and } r \neq t \neq n \tag{15}$$

where, $B_{S,n}^o$ denotes the position of the horse o from a group n, $B_{S,r}^x$ signifies the location of the foal x from the group r, and $B_{S,t}^y$ represents the position of the foal y from a group t.

v) *Group leadership*

The group member is instructed by the group stallion to move towards the waterhole in this phase. The stallion fights along with other stallions to compete for the waterhole, and the dominant group uses the waterhole and the other group doesn't have access to water until the leading group leaves the place. This behavior is expressed as,

$$G_{r+1,E} = \begin{cases} 2F \times \cos(2\pi HF) \times (IJ - G_{r,E}) + IJ \ \mathit{if}\ u_1 > 0.5 \\ 2F \times \cos(2\pi HF) \times (IJ - G_{r,E}) - IJ \ \mathit{if}\ u_1 \geq 0.5 \end{cases} \tag{16}$$

where, $G_{r,E}$ and $G_{r+1,E}$ are existing leader position (stallion) in the rth group in the current and next iteration, IJ represents waterhole location, u_1 is a random number ranging within [0,1].

vi) *Leader Exchange and Selection*

To attain an optimal fitness value, the group leader is wisely selected by the optimal leader among all leaders and is determined as,

$$G_{r,E} = \begin{cases} B_{r,E} \ \mathit{if} \cos t(B_{r,E}) < \cos t(G_{r,E}) \\ G_{r,E} \ \mathit{if} \cos t(B_{r,E}) > \cos t(G_{r,E}) \end{cases} \tag{17}$$

vii) *Feasibility Evaluation*

The position of the member is updated by the WHO algorithm and this algorithm is applied to the next iteration and the position of the leader is updated until it reaches the last iteration. When the updated position is better than other positions, then that position is considered as newly updated position.

viii) *Termination*

At last, the above steps are repeated and the best result is recorded. The execution steps and pseudo code are demonstrated in Algorithm 1.

WHO algorithm's pseudo code
Primary horse population arbitrary initialization
Initialization of parameter
Evaluate fitness
Generate foal sets and choose stallions.
Find the Optimal horse.
When the ending condition is not met
Measure Time Domain Reflectometry (TDR)
For each stallion
Determine adaptive mechanism (F)
To number of foals in several groups
If rand > Crossover Percentage (PC)
Update foal position using equation (15)
Else
Update foal position using equation (16)
End
End
Update leader position using equation (17)
Exchange leader using equation (18)
Foal arrangement for a group by cost
Least cost foal election
If $cost(B_{r,E}) > cost(G_{r,E})$
Swap stallion and foal position.
End
End
Upgrade optimal
End

5 Results and Discussion

The WHO_QDCNN proposed for malware variant identification in this work is estimated for its efficiency with several metrics and a brief explanation of this is provided as follows:

5.1 Experimental Setup

The implementation of the proposed WHO_QDCNN is executed on a PC with Windows 10 OS, 8 GB RAM, more than 100 GB ROM, and 1.7 GHz CPU by utilizing the PYTHON Version: 3.9.11. The parameters of the WHO_QDCNN are described in Table 1.

Table 1. Parameter details

Parameter	Value
Epoch	50
Optimizer	WHO
Batch size	16
Activation function	softmax
Dropout	0.2
Kernel size	(3,3)
Loss	categorical cross entropy
Population size	40
Dimension	10
Lower bound	0
Upper bound	1
Maximum Iteration	500

5.2 Description of Dataset

Malware capture facility dataset: The dataset utilized by WHO_QDCNN is the malware capture facility dataset [11]. This dataset is used for analysing and capturing long-lived and real malware traffic. The dataset contains information about the infection date, malware, capture name, and capture URL.

IoT-23 Dataset: The IoT-23 dataset [16] is a labeled collection of IoT network traffic, developed by the Stratosphere Laboratory, AIC group at CTU University, Czech Republic, with support from Avast Software. It contains 23 network traffic captures, 20 from infected IoT devices running various malware samples on Raspberry Pi, and 3 from benign IoT devices. The dataset is designed to facilitate machine learning-based research in IoT malware detection by providing real, diverse, and labeled network traffic collected in a controlled environment.

5.3 Explainability Analysis

Figure 4 shows the explainability analysis using SHapley Additive exPlanations (SHAP). SHAP is a popular explainable AI (XAI) technique used to understand the predictions made by machine learning models. Figure 4a) shows the mean plot, which shows the average impact of each feature on the model's output. Here, id.orig_h has the highest influence, suggesting the originating host significantly affects the malware classification. The maximum plot is shown in Fig. 4b) and it shows the highest individual contribution (positive or negative) to each feature made to any single prediction. In this plot, id.orig_h has the strongest maximum impact (+1.23), indicating it can heavily influence the model's decision in certain cases. Features like RA, RD, rcode_name, and query show no maximum effect on any prediction. Figure 4c) depicts the summary plot of the model, which helps to visualize both the importance and effect of each feature in a simple, interpretable way. Features like id.orig_h, ctype_name, and qclass_name have the most impact, with id.orig_h being the strongest contributor. Red dots represent high feature values, while blue dots indicate low values.

The heatmap plot is provided in Fig. 4d) and it provides a clear, instance-wise view of how feature values affect model decisions. In this plot, each row represents a feature, and each column represents a specific input instance. Red areas indicate a positive impact on the prediction, while blue areas indicate a negative impact. Figure 4e) represents the average impact plot, which shows the contribution of each feature to a single prediction. The SHAP average impact plot above shows the contribution of each feature to a single prediction. Here, id.orig_h has the highest positive impact, pushing the model's output toward predicting malware. In contrast, features like qtype_name and proto have negative impacts. The waterfall plot is described in Fig. 4f) and it clearly explains how the model combined feature effects to make its final decision. It starts from a base value and adds or subtracts the effect of each feature to reach the final output. In this case, id.orig_h had the highest positive impact, increasing the prediction score, while features like qtype_name, proto, and id.resp_h slightly reduced it. Figure 4g) implies the decision plot, which helps trace how feature impacts accumulate to shape each prediction. Here, each line represents a prediction path, starting from the base value and moving toward the final output. The thickness of the paths shows the strength of influence, and the color indicates feature values.

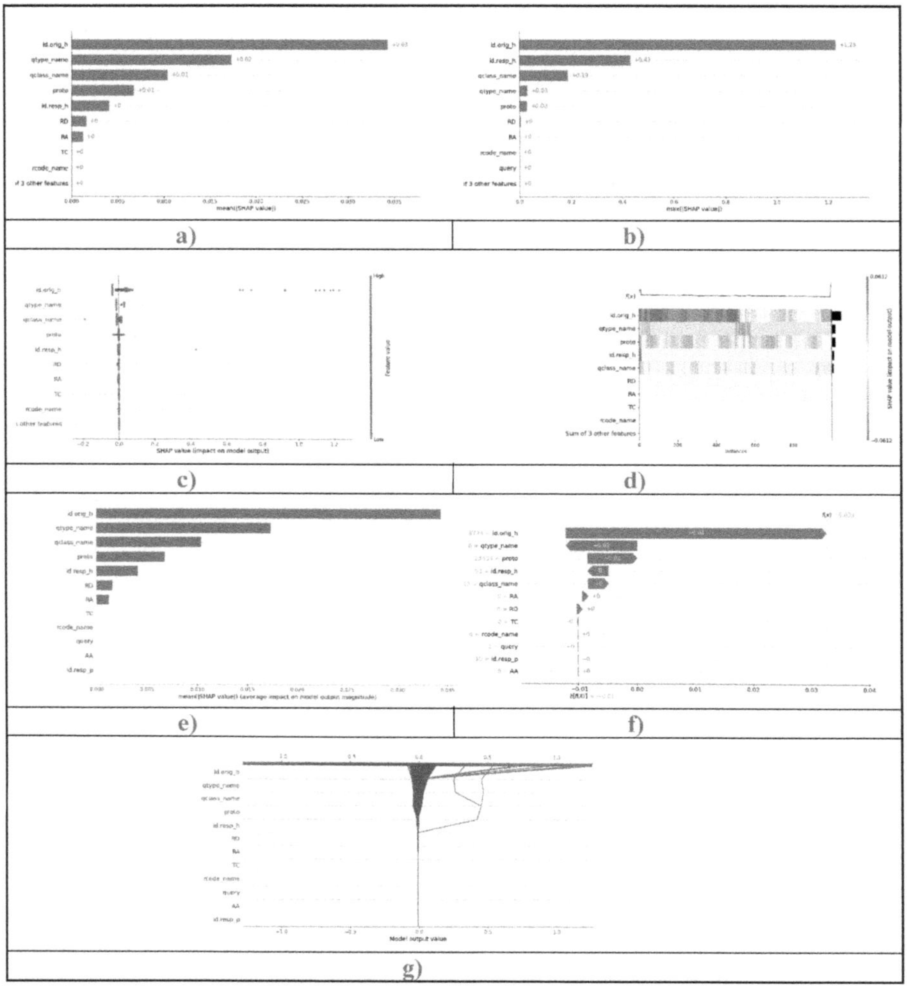

Fig. 4. Explainability analysis using SHAP a) Mean plot, b) Maximum plot, c) Summary plot, d) Heatmap plot, e) Average impact plot, f) Waterfall plot, and g) Decision plot

5.4 Evaluation of Metrics

WHO_QDCNN for malware variant traffic identification is analysed for its effectiveness considering three metrics, namely accuracy, precision, and recall.

5.4.1 Accuracy

Accuracy is used to measure the efficacy of the WHO_QDCNN in detecting malware and normal events correctly, and it is given by,

$$Accuracy = \frac{TP + TN}{TP + TN + FP + FN} \tag{18}$$

where, *TP* represents true positive, *TN* indicates true negative, *FP* denotes false positive, and *FN* is false negative.

5.4.2 Precision

The ratio of true positive predictions to the total number of positive predictions is called precision and it is expressed as,

$$\Pr ecision = \frac{TP}{TP + FP} \tag{19}$$

5.4.3 Recall

The measurement of the ratio of true positive value prediction with the total number of original positive occurrences is called recall and is measured by,

$$\text{Re}call = \frac{TP}{TP + FN} \tag{20}$$

5.5 Comparative Techniques

WHO_QDCNN is evaluated by comparing it with various methods, such as DSAN-AT [1], EBF [2], CNN and AE [3], LSTM-based DL [4], synthetic data generated via Generative Adversarial Networks (SYN-GAN) [17], and Nuclear Reactor Optimization-based Support Vector Machine (NRO_SVM) [18].

5.6 Comparative Analysis

The malware variant traffic identification is estimated by WHO_QDCNN with metrics, such as recall, precision, and accuracy considering two datasets and it is discussed below:

5.6.1 Evaluation Using Malware Capture Facility Dataset

a) **Based on the learning set**

The evaluation of WHO_QDCNN regarding the Malware capture facility dataset and the learning set is demonstrated in Fig. 5. Figure 5 a) reveals the estimation of accuracy for WHO_QDCNN. With a learning set of 90%, by traditional methods, such as DSAN-AT, EBF, CNN and AE, LSTM-based DL, SYN-GAN, and NRO_SVM measured accuracy of 78.17%, 83.34%, 87.33%, 89.78%,90.43%, and 91.54%, and accuracy recorded by proposed WHO_QDCNN is 92.88%. Here, the accuracy of WHO_QDCNN is 6.27% better than LSTM-based DL. The estimation of precision for WHO_QDCNN is exploited in Fig. 5 b). The precision recorded with a learning set of 90% by existing methods, namely DSAN-AT is 74.03%, EBF is 77.03%, CNN and AE is 81.23%, LSTM-based DL is 84.15%, SYN-GAN is 86.67%, and NRO_SVM is 88.77%, and the proposed WHO_QDCNN is 89.22%. In this, the precision of WHO_QDCNN is 5.16% better than CNN and AE. In Fig. 5 c), the evaluation based on the recall of WHO_QDCNN is represented. Moreover, with a learning set of 90% the methods, including DSAN-AT, EBF, CNN and AE, LSTM-based DL, SYN-GAN, and NRO_ SVM figured recall of 78.20%, 81.97%, 84.21%, 87.14%, 89.77%, and 91.88%, and proposed WHO_QDCNN is 93.88%. The recall of WHO_QDCNN improved by 8.00% to LSTM-based DL respectively.

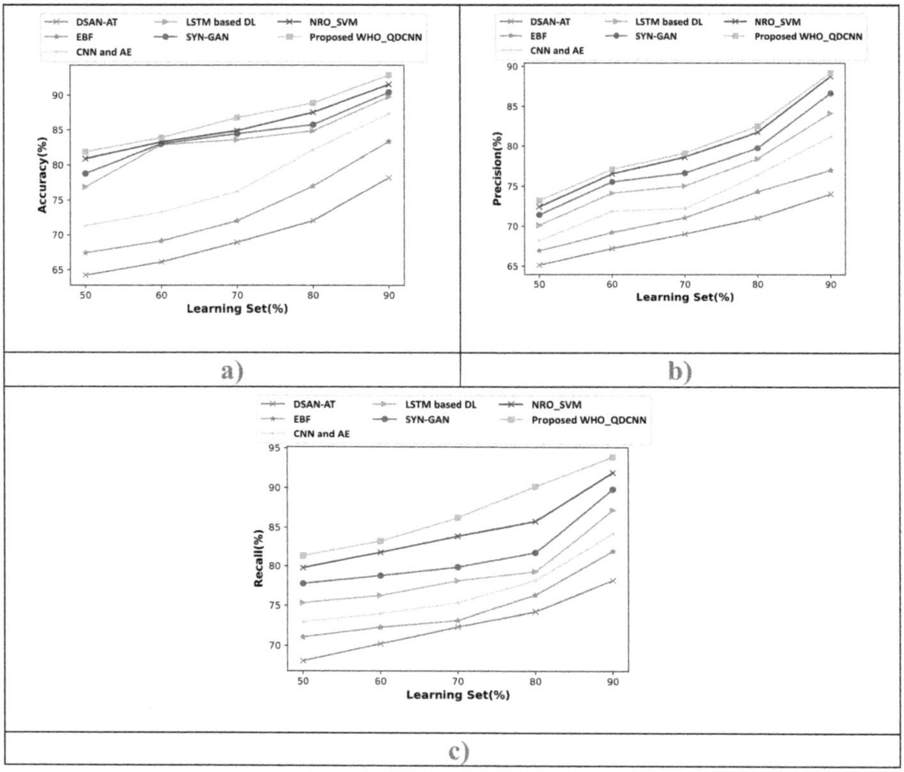

Fig. 5. Estimation of WHO_QDCNN using Malware capture facility dataset a) accuracy, b) precision, and c) recall with learning set

b) **Based on k-group**

Figure 6 reveals the estimation of WHO_QDCNN with respect to the Malware capture facility dataset based on k-group. In Fig. 6 a) the evaluation of accuracy for WHO_QDCNN is figured. The accuracy figured by existing methods, namely DSAN-AT, EBF, CNN and AE, LSTM-based DL, SYN-GAN, and NRO_SVM with k-group value of 9 is 77.22%, 79.04%, 88.78%, 91.77%, 92.12%, and 92.78%, and the proposed WHO_QDCNN measured accuracy of 93.77%. Now, the accuracy of WHO_QDCNN is 5.95% better than DSAN-AT. Figure 6 b) reveals the estimation of precision for WHO_QDCNN. Additionally, the precision attained with the k-group value of 9 by techniques, like DSAN-AT, EBF, CNN and AE, LSTM-based DL, SYN-GAN, and NRO_SVM is 78.06%, 81.33%, 83.27%, 86.94%, 87.35%, and 89.33%, and the proposed WHO_QDCNN is 90.05%. Here, the precision of WHO_QDCNN is improved by 5.28% than LSTM-based DL. The estimation of recall for WHO_QDCNN is shown in Fig. 6 c). The methods, such as DSAN-AT, EBF, CNN and AE, LSTM-based DL, SYN-GAN, NRO_SVM and WHO-QDCNN measured recall of 79.08%, 81.25%, 84.37%, 87.22%, 88.78%, 91.77%, and 94.66% with k-group of 9. Also, the recall of WHO_QDCNN is 5.74% more than CNN and AE.

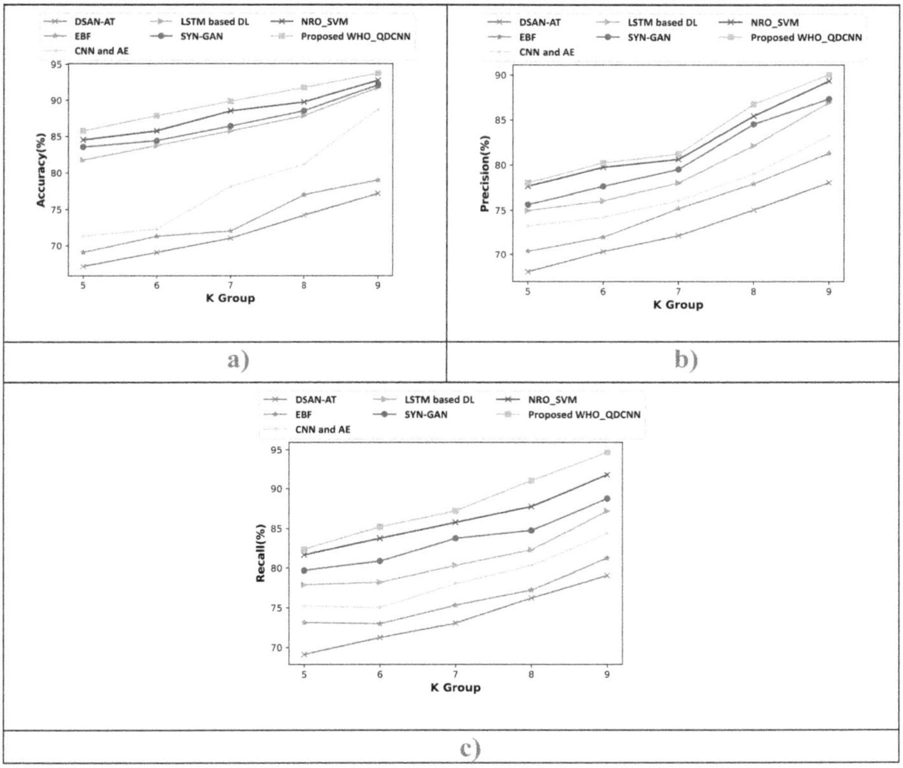

Fig. 6. Analysis of WHO_QDCNN based on Malware capture facility dataset a) accuracy, b) precision, and c) recall with K-group

5.6.2 Evaluation Using IoT-23 Dataset

The evaluation of WHO_QDCNN using the IoT-23 Dataset regarding the learning set is demonstrated in Fig. 7. Figure 7 a) reveals the estimation of accuracy for WHO_QDCNN. With a learning set of 90%, by traditional methods, such as DSAN-AT, EBF, CNN and AE, LSTM-based DL, SYN-GAN, and NRO_SV measured accuracy of 84.88%, 87.88%, 88.78%, 89.77%, 90.36%, and 90.89%, and accuracy recorded by proposed WHO_QDCNN is 91.76%. The estimation of precision for WHO_QDCNN is exploited in Fig. 7 b). The precision recorded with a learning set of 90% by existing methods, namely DSAN-AT is 81.77%, EBF is 82.78%, CNN and AE is 83.88%, LSTM-based DL is 86.77%, SYN-GAN is 87.33%, and NRO_ SVM is 87.79%, and the proposed WHO_QDCNN is 88.77%. In Fig. 7 c), the evaluation based on the recall of WHO_QDCNN is represented. Moreover, with a learning set of 90% the methods, including DSAN-AT, EBF, CNN and AE, LSTM-based DL, SYN-GAN, and NRO_ SVM figured recall of 85.88%, 86.87%, 87.88%, 89.78%, 90.35%, and 91.43%, and proposed WHO_QDCNN is 92.88%.

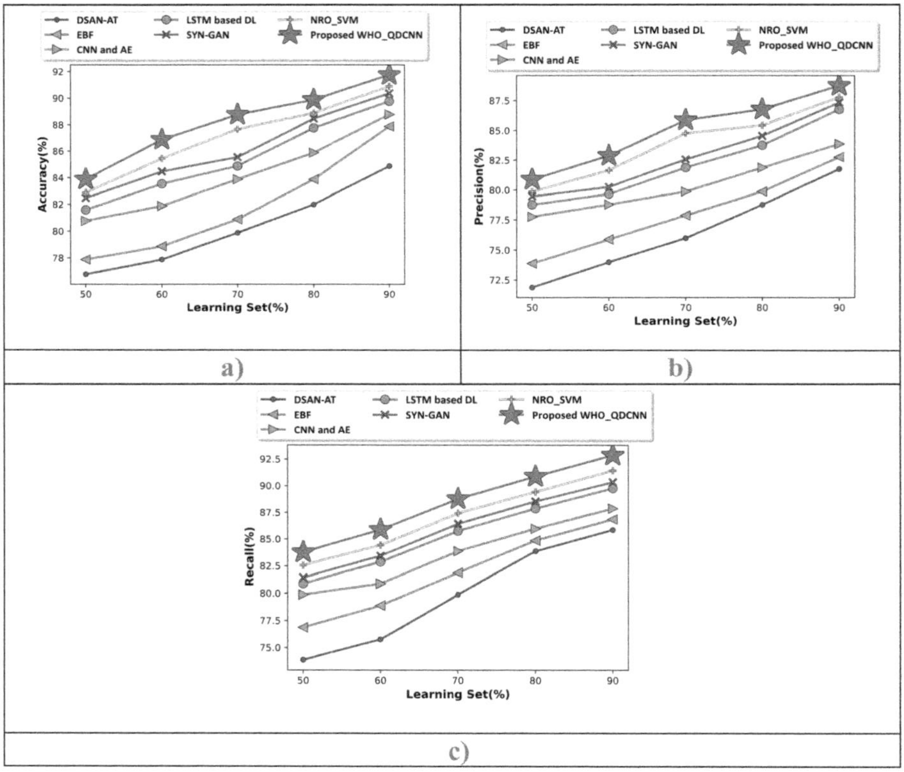

Fig. 7. Estimation of WHO_QDCNN using IoT-23 Dataset a) accuracy, b) precision, and c) recall with learning set

5.7 Algorithmic Analysis

The algorithmic analysis of this research is provided in Fig. 8. The performance of the proposed WHO_QDCNN is compared with other optimization algorithms, such as Particle swarm optimization (PSO)_QDCNN [19], Genetic algorithm (GA)_QDCNN [20], Coati Optimization Algorithm (COA)_QDCNN [21], and Kookaburra Optimization Algorithm (KOA)_QDCNN [22]. The algorithmic evaluation using the Malware capture facility dataset is depicted in Fig. 8a). When the population is 40, the accuracy obtained by the PSO_QDCNN, GA_QDCNN, COA_QDCNN, KOA_QDCNN, and proposed WHO_QDCNN is 87.77%, 88.77%, 89.88%, 90.77%, and 92.88%. Figure 8b) shows the algorithmic evaluation using the IoT-23 Dataset. When the population is 40, the accuracy obtained by the PSO_QDCNN is 85.67%, GA_QDCNN is 87.78%, COA_QDCNN is 88.77%, KOA_QDCNN is 89.77%, and proposed WHO_QDCNN is 91.76%.

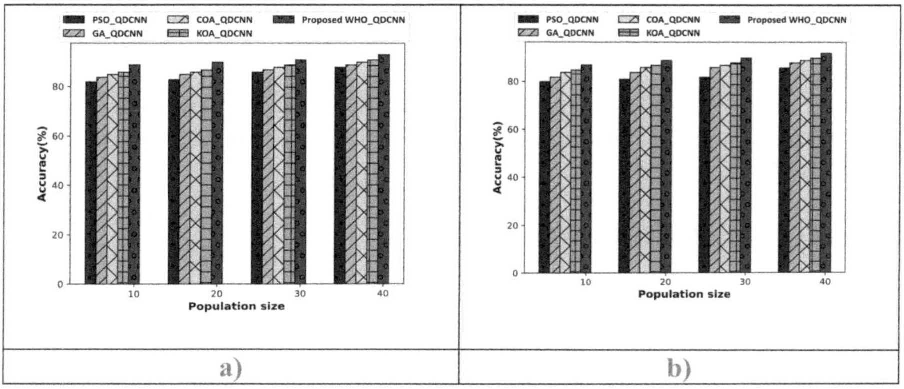

Fig. 8. Algorithmic analysis using a) Malware capture facility dataset, and b) IoT-23 Dataset

5.8 Analysis Based on Energy Consumption and Latency

The analysis based on energy consumption and latency using the Malware capture facility dataset is provided in Fig. 9. The performance of the proposed WHO is compared with the Secretary Bird Optimization Algorithm (SBOA) [23], Greylag Goose Optimization (GGO) [24], and Lyrebird Optimization Algorithm (LOA) [25] in terms of energy consumption and latency. The energy consumption-based evaluation is given in Fig. 9a). When the iteration is 80, the energy consumption of the SBOA, GGO, LOA, and proposed WHO is 8.00J, 6.58J, 3.99J, and 2.99J. This shows the energy consumption of the proposed model is minimum. The latency-based evaluation is provided in Fig. 9b). When the iteration is 80, the latency obtained by the SBOA is 0.637s, GGO is 0.499s, LOA is 0.316s, and proposed WHO is 0.188s. The latency of the proposed method is less than other algorithms.

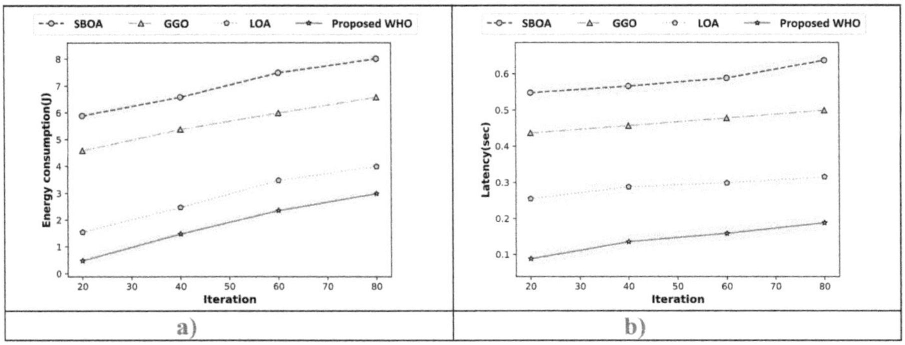

Fig. 9. Comparative analysis based on Malware capture facility dataset a) Energy consumption and b) Latency

5.9 Analysis Based on Attacks Detection

The comparative analysis based on various attack detection, such as Replay, Poisoning, and Evasion is specified in Fig. 10. The attack detection using the Malware capture facility dataset is given in Fig. 10a). The detection rate of the replay attack is 82.89%, 85.88%, 88.88%, 89.78%, 90.38%, 90.78%, and 91.78% for DSAN-AT, EBF, CNN and AE, LSTM-based DL, SYN-GAN, NRO_SVM, and proposed WHO_QDCNN. The poisoning attack detection rate is 84.99% for DSAN-AT, 86.89% for EBF, 89.88% for CNN and AE, 91.77% for LSTM-based DL, 92.36% for SYN-GAN, 92.65% for NRO_SVM, and 93.47% for proposed WHO_QDCNN. The detection rate of the DSAN-AT, EBF, CNN and AE, LSTM-based DL, SYN-GAN, NRO_SVM, and proposed WHO_QDCNN are 87.77%, 89.99%, 91.88%, 93.77%, 94.17%, 94.57%, and 95.76% for Evasion attack. Figure 10b) depicts the attack detection using the IoT-23 Dataset. The detection rate of the replay attack is 81.78%, 83.88%, 86.88%, 88.77%, 89.37%, 89.78%, and 90.88% for DSAN-AT, EBF, CNN and AE, LSTM-based DL, SYN-GAN, NRO_SVM, and proposed WHO_QDCNN. The poisoning attack detection rate is 82.88% for DSAN-AT, 85.88% for EBF, 88.89% for CNN and AE, 89.67% for LSTM-based DL, 90.27% for SYN-GAN, 90.66% for NRO_SVM, and 91.77% for proposed WHO_QDCNN. The detection rate of the DSAN-AT, EBF, CNN and AE, LSTM-based DL, SYN-GAN, NRO_SVM, and proposed WHO_QDCNN are 85.77%, 88.88%, 90.88%, 91.77%, 92.37%, 92.77%, and 93.77% for Evasion attack.

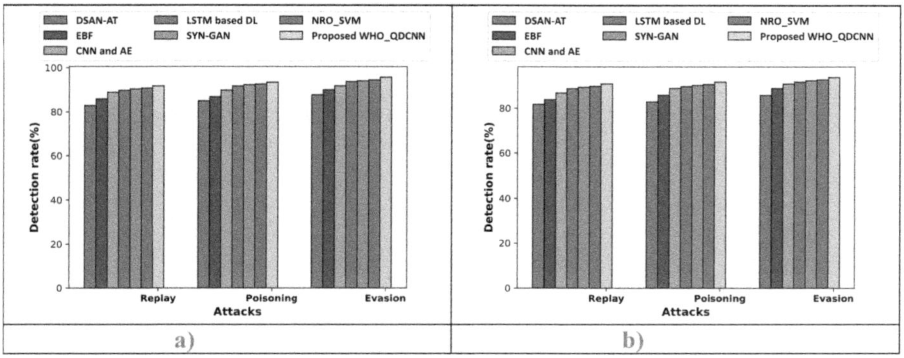

Fig. 10. Comparative analysis based on various attack detection using a) Malware capture facility dataset, and b) IoT-23 Dataset

5.10 Comparative Discussion

The comparative description of WHO_QDCNN concerning metrics, namely precision, recall, and accuracy by learning set and k-group is elaborated in Table 2. The highest accuracy, precision, and recall obtained by the proposed model is 93.77%, 90.05%, and 94.66%by varying k-fold using the Malware capture facility dataset. The application of WHO for tuning the QDCNN enabled the determination of the ideal hyperparameters of the QDCNN, thereby avoiding local optima leading to superior performance.

Table 2. Comparative Discussion

Analysis based	*Metrics*	*Methods*						
		DSAN-AT	*EBF*	*CNN and AE*	*LSTM based DL*	*SYN-GAN*	*NRO_SVM*	*Proposed WHO_QCNN*
		Malware capture facility dataset						
Learning set is 90%	***Accuracy (%)***	78.17	83.34	87.33	89.78	90.43	91.54	92.88
	Precision (%)	74.03	77.03	81.23	84.15	86.67	88.77	89.22
	Recall (%)	78.20	81.97	84.21	87.14	89.77	91.88	93.88
k-value is 9	***Accuracy (%)***	77.22	79.04	88.78	91.77	92.12	92.78	93.77
	Precision (%)	78.06	81.33	83.27	86.94	87.35	89.33	90.05
	Recall (%)	79.08	81.25	84.37	87.22	88.78	91.77	94.66
		IoT-23 Dataset						
Learning set is 90%	***Accuracy (%)***	84.88	87.88	88.78	89.77	90.36	90.89	91.76

(continued)

Table 2. (*continued*)

Analysis based	*Metrics*	*Methods*						
		DSAN-AT	*EBF*	*CNN and AE*	*LSTM based DL*	*SYN-GAN*	*NRO_SVM*	*Proposed WHO_QCNN*
	Precision (%)	81.77	82.78	83.88	86.77	87.33	87.79	88.77
	Recall (%)	85.88	86.87	87.88	89.78	90.35	91.43	92.88

5.10.1 Analysis Based on Time and Memory

Table 3 compares the computational efficiency and memory usage of various models on two datasets. The proposed WHO_QCNN model consistently demonstrates the lowest computational time and least memory usage across both the Malware Capture Facility and IoT-23 datasets. On the Malware dataset, WHO_QCNN requires only 0.258 s and 5.77 MB, while on the IoT-23 dataset, it uses 1.357 s and 7.58 MB. These results indicate that WHO_QCNN is the most efficient among the compared models, including DSAN-AT, EBF, CNN-AE, LSTM, SYN-GAN, and NRO_SVM.

Table 3. Computational time and memory

Metrics	*DSAN-AT*	*EBF*	*CNN and AE*	*LSTM based DL*	*SYN-GAN*	*NRO_SVM*	*Proposed WHO_QCNN*
	Malware capture facility dataset						
Computational time (Sec)	5.766	3.579	1.590	0.977	0.654	0.443	0.258
Memory (MB)	15.88	13.78	10.88	9.88	8.58	7.55	5.77
	IoT-23 Dataset						
Computational time (Sec)	6.677	4.579	2.688	1.988	1.888	1.664	1.357
Memory (MB)	20.88	18.88	13.88	10.77	9.67	8.66	7.58

6 Conclusion

Nowadays, malware is generated by cybercriminals by exploiting the vulnerability of IoT devices to leak the private data of IoT users. In order to prevent severe losses in IoT edge gateways, the identification of malware variant traffic is necessary. However, most of the techniques suffer from slow convergence and fail to accurately detect this malware Therefore, an efficient WHO_QDCNN is established for malware variant traffic identification. Initially, the IoT edge gateway system model is considered and log file collection is carried out. Then, the collected data is subjected to feature extraction, where features, such as network-oriented and traffic-oriented features are extracted. Then, the features are fed into data augmentation, which is carried out by the oversampling method. At last, the malware variant traffic identification is done by utilizing the proposed WHO_QDCNN. In addition, WHO_QDCNN achieved a good accuracy of 93.77%, recall of 90.05%, and precision of 94.66% for malware variant traffic identification. In the future, hybrid deep learning methods can be devised to enhance the detection performance. Further, the application of transfer learning can be considered to cope with unseen malware identification issues.

Ethical Considerations This study aims to enhance malware detection in IoT environments using publicly available datasets. No components related to malware generation are included, and the model is intended only for defensive cybersecurity applications in line with responsible AI practices.

References

1. Hu, X., Zhu, C., Cheng, G., Li, R., Wu, H., Gong, J.: A deep subdomain adaptation network with attention mechanism for malware variant traffic identification at an iot edge gateway. IEEE Internet Things J. **10**(5), 3814–3826 (2022)
2. Indrasiri, P.L., Lee, E., Rupapara, V., Rustam, F., Ashraf, I.: Malicious traffic detection in IoT and local networks using stacked ensemble classifier. Computers, Materials and Continua **71**(1), 489–515 (2022)
3. He, M., Wang, X., Zhou, J., Xi, Y., Jin, L., Wang, X.: Deep-feature-based autoencoder network for few-shot malicious traffic detection. Security and Communication Networks, pp.1–13 (2021)
4. Ali, S., Abusabha, O., Ali, F., Imran, M., Abuhmed, T.: Effective multitask deep learning for iot malware detection and identification using behavioral traffic analysis. IEEE Transactions on Network and Service Management (2022)
5. Khan, M.A., et al.: A deep learning-based intrusion detection system for MQTT enabled IoT. Sensors **21**(21), 7016 (2021)
6. Javaid, A., Niyaz, Q., Sun, W., Alam, M.: A deep learning approach for network intrusion detection system. In: Proceedings of the 9th EAI International Conference on Bio-inspired Information and Communications Technologies (formerly BIONETICS), pp. 21–26 (2016)
7. Liu, H., Lang, B.: Machine learning and deep learning methods for intrusion detection systems: a survey. Applied Sciences **9**(20), 4396 (2019)
8. Mohammed, R., Rawashdeh, J., Abdullah, M.: Machine learning with oversampling and undersampling techniques: overview study and experimental results. In: Proceedings of 11th international conference on information and communication systems (ICICS), pp. 243–248, IEEE (2020)
9. Chen, Y.: QDCNN: Quantum Dilated Convolutional Neural Network. arXiv preprint arXiv: 2110.15667 (2021)
10. Motwakel, A., et al.: Wild horse optimization with deep learning-driven short-term load forecasting scheme for smart grids. Sustainability **15**(2), 1524 (2023)
11. The malware capture facility dataset is taken from "https://mcfp.felk.cvut.cz/publicDatasets/datasets.html". Accessed on May 2024

12. Soe, Y.N., Feng, Y., Santosa, P.I., Hartanto, R., Sakurai, K.: Implementing lightweight iot-ids on raspberry pi using correlation-based feature selection and its performance evaluation. In: Advanced Information Networking and Applications: Proceedings of the 33rd International Conference on Advanced Information Networking and Applications (AINA-2019), **33**, pp. 458–469, Springer International Publishing (2020)
13. Alrzini, J.R.S., Pennington, D.: A review of polymorphic malware detection techniques. International Journal of Advanced Research in Engineering and Technology **11**(12), 1238–1247 (2020)
14. Beaugnon, A., Chifflier, P.: Machine learning for computer security detection systems: practical feedback and solutions. In: Proceedings of the 2018 Intelligence Artificielle et Cybersécurité/Artificial Intelligence and Cybersecurity (C&ESAR), Rennes, France, pp.19–21 (2018)
15. Sadeghzadeh, A.M., Shiravi, S., Jalili, R.: Adversarial network traffic: towards evaluating the robustness of deep-learning-based network traffic classification. IEEE Trans. Netw. Serv. Manage. **18**(2), 1962–1976 (2021)
16. IoT-23 Dataset is taken from "https://www.kaggle.com/datasets/engraqeel/iot23preprocesseddata,". Accessed on June 2025
17. Rahman, S., Pal, S., Mittal, S., Chawla, T., Karmakar, C.: SYN-GAN: a robust intrusion detection system using GAN-based synthetic data for IoT security. Internet of Things **26**, 101212 (2024)
18. Ahmad, I., Wan, Z., Ahmad, A., Ullah, S.S.: A hybrid optimization model for efficient detection and classification of malware in the internet of things. Mathematics **12**(10), 1437 (2024)
19. Wang, D., Tan, D., Liu, L.: Particle swarm optimization algorithm: an overview. Soft. Comput. **22**, 387–408 (2018)
20. Mirjalili, S.: Genetic algorithm. Evolutionary Algorithms and Neural Networks, pp. 43–55 (2018)
21. Dehghani, M., Montazeri, Z., Trojovská, E., Trojovský, P.: Coati optimization algorithm: a new bio-inspired metaheuristic algorithm for solving optimization problems. Knowl.-Based Syst. **259**, 110011 (2023)
22. Dehghani, M., Montazeri, Z., Gulnara, B., Malik, O.P., Dhiman, G., Ahmed, A.E.M.: Kookaburra Optimization Algorithm: A New Bio-Inspired Metaheuristic Algorithm for Solving Optimization Problems **8**(6), 470 (2023)
23. Fu, Y., Liu, D., Chen, J., He, L.: Secretary bird optimization algorithm: a new metaheuristic for solving global optimization problems. Artificial Intelligence Review **57** (2024)
24. El-kenawy, E.-S., Khodadadi, N., Mirjalili, S., Abdelhamid, A.A., Eid, M.M., Ibrahim, A.: Greylag goose optimization: nature-inspired optimization algorithm. Expert Syst. Appl. **238**, 122147 (2024)
25. Dehghani, M., Bektemyssova, G., Montazeri, Z., Shaikemelev, G., Malik, O.P., Dhiman, G.: Lyrebird optimization algorithm: a new bio-inspired metaheuristic algorithm for solving optimization problems. Biomimetics **8**(6), 507 (2023)

Author Index

R. Geetha et al. (Eds.): AAIMB 2024, CCIS 2654, pp. 423–424, 2026.
https://doi.org/10.1007/978-3-032-13029-7

www.ingramcontent.com/pod-product-compliance
Ingram Content Group UK Ltd.
Pitfield, Milton Keynes, MK11 3LW, UK
UKHW020648021026
476200UK00002B/85
* 9 7 8 3 0 3 2 1 3 0 2 8 0 *